Light Propagation in Periodic Media

OPTICAL ENGINEERING

1. Electron and Ion Microscopy and Microanalysis: Principles and Applications, *Lawrence E. Murr*
2. Acousto-Optic Signal Processing: Theory and Implementation, *edited by Norman J. Berg and John N. Lee*
3. Electro-Optic and Acousto-Optic Scanning and Deflection, *Milton Gottlieb, Clive L. M. Ireland, and John Martin Ley*
4. Single-Mode Fiber Optics: Principles and Applications, *Luc B. Jeunhomme*
5. Pulse Code Formats for Fiber Optical Data Communication: Basic Principles and Applications, *David J. Morris*
6. Optical Materials: An Introduction to Selection and Application, *Solomon Musikant*
7. Infrared Methods for Gaseous Measurements: Theory and Practice, *edited by Joda Wormhoudt*
8. Laser Beam Scanning: Opto-Mechanical Devices, Systems, and Data Storage Optics, *edited by Gerald F. Marshall*
9. Opto-Mechanical Systems Design, *Paul R. Yoder, Jr.*
10. Optical Fiber Splices and Connectors: Theory and Methods, *Calvin M. Miller with Stephen C. Mettler and Ian A. White*
11. Laser Spectroscopy and Its Applications, *edited by Leon J. Radziemski, Richard W. Solarz, and Jeffrey A. Paisner*
12. Infrared Optoelectronics: Devices and Applications, *William Nunley and J. Scott Bechtel*
13. Integrated Optical Circuits and Components: Design and Applications, *edited by Lynn D. Hutcheson*
14. Handbook of Molecular Lasers, *edited by Peter K. Cheo*
15. Handbook of Optical Fibers and Cables, *Hiroshi Murata*
16. Acousto-Optics, *Adrian Korpel*
17. Procedures in Applied Optics, *John Strong*
18. Handbook of Solid-State Lasers, *edited by Peter K. Cheo*
19. Optical Computing: Digital and Symbolic, *edited by Raymond Arrathoon*
20. Laser Applications in Physical Chemistry, *edited by D. K. Evans*
21. Laser-Induced Plasmas and Applications, *edited by Leon J. Radziemski and David A. Cremers*
22. Infrared Technology Fundamentals, *Irving J. Spiro and Monroe Schlessinger*
23. Single-Mode Fiber Optics: Principles and Applications, Second Edition, Revised and Expanded, *Luc B. Jeunhomme*
24. Image Analysis Applications, *edited by Rangachar Kasturi and Mohan M. Trivedi*
25. Photoconductivity: Art, Science, and Technology, *N. V. Joshi*
26. Principles of Optical Circuit Engineering, *Mark A. Mentzer*
27. Lens Design, *Milton Laikin*
28. Optical Components, Systems, and Measurement Techniques, *Rajpal S. Sirohi and M. P. Kothiyal*
29. Electron and Ion Microscopy and Microanalysis: Principles and Applications, Second Edition, Revised and Expanded, *Lawrence E. Murr*

30. Handbook of Infrared Optical Materials, *edited by Paul Klocek*
31. Optical Scanning, *edited by Gerald F. Marshall*
32. Polymers for Lightwave and Integrated Optics: Technology and Applications, *edited by Lawrence A. Hornak*
33. Electro-Optical Displays, *edited by Mohammad A. Karim*
34. Mathematical Morphology in Image Processing, *edited by Edward R. Dougherty*
35. Opto-Mechanical Systems Design: Second Edition, Revised and Expanded, *Paul R. Yoder, Jr.*
36. Polarized Light: Fundamentals and Applications, *Edward Collett*
37. Rare Earth Doped Fiber Lasers and Amplifiers, *edited by Michel J. F. Digonnet*
38. Speckle Metrology, *edited by Rajpal S. Sirohi*
39. Organic Photoreceptors for Imaging Systems, *Paul M. Borsenberger and David S. Weiss*
40. Photonic Switching and Interconnects, *edited by Abdellatif Marrakchi*
41. Design and Fabrication of Acousto-Optic Devices, *edited by Akis P. Goutzoulis and Dennis R. Pape*
42. Digital Image Processing Methods, *edited by Edward R. Dougherty*
43. Visual Science and Engineering: Models and Applications, *edited by D. H. Kelly*
44. Handbook of Lens Design, *Daniel Malacara and Zacarias Malacara*
45. Photonic Devices and Systems, *edited by Robert G. Hunsperger*
46. Infrared Technology Fundamentals: Second Edition, Revised and Expanded, *edited by Monroe Schlessinger*
47. Spatial Light Modulator Technology: Materials, Devices, and Applications, *edited by Uzi Efron*
48. Lens Design: Second Edition, Revised and Expanded, *Milton Laikin*
49. Thin Films for Optical Systems, *edited by François R. Flory*
50. Tunable Laser Applications, *edited by F. J. Duarte*
51. Acousto-Optic Signal Processing: Theory and Implementation, Second Edition, *edited by Norman J. Berg and John M. Pellegrino*
52. Handbook of Nonlinear Optics, *Richard L. Sutherland*
53. Handbook of Optical Fibers and Cables: Second Edition, *Hiroshi Murata*
54. Optical Storage and Retrieval: Memory, Neural Networks, and Fractals, *edited by Francis T. S. Yu and Suganda Jutamulia*
55. Devices for Optoelectronics, *Wallace B. Leigh*
56. Practical Design and Production of Optical Thin Films, *Ronald R. Willey*
57. Acousto-Optics: Second Edition, *Adrian Korpel*
58. Diffraction Gratings and Applications, *Erwin G. Loewen and Evgeny Popov*
59. Organic Photoreceptors for Xerography, *Paul M. Borsenberger and David S. Weiss*
60. Characterization Techniques and Tabulations for Organic Nonlinear Optical Materials, *edited by Mark Kuzyk and Carl Dirk*

61. Interferogram Analysis for Optical Testing, *Daniel Malacara, Manuel Servín, and Zacarias Malacara*
62. Computational Modeling of Vision: The Role of Combination, *William R. Uttal, Ramakrishna Kakarala, Sriram Dayanand, Thomas Shepherd, Jagadeesh Kalki, Charles F. Lunskis, Jr., and Ning Liu*
63. Microoptics Technology: Fabrication and Applications of Lens Arrays and Devices, *Nicholas F. Borrelli*
64. Visual Information Representation, Communication, and Image Processing, *Chang Wen Chen and Ya-Qin Zhang*
65. Optical Methods of Measurement: Wholefield Techniques, *Rajpal S. Sirohi and Fook Siong Chau*
66. Integrated Optical Circuits and Components: Design and Applications, *edited by Edmond J. Murphy*
67. Adaptive Optics Engineering Handbook, *edited by Robert K. Tyson*
68. Entropy and Information Optics, *Francis T. S. Yu*
69. Computational Methods for Electromagnetic and Optical Systems, *John M. Jarem and Partha P. Banerjee*
70. Laser Beam Shaping: Theory and Techniques, *edited by Fred M. Dickey and Scott C. Holswade*
71. Rare-Earth-Doped Fiber Lasers and Amplifiers: Second Edition, Revised and Expanded, *edited by Michel J. F. Digonnet*
72. Lens Design: Third Edition, Revised and Expanded, *Milton Laikin*
73. Handbook of Optical Engineering, *edited by Daniel Malacara and Brian J. Thompson*
74. Handbook of Imaging Materials, *edited by Arthur S. Diamond and David S. Weiss*
75. Handbook of Image Quality: Characterization and Prediction, *Brian W. Keelan*
76. Fiber Optic Sensors, *edited by Francis T. S. Yu and Shizhuo Yin*
77. Optical Switching/Networking and Computing for Multimedia Systems, *edited by Mohsen Guizani and Abdella Battou*
78. Image Recognition and Classification: Algorithms, Systems, and Applications, *edited by Bahram Javidi*
79. Practical Design and Production of Optical Thin Films: Second Edition, Revised and Expanded, *Ronald R. Willey*
80. Ultrafast Lasers: Technology and Applications, *edited by Martin E. Fermann, Almantas Galvanauskas, and Gregg Sucha*
81. Light Propagation in Periodic Media: Differential Theory and Design, *Michel Nevière and Evgeny Popov*

Additional Volumes in Preparation

Optical Remote Sensing: Science and Technology, *Walter Egan*

Light Propagation in Periodic Media

Differential Theory and Design

Michel Nevière and Evgeny Popov

Université de Provence
Marseille, France

MARCEL DEKKER, INC. NEW YORK • BASEL

Library of Congress Cataloging-in-Publication Data
A catalog record for this book is available from the Library of Congress.

ISBN: 0-8247-0893-8

This book is printed on acid-free paper.

Headquarters
Marcel Dekker, Inc.
270 Madison Avenue, New York, NY 10016
tel: 212-696-9000; fax: 212-685-4540

Eastern Hemisphere Distribution
Marcel Dekker AG
Hutgasse 4, Postfach 812, CH-4001 Basel, Switzerland
tel: 41-61-260-6300; fax: 41-61-260-6333

World Wide Web
http://www.dekker.com

The publisher offers discounts on this book when ordered in bulk quantities. For more information, write to Special Sales/Professional Marketing at the headquarters address above.

Current printing (last digit):
10 9 8 7 6 5 4 3 2 1

PRINTED IN THE UNITED STATES OF AMERICA

Preface

Periodic media are common in nature. Solids are hard because their atoms are kept more or less fixed at the nodes of a three-dimensional mesh. This arrangement is responsible not only for their mechanical properties, but also for interference phenomena that lead to strange effects. Probably the most remarkable among these is the band gap in electron energy levels, the discovery of which opened the highway to semiconductor technology and modern information processing in the mid-twentieth century.

There is no doubt that periodicity on a microscopic level can produce other interesting and important effects not only in electron behavior, but also in optics. Diffraction gratings are the most famous example of periodic optical devices. Their ability to redirect light according to its spectral content is due to their periodicity and has been widely used in spectroscopy for two centuries. More complicated periodic devices, known as photonic crystals, designed with two- or three-dimensional periodicities, have been recently developed. Like common crystals, their periodicity on the scale of light wavelength leads to the formation of band gaps, but for photons instead of electrons–hence their high potential for light processing and the great interest they hold for immediate applications.

Light behavior in these media is subject to Maxwell equations. From a theoretical and practical point of view, it is quite critical that they constitute a multiparametric problem, with many degrees of freedom for the optogeometrical parameters. This requires numerical treatment for modelization and optimization, even when remaining in the more simple case of diffraction gratings. The efforts to do this in a manner appropriate and useful for optical engineers began in the late nineteenth century. Since the late 1960s, several different approaches have been established. The most widely used theories are the integral method, the modal method and the differential method. While a great number of fundamental and applied papers have been published during the last 30 years dealing with theoretical and numerical modeling of light propagation in periodic media, the only detailed monograph on this problem is dated 1980, and a lot of progress has been made since that time.

Our book aims to present the development of modeling the propagation and diffraction of light in periodic media during the last 20 years. We pay special attention to the class of differential methods, due to their relative simplicity for teaching to the engineering community. They also have the virtue of easy adaptation to different geometries, anisotropy and nonlinearity. Applications studied concern various periodic media – stacks of gratings, crossed gratings, and photonic crystals, as well as some quasiperiodic and nonperiodic systems – separate cylindrical objects, inhomogeneous layers with volume incorporations, etc. The spectral region covered extends from X-rays to millimeter waves.

The book is designed for scientists studying light propagation in periodic media and optical engineers of conventional, night vision, fiber and communication optics, and optical detectors, but also for all grating users - astronomers, synchrotron engineers, spectroscopists, etc. The book can be used in graduate courses and by Ph.D. students in optics, physics, optical engineering, electromagnetism, photonics and optical communications.

The book starts with simpler problems and gradually moves to more complicated ones. After discussing the general properties of light diffracted by periodic media in Chapter I, the classical differential theory of gratings is presented in Chapter II. This presentation is useful in helping beginners in diffraction theory to easily understand the basic principles. Numerical instabilities that appear during the integration process are analyzed in detail in Chapter III, where an algorithm is proposed to get rid of these difficulties. Another key step in the theory is the improvement of the convergence rate of the Fourier series of the field diffracted by metallic objects. This problem plagued the performance of differential theories during the last quarter of the twentieth century. The book presents a new theory, called the Fast Fourier Factorization (FFF) approach, which drastically ameliorates the convergence of the truncated series. Chapter IV presents the approach and its applications to classical diffraction gratings made of isotropic materials in in-plane and conical diffraction. The FFF method is generalized in Chapter V to anisotropic media with one-, two- or three-dimensional periodicity. To that aim, a new formulation of the constitutive relations between the vectors of electric field $\vec{E}$ and the induction $\vec{D}$ is proposed in a truncated Fourier space. The same relations apply between $\vec{H}$ and $\vec{B}$ in the case of magnetic media. As a result, a specific form of Maxwell equations in a truncated Fourier space is obtained in the general case, and then applied to specific structures in what follows. These are the lamellar gratings in Chapter VI, anisotropic gratings in Chapter VIII and crossed gratings in Chapter IX. Chapter VII briefly presents numerical algorithms based on transformation of the coordinate system to avoid dealing with Fourier representation of discontinuous functions.

Chapters X to XIV are devoted to applications of the differential theory to specific structures and their properties. Chapter X deals with photonic crystals and, in particular, their ability to efficiently diffract light with blazing properties unachieved by classical gratings. Chapter XI gives a detailed study of X-ray gratings, bare or multilayered, with several examples of particular designs. Transmission grating properties are presented in Chapter XII, while guided and surface wave phenomena are presented in Chapter XIII. Chapter XIV presents an extension of the diffraction theory to nonperiodic, random and isolated objects. Finally, Chapter XV generalizes the FFF approach to optically nonlinear media, numerically analyzed in truncated Fourier space by using iterative methods with a linearization of the nonlinear set of equations.

The book is the result of more than 30 years of research on the differential theory of light diffraction by periodic media, made in the Laboratoire d'Optique Electromagnetique (now called the Institut Fresnel) in Marseille, France, and joined later by the grating group at the Institute of Solid State Physics, Sofia, Bulgaria. The research involved a significant amount of collaboration with other research laboratories and grating manufacturers throughout the world.

Special thanks are due to Prof. Lifeng Li for his careful reading of the manuscript. We highly appreciated his useful remarks and critiques. We are also glad to thank Mrs. Galia Popov for the great job she did typing the manuscript with its many equations.

Michel Nevière
Evgeny Popov

Contents

Chapter I
General Properties

I.1 Introduction

I.1.1 Types of diffracting periodic media

Periodic media can be found everywhere in nature. Crystals are made of a huge amount of ordered atoms and there is no doubt that their order induces special properties, mechanical frigidness being the most obvious. The solid-state physics of the first half of the 20th century took into account periodicity in terms of Bloch waves revealing interesting properties. Among them, the band gaps happened to open an impressive domain of applications in electronic signal processing.

The aim of this book is to study the larger-scale periodicity, when the structure is periodic on the scale of light wavelength (usually much larger than the atomic distances). This leads to many interesting and important phenomena connected with light propagation, scattering, and diffraction, which in general are not concerned by quantum phenomena. However, the experience acquired in quantum mechanics and solid-state physics can be of great help in studying light behavior in periodic media.

At a macroscopic level, the most famous examples of periodic media are diffraction gratings. Since their invention two centuries ago [I.1], gratings are known to behave differently from rough surfaces in the sense that they diffract an incident plane wave into privileged directions called *spectral orders* instead or scattering it over the entire space. This property is a direct consequence of the periodicity and turns out to depend on the wavelength, making gratings famous in many spectroscopic applications [I.2, I.3]. The development of high-quality ruling engines and a reliable replication process greatly increased the use of gratings instead of prisms. Holographic recording in the 1960s and, more recently, the ion-etching technique has opened the road to common-use applications. Among them are many domains of physics: spectroscopy, acoustics, solid-state physics, X-ray instrumentations, lasers, but also wider scientific and technological applications in optical communications, information processing, night vision, astronomy and space science, together with many every-day uses in CD players, safety features in credit cards and banknotes, displays and advertisements.

Gratings exist nowadays in many forms, not only as a simple modulated surface. They can be ruled (or holographically or lithographically recorded) on dielectric waveguides (planar, channel or fiber) for input-output coupling or

Bragg interaction. They can be combined with stacks of dielectric layers for hostile-environment protection or to increase the efficiency and reduce absorption losses, as done in high-power laser pulse compression. Gratings made of hundreds of modulated dielectric and metallic layers are in regular use in the X-ray domain.

Crossed gratings emerge when using two-dimensional refractive index or surface modulation. Another complexity is the use of anisotropic or chiral materials. Nonlinear materials can be combined with gratings to excite guided or surface waves in order to enhance various nonlinear effects [I.4]. All this points out how complex the theoretical means need to be to model so many different grating configurations.

On top of the complexity of the grating problems, during the last ten years new devices have been created for light control, known now under the name of photonic crystals [I.5]. They reproduce and use on the scale of wavelength of light what nature has done on the atomic level. The crystal atoms are replaced by dielectric or metallic particles arranged in 1-, 2-, or 3-dimensional periodic lattice. As in the case of ordinary crystals, where electronic band gaps are formed, the photonic crystals create band gaps for photons, paving the way to precise control of light propagation and emission. Theoretical analysis of light diffraction in photonic crystals is another topic discussed in this book.

A separate object can also be regarded as a periodic surface. Although this approach is not obvious, to accept it is necessary to describe it in a system of spherical coordinates (r, θ, φ). Then the device surface, as well as the diffracted field will be periodical with respect to θ and φ with a period of 2π. In the same way, cylindrical objects are periodical with respect to the polar angle in a cylindrical coordinate system.

The periodicity naturally imposes Fourier transformation of the optogeometrical parameters and the electromagnetic fields. Moreover, discrete Fourier analysis can be applied using the same theoretical methods developed for periodic objects to study non-periodic media, such as isolated bump(s) on a flat or corrugated or rough surface, nonperiodic inhomogeneities inside or over dielectric and absorbing layers, etc. For these non-periodic cases, the discrete Fourier transform substitutes numerically the Fourier series required to describe periodic objects.

Many useful devices in conventional, microwave and X-ray optics can be successfully treated using the methods developed for periodic objects. Such are the Fresnel planar lenses and the diffractive optics elements with focusing and beam-shaping properties, which can be considered to work in the "adiabatic" region, when the grating groove parameters over the device surface or volume vary much more slowly than the wavelength of light.

I.1.2 Brief overlook of grating theories

Among all the objects that can scatter and diffract light, diffraction gratings have attracted the greatest attention. The first observations of light diffraction through a fine silk handkerchief by Francis Hopkinson in 1785 and the more scientific investigation of this phenomenon by David Rittenhouse [I.1] immediately pointed out the usefulness of the gratings in spectral analysis. Thomas Young [I.6] was the first to discover the sine relationship for the directions of different diffraction orders, but only to Fraunhofer we due the serious studies of grating properties and their physical explanation [I.3, I.7, I.8].

To understand why light goes into different directions when reflected by a grating and what are these directions, it is necessary to analyze the difference between a flat and a corrugated reflection surface from the point of view of light striking the surface. For a flat surface, Snell's laws require a preservation of the component k_x

$$k_x = k \sin\theta_i \equiv \frac{2\pi}{\lambda}\sin\theta_i \tag{I.1}$$

of the light wavevector $\vec{k}$ along the flat surface, due to the invariance of the optogeometrical properties along the interface. Here λ is the light wavelength and θ_i is the angle of incidence with respect to the normal to the mean plane of the grating. When the surface is corrugated, say, in the x-direction with a period d, the properties are no longer invariant in x, but periodical. In the k-space, the periodicity is expressed in terms of the so-called grating number K, defined in a similar manner as the wavenumber k:

$$K = \frac{2\pi}{d}. \tag{I.2}$$

Instead of Eq. (I.1), the x-component of the wavevector of the reflected wave can take several possible values

$$k_{x,m} = \frac{2\pi}{\lambda}\sin\theta_i + m\frac{2\pi}{d} \tag{I.3}$$

for different integer values of m. This equation determines a set of discrete directions of propagation, called diffraction orders. As long as the wavelength is not changed, the wavenumber is preserved and for only a limited number of m the component k_y of the orders along the y-axis will be real and the orders will be propagating. For m outside that set, k_y is imaginary and the corresponding diffraction orders are evanescent when going away from the grating surface:

$$k_{y,m} = \sqrt{k^2 - k_x^2} \quad \text{real : propagating orders}$$
$$k_{y,m} = \sqrt{k^2 - k_x^2} \quad \text{imaginary : evanescent orders} \tag{I.4}$$

Eq. (I.3) can be written in a simpler form, well known to the spectroscopic and grating community as the famous *grating equation for the diffraction angle of the m-th order*:

$$\sin\theta_m = \sin\theta_i + m\frac{\lambda}{d}\,, \tag{I.5}$$

wherefrom come the spatial dispersion of the nonspecular ($m \neq 0$) diffraction orders and the sine law of the angles of diffraction θ_m of the consecutive orders.

The dispersion of the propagating orders was so important to the development of spectroscopy that many famous scientists (T. Young, J. Fraunhofer, A. Ångström, H. Rowland, R. Wood, Lord Rayleigh, C. Raman, A. Michelson, etc.) spent an important part of their life in studying and developing grating theory, technology and apparatus [I.3]. The evanescent orders, although non-measurable in the far field, play a very important role in surface enhanced Raman scattering, nonlinear effects, guided wave coupling, near-field microscopy, etc.

From a grating user point of view, the grating equation is very important, because it determines the directions of propagation of the diffraction orders. However, this is never sufficient, because people want not only to know where the light goes but also how much of it goes into specific directions, i.e., what is the grating efficiency to diffract light. Taken the grating importance, it is not surprising that this question attracted the biggest theoretical efforts [I.4, I.9] in electromagnetic optics since the early work of Lord Rayleigh [I.10]. At the beginning of the 20th century, Rayleigh established that outside the groove region, the electromagnetic field near the grating can be represented as series of propagating and evanescent waves, these series being known as Rayleigh expansion. While evident above the top of the grooves, the validity of this expansion inside the grooves was under doubt for him and for many other grating theoreticians. After years of intensive debates, the answer was given in 1966 [I.11] that, in general, the Rayleigh expansion is not valid to represent the field inside the modulated region, except for a limited type of shallow grooves. Nevertheless, the theory based on the Rayleigh assumption is very easy for understanding and programming and has attracted the attention of many scientists. The same is also true for an amount of approximate methods developed later. Although simple, these methods have limited regions of validity and application.

On the other hand, there exist several electromagnetic theories developed since 1960 that are considered as "rigorous" ones, in the sense that they provide numerical solutions of the Maxwell equations and boundary conditions without any theoretical approximation, except for those introduced by the computation process (round-off errors, truncation of the number of equations, of the number of field harmonics, of the computer word length, etc.). There are several such rigorous theories: the integral method, the classical differential method, the method of fictitious sources, the finite-element method, the modal methods (there are several), the method of Green functions, etc. An interested reader can find recent reviews on the subject [I.12, I.13]. In general, the methods start with Maxwell equations written in integral or differential form, project them on some set of functions, use various techniques to simplify the equations and to avoid singularities and instabilities, and arrive at a linear algebraic system of equations for the diffraction order amplitudes.

The integral method was the first rigorous grating theory. It was developed by several authors around 1966 [I.14a-d] for perfectly conducting gratings and generalized in 1972 for finite conductivity [I.15]. The basic principle of the integral method can be understood more easily for a perfectly conducting substrate. When an incident plane wave falls on the grating surface, it induces a surface current $j_S(M')$ at each point M' of the surface. When propagating along the grating, the surface current radiates a diffracted field E(P) at a given point P above the surface, which is related to j_S by the Kirchhoff-Huygens formula using the Green function G technique:

$$E(P) = \int_{\substack{\text{one}\\ \text{grating}\\ \text{period}}} G(P,M')\ \varphi(M')\ ds'$$

where $\varphi(M')$ is proportional to $j_S(M')$ and the integration is carried over the curvilinear coordinate s' along the grating profile. However, the problem is more complex than just a single integration, because the surface current is not produced only by the incident wave $E^{(i)}$, but also results from the diffracted field, radiated from the other points of the grating surface. Thus $j_S(M')$ depends both on $E^{(i)}$ and on the current $j_S(M'')$ at each other point M'' on the profile and the above equation is then transformed into an integral equation for $\varphi(M')$. The situation becomes even more complicated for finitely conducting gratings.

The key point of the theory is the numerical resolution of an integral equation with singular kernel, due to the divergence of the Green functions when the points M' and M'' on the profile become too close to each other. Fortunately, the singularities can be eliminated analytically and the integral equation is reduced to a linear set of algebraic equations by discretization along the profile.

Several other methods use similar projection of the integral equations on discretized points or segments along the profile (one-dimensional finite-element technique, Green function method), on separated points outside the interface (method of fictitious sources) or on a 2-D segmental mesh (two-dimensional finite-element method).

Instead of using the integral presentation of Maxwell equations or their solutions, several other methods directly use the differential form of the equations to study the grating diffraction. The classical modal method [I.16] deals with a step-like (also called lamellar, rectangular or laminar) profile and searches the solution of the Maxwell equations and boundary conditions inside the grooves in form of "modes" with the field given in the form:

$$F_m(x,y) = u_m(x)\exp(i\rho_m y) \tag{I.6}$$

so that the function u(x) has a different form inside the grooves and the lamellae, determined by the refractive index of the media and the boundary conditions along the groove walls. Due to the periodicity along x, there is only a discrete set of values for ρ_m, called modal constants. The total field, which must satisfy the boundary conditions on the interfaces between the corrugated and the outer homogeneous media, is searched as a linear combination of the modes with unknown amplitudes, determined again by solving a linear algebraic system. Although simple for understanding and programming, this method is limited to lamellar profiles or to a stack of several such profiles, and poses difficulties to locate the modal constants for highly conducting materials. Several different techniques have been proposed for that sake [I.17].

The differential theory uses similar representation of the field, but contrary to Eq. (I.6) the exponentials are along x and the unknown functions, along y:

$$F(x,y) = \exp(ik\sin\theta_i x)\sum_m F_m(y)\exp(i\frac{2\pi}{d}x)\,, \tag{I.7}$$

which, obviously, is nothing but a Fourier series representation of the field possible due to the periodicity in x (see later Section I.2.4). By comparing Eq. (I.7) and (I.5), it is evident that, outside the groove region, the series reduce to Rayleigh expansion (Section I.2.5). Inside the modulated region the Fourier components are unknown functions of y only, which reduces the Maxwell equations to a set of coupled ordinary differential equations. This set can be solved using numerical integration techniques; then the solution is matched to the Rayleigh expansion outside the modulated region to determine the field everywhere, thus the grating efficiency.

For rectangular grooves, the set of differential equations has constant coefficients and its solution can be found by using an eigenvalue numerical technique. This possibility, first presented in the early works of Burckhardt [I.18a] and Peng et al. in 1975 [I.18b] and independently developed in Russia in 1977 [I.19], has been widely studied in United States since the early 1980s and popularized under the name of rigorous coupled wave theory (RCW) [I.20].

Several other differential approaches use coordinate transformation to simplify the boundary conditions at the modulated interface. Such are the conformal mapping technique [I.21] and the more popular coordinate transformation method (C-method) [I.22]. The latter again uses an eigenvalue technique.

In what follows we will concentrate our study mainly on the differential method, which requires a numerical integration of a set of ordinary differential equations for the unknown Fourier components of the field, Eq. (I.7). The choice is determined by the versatility of the method, its adaptability to various kinds of groove profiles and media types: linear, nonlinear, anisotropic, magnetooptic, etc. In addition, the differential theory is easy to present to students, researchers and engineers, because it does not involve any sophisticated mathematical background, only linear algebra and ordinary differential equations. Anyway, one has to keep in mind that recent development of grating theories has proved their complementarity, rather than the superiority of one approach over the others (as shown in the next section), so that sophisticated and extensive grating researches must be based on a diversity of theoretical tools.

I.1.3 The grating theory user guide

In order to understand the complexity of the grating problem, it is necessary to realize that nowadays gratings are not restricted to *bare metallic* corrugated surfaces, as used in reflection for spectroscopic purposes. Such gratings are usually produced by diamond ruling. Holographic recording, beam-etching and lithographic methods produce, in general, symmetrical profiles with sinusoidal, lamellar (rectangular) or trapezoidal grooves. Electron microscopy reveals that for very high groove frequencies (e.g., 6000 gr/mm), as well as very low ones (e.g., echelles, widely used in astronomy), the real profiles have much more complicated forms, different from the classical four types mentioned above.

In far-infrared and millimeter range of wavelength the metal can be assumed with infinite conductivity. In the visible region and for shorter wavelengths the finite conductivity complicates the grating response and requires different theoretical methods.

Transmission gratings are frequently used as beam splitters and as Carpenter gratings or grisms to transform a telescope or a camera into a spectroscope. Under special conditions the diffraction compensates the refraction and a dispersive diffraction order can propagate in the initial direction. Transmission gratings require a transparent (lossless) grating material, contrary to reflection gratings.

Grating *couplers* combine a dielectric grating and an optical waveguide. They are used in integrated optics to couple an incident beam into a slab or vice versa.

Dielectric coated gratings are mainly used in vacuum UV with a dielectric layer (typically MgF_2) to prevent aluminum from oxidation. Similar technique is used in the visible region to protect silver gratings from tarnishing.

Multilayer dielectric gratings consist of a dielectric grating replicated on the top or at the bottom of a stack of plane dielectric layers with alternating high and low refractive indices and a quarter of wavelength optical thickness. They are used in high-power laser optics to achieve peak absolute efficiency, typically more than 96%, at a single wavelength.

Another kind of multilayer gratings, consisting of a *stack of modulated interfaces*, is used in XUV and X-ray domain. Their theoretical treatment requires to distinguish between two main cases: profiles without and with an interpenetration. There is no interpenetration when the bottom of the upper profile lies higher than the top of the lower profile. A grating coupler with a *double corrugation* is an example of a "thin" stack of identical profiles without interpenetration. It can be optimized to couple as high as 80% of the incident beam into the waveguide. In a more complicated case, the high diffraction efficiency (45%) in XUV region (wavelength $\lambda \sim 15$ nm) is obtained by combining profiles with and without interpenetration. The case of identical profiles is called *multilayer coated grating* and the layers are usually thinner than the profile depth, i.e., there is interpenetration. Such gratings are used in the two extremities of the spectrum. Another alternative of highly efficient gratings for X-rays are the *Bragg multilayer gratings*.

Echelles are echelette gratings used in high orders, typically 10 to 500. They can be bare or multilayer coated depending on the spectral domain. The low wavelength-to-period (λ/d) ratio requires special theoretical methods.

Multilevel gratings are used in diffractive optics to produce multiple beams from a single incident ray. Usually, they also work under extremely low λ/d ratios.

In contrast, *Dammann gratings* consist of several bumps with different widths within a single period. Not only do they also work in a low λ/d region, but their profile is more complicated than the usual profiles to be introduced in grating theories. Dammann gratings have also become a common element in diffractive optics.

Gratings with varying period or groove shape are used in integrated optics and in X-ray domain for focusing and beam shaping. If only the groove period is varied, we obtain a *linear zone plate*. When etched inside a planar multilayer structure, it is called a *Bragg-Fresnel linear zone plate*, used in X-ray microscopy and spectroscopy.

There are also some gratings with discontinuous profiles, e.g., *grid* or *rod gratings* made of dielectric or metallic rods, the latter used as frequency filters in infrared, selective absorbers of solar energy, or in radar furtivity. In photolithography, chromium masks are made of rectangular rods (strips, sometimes called *Ronchi gratings*) and are used to cast their shadow onto a photoresist layer, which represents a combination of a rod grating and several plane layers.

Phase or *volume* gratings do not contain corrugated interfaces, but rather a plane layer with a periodic modulation of the refractive index. The result is that lamellar and rectangular rod gratings belong to both relief and phase grating types.

The next complication appears when going to two-dimensional geometry with a modulation of the index or corrugation of the surface made in two directions. The result is called a *crossed grating*. Such devices can be useful in solar absorption, beam splitting and memory storage. Their other name, *bi-grating*, is often confused with a one-dimensional grating having two different periods of modulation. A further step in increasing the complexity is periodicity in three dimensions, which brings us to 3-D *photonic crystals*.

Another problem that can be treated numerically using grating theories comes from the necessity to interpret the images of the photon scanning tunnelling microscope. When a grating surface is studied, it is necessary to analyze the near-field picture as diffracted by a periodic surface in presence of a single object, namely the tip of the scanning device (optical fiber). This structure can be modelled by substituting the single tip with a collection of periodic tips with a distance much larger than the grating period in order to avoid parasitic coupling. The geometry is then reduced to two gratings with different periods, one being a multiple of the other.

The diversity of grating problems and existing theories is nowadays so high that it is not obvious for an engineer or a scientist to know what method is the best suited to resolve the problem he encounters in his work. In order to try to help him in this aim, we first distuinguish between two kinds of grating problems. The first kind will be called "simple grating problems" and will concern relief gratings made with homogeneous isotropic materials, used in linear optics in the resonance domain. The modulated region will be limited to one or a couple (less than 3) nonseparable interfaces, plus a stack of plane interfaces, plus other separable modulated interfaces, if any. The second kind will be called "exotic situations" and will include bare echelles, multilevel and Dammann gratings ($\lambda/d \ll 1$), phase gratings, multilayer coated gratings and echelles, grating ruled on anisotropic materials, gratings used in nonlinear optics, crossed gratings, Bragg gratings and any kind of X-ray gratings. Photonic crystals are considered in a separate table.

Concerning simple grating problems, Table I.1 lists the relative merit of the theories confronted to the various types of grating problems. The criteria of simplicity, short computation time and universality (groove shape, TE and TM polarization, etc.) are added to the performance of computation in view of attributing the stars. The general impression is that if we except the low modulated case, a grating with high (but not infinite) conductivity will be analyzed through the integral theory or through the coordinate transformation method. For dielectric gratings, the differential theory will be preferred. The same criteria are used in order to construct Table I.2, which gives an idea of the versatility and universality of the theories when confronted to new problems. Table I.3 summarizes the experience in applying diffraction theories to one-, two- or three-dimensional photonic crystals.

Table I.1. Ability of Grating Theories to Deal with Common Gratings

Method / Grating types	Rayleigh	Differential	RCW	Modal	Integral	Coordinate transform	Conformal mapping
perfectly conducting gratings	*			***(*)	*****	*****	****
dielectric gratings	*	*****	****	****	****	*****	
real metals	*	***	*	***(*)	*****	*****	
deep dielectric gratings		*****	****	***	*****	*****	
deep metallic gratings		***	**	***	*****	*****	
binary, lamellar gratings		***	*****	*****	**		
dielectric coated gratings	*	***			***	*****	***
multilayer dielectric gratings	*	*****	*	*	***	***	
low modulated gratings	****	*****	*	*	*****	*****	

In order to illustrate how these tables can be used, let us go back to the grating problems described in the beginning of this section, and give for each of them our preference, keeping in mind that in most cases several theories could work. The metallic grating would be studied with the integral method; the transmission gratings with the differential theory as well as the dielectric coated grating for VUV region. The multilayer dielectric grating would be studied with the differential method, the lamellar grating couplers with the RCW one. The stack of gratings, as well as Bragg gratings and Bragg-Fresnel zone plates, would be studied with the differential method; the multilevel grating, with the integral one; the Dammann grating with the differential theory; the crossed grating with the coordinate transformation.

Table I.2. Ability of Grating Theories to Deal with Exotic Gratings

Grating types	Method	Rayleigh	Differential	RCW	Modal	Integral	Coordinate transform
echelles, echelettes	bare metallic echelle		*			*****	
	bare dielectric echelle		*****			*****	
	multilayer X-ray echelle		*****				
phase gratings, holograms			*****	*****			
multilayer	multilevel gratings		***	*****	****	*****	
	multilayer coated gratings		*****			**	*****
	etched multilayers (Bragg gratings)		*****	*****	*****	**	
Dammann gratings			****	*****	*****	****	
anisotropic, chiral, etc. media		****	*****		*	*	**
Bragg-Fresnel zones			*****	****	****	***	
crossed gratings		*	***	**	**	****	*****
nonlinear media	Second-harmonic generation (SHG) in dielectrics	*	****	***	***		****
	SHG in metals		***	**	**	*****	****
	Kerr effect		*****	****	***	****	****
X-ray domain		****(*)	*****	****		*	****

On top of that, phase gratings and holograms would be studied with the differential technique and grid gratings with the differential or integral theory depending on the fact the rods are dielectric or metallic. These examples give an idea of the choice that should be done for various problems encountered in linear optics.

Table I.3. Ability of Grating Theories to Deal with Photonic Crystals

Types	Method	Rayleigh	Differential	RCW	Modal	Integral	Coordinate transform
1-D	isotropic multilayer mirrors	*****	*	*	*	*	*
		analytic solution					
	anisotropic multilayer mirrors	*****	*****	*****	*****	*	*
2-D	metal, arbitrary form	*	***	*	*	*****	
	dielectric, arbitrary form	*	*****	***	***	*****	
	rectangular form	*	***	*****	*****	*****	
3-D	cubes, bricks	*	*****	*****	*****	*****	
	spheres	*	*****	*	*	*****	
	arbitrary forms	*	*****	*	*	*****	

I.1.4 The early years and the maturity of the differential theory of gratings

The late 1960s gave birth to both the integral and the differential theory of gratings [I.23, I.24], with the beginning of the scientific applications of the early computers. At that time, great theoretical efforts were made in France under the Atomic Energy Commission in nuclear physics and, in particular, in the study of diffusion of particles in nuclear potential [I.25]. Computers allowed to calculate the differential elastic cross-sections using integration by separation of variables of the radial equations derived from the Schrödinger equation. Despite the fact that this first effort [I.25] was called the "optical model," computers were not in regular use for optical community. However, the similarity between the Schrödinger and the Helmholtz equations raised hopes that the method having success in nuclear physics could also be successful in optics. The first attempts [I.24] led to results with accuracy too low to be used in the experiments and to be reproduced by forecoming studies. To better understand the difficulties, one of the authors, M. Nevière, returned back to perfectly conducting gratings [I.21] and introduced a conformal mapping that maps the grating surface to a plane. As soon as 1971, by using an accurate integration algorithm (the Numerov's algorithm), accurate and reliable results were obtained for the diffraction efficiencies of perfectly conducting gratings. The following efforts to extend the approach to gratings with zero [I.26] and finite [I.27] conductivity resulted in what can be called now the classical differential theory. The conformal mapping was abandoned because it introduced complications not necessary for solving the problem in the case of dielectric or real metallic gratings; the approach was significantly simplified and became much easily adaptable to various groove profiles, not to mention the significant reduction of the programming and computation times.

In its classical formulation, the differential method turned out to be very successful for dielectric gratings in both TE and TM polarizations, with, respectively, the electric field vector parallel or perpendicular to the grooves. Metallic gratings could also be treated, but if the grooves became deep enough, accurate results were only obtained in TE polarization. The simplicity of the method ensured its large applications and it became in regular use from X-rays [I.28] to millimeter waves.

On metallic gratings with high reflectivity, when used in TM polarization, a poor convergence was observed with respect to the number of spectral orders (the total number of Fourier components in the sum in Eq. (I.7) retained in the computations). An important remark is due here: When working with computers, one has to truncate the otherwise mathematically infinite series of the field expansion. The number of the terms retained in the series is called *truncation parameter*.

To worsen the problem, numerical instabilities appeared above some value of the truncation parameter for some groove depth and metal conductivity, which made the convergence test when increasing the truncation parameter impossible, and giving results violating energy balance and other accuracy criteria. This imposed severe limitations on the use of the differential method for metallic gratings in TM polarization. When first observed in 1974 [I.27], these instabilities were believed to be linked with the numerical integration process that is used in the differential theory and a lot of unpublished efforts were devoted to get rid of them by changing the integration algorithm, modifying the set of equations to be integrated and the sense of integration, using asymptotic expansion of the field, etc., but all in vain. Small progress was made in 1980 [I.29] by using two simultaneous counter-running integration processes, enlarging the numerical stability to twice deeper gratings than before. However, this is not enough to eliminate the problem, and the differential theory was believed to be severely limited in its application range.

This serious drawback had as a happy consequence the development of alternative approaches [I.20, I.22]. Since the numerical instabilities of the differential method were attributed to the numerical integration process, a special technique for rectangular profiles was developed, known as rigorous coupled wave theory [I.20], which avoided the numerical integration by using an eigenvalue technique. The RCW method was later extended to arbitrary profiles represented through a staircase approximation. However, it was observed that this method also suffers from bad convergence for arbitrary-shaped metallic gratings in TM polarization; the reasons for this behavior have only recently been found [I.30] and they limit the method to special profiles, rectangular or made by few rectangular steps, as is the situation with the classical modal method [I.16].

The eigenvalue-eigenvector technique is also utilized by another approach [I.22], which uses a non-conformal mapping of the grating surface onto a plane. The method is used with success for deep gratings, independently on the polarization and refractive index, but is not well adapted to profiles with sharp edges and discontinuities, although improvements have been found recently.

In the meantime, the search of improvement of the classical differential method continued, together with the search of the origin of its problems. Some authors even claimed that the propagation equations that are the basis of the theory could be wrong; fortunately, their arguments appeared to be wrong [I.31]. In fact, it was difficult to understand and remove the numerical instabilities, because they have two independent origins: (1) the numerical contamination due to growing (during the integration process) exponential terms associated with evanescent orders and (2) the poor convergence of the Fourier series of some of the field components, which are discontinuous in TM polarization. The poor convergence requires a larger set of Fourier components, thus higher evanescent orders, which aggravates the contamination due to the growing exponentials. The problem due to these exponentials is not limited to the TM polarization and metallic gratings, but in TE polarization and/or dielectric gratings the required number of Fourier components of the field is smaller and the problem due to the exponentials starts to be observed for much deeper grooves (i.e., longer integration distances).

After ten years of studies by many researchers, the first breakthrough was observed in 1990 concerning the first reason, the growing exponential terms during the integration. G. Tayeb [I.32] proposed to use an orthonormalization procedure during the integration process to avoid shrinking the dimensions of the Fourier space during the integration. Several other different approaches were proposed at that time [I.33-I.40], based on dividing the field components "propagating" upward and downward and treating them differently. These approaches will be discussed in detail further on. Among them, the so-called "S-matrix propagation algorithm" [I.38] seems to be the easiest one to implement and gives stable results when combined with the differential method [I.41].

The success of these algorithms when applied to both the classical differential method or to the RCW theory (and also to a multilayer stack of gratings treated by other numerical methods) was to completely eliminate the numerical contamination due to the growing exponential terms, even when the integration was made along thousands of wavelengths (in thick holograms or photonic crystals, for example). However, this only sharpens the second numerical problem, the slow convergence of the Fourier series for metallic gratings in TM polarization.

The second breakthrough started with the empirical discovery [I.42, I.43] of how to improve the convergence of the Fourier series for lamellar gratings by reformulation of the propagation equations, which constituted a great progress for the RCW method. However, when applied to sinusoidal profiles, this new formulation did not work better than the old one. In addition, it was not clear why the new one improves the convergence for lamellar gratings in TM polarization. The explanation was found by L. Li [I.44] in the convergence behavior of the *truncated* Fourier series of the product of two continuous or discontinuous functions. As a consequence of his work, the conclusion was that although the propagating equations of the classical differential theory were exact in the coordinate space and in the discrete Fourier space of infinite dimensions, they become ill-suited in a truncated Fourier space, a truncation indispensable during the numerical treatment. Li proposed appropriate factorization rules that were used to improve the equations when being truncated.

Using these factorization rules, in 2000 we established [I.45] a new set of propagation equations for arbitrary grating profiles, which ameliorated significantly the convergence rate of the differential method in TM polarization. The method is nowadays known as *the fast Fourier factorization* (FFF) method and has recently been extended [I.46] to arbitrary shaped periodic (with 1-, 2-, or 3-D periodicity) anisotropic and magnetic media. This last effort enabled the differential theory to be able to tackle a wide class of diffraction problems.

I.2 General properties of electromagnetic field in periodic media

Grating properties are taught in universities in the frame of scalar optics. Gratings are assumed as periodic sets of slits or small mirrors and their diffraction properties are analyzed with the Kirchhoff diffraction theory [I.47] in the Fraunhofer approximation. The directions of the diffraction orders are given by the Fraunhofer equation (I.5) and they correspond to the direction of interference maxima, represented by the so-called "interference function." The normalized diffraction intensity is then a product of the interference function by the normalized intensity function I_0 of the slit. According to the scalar theory, the intensity function varies with the angle of incidence and the wavelength much slowly than the interference function. Thus the diffraction intensity is found to consist of sharp peaks (due to the interference function), modulated by the intensity function, which depends on the groove shape. Such an approach cannot predict the complex behavior of grating efficiencies observed experimentally and, in particular, the grating anomalies known since R. Wood's observation a century ago [I.48]. Neither it could account for the variety of grating properties in the so-called "resonance domain," when the grating pitch is small enough to be comparable to the wavelength of light. Such gratings are widely accessible nowadays due to the spectacular progress in grating manufacturing since the middle of the 20th century, which became possible with the use of interferometric control and servo-control systems [I.49]. An evident influence of the polarization in the resonance domain was observed [I.50], although not for the first time, since Wood's anomalies showed strong polarization dependence half a century before [I.48]. However, it was necessary to widely expand the practical use of gratings after the 1950s to get opticians convinced of the necessity of electromagnetic, rather than scalar study of gratings, although Lord Rayleigh already tried to do this 50 years earlier. It was time to go back to Maxwell equations and their boundary conditions, instead of using the physical optics diffraction approximations.

I.2.1 Time-harmonic Maxwell equations

Most of the early and recent grating problems are linear, without time-frequency conversion, so that the different wavelengths diffract independently of each other. Nevertheless, the spacial part of Maxwell equations contains partial derivatives. Moreover, some of the electromagnetic field vector components are discontinuous across the grating surface, so that the first difficulty comes from the necessity of interpreting derivatives of discontinuous functions. Mathematical theory of distributions probably gives the most elegant way in treating this problem by presenting Maxwell equations in the sense of distributions [I.51]; such a representation will automatically include the boundary conditions on the surfaces of discontinuity. However, the distribution theory is far from being popular with the grating community. There are two basic approaches to avoid the distribution theory: the first one is to consider the field components separately in each media and then to apply the boundary conditions on the interfaces (e.g., the classical modal method); the second approach deals with the Fourier transform of the field components. This transform is represented by integrals over the groove period, defined as well for discontinuous functions as for continuous ones, although more slowly convergent.

The Cartesian coordinate system is suitable for most of the periodic structures, with any point M represented by its coordinate vector $\vec{r} = (x, y, z)$. The directions of the three axis are determined by their unit vectors $\vec{x}$, $\vec{y}$, and $\vec{z}$. Considering only time-harmonic fields with time dependence of the kind $\exp(-i\omega t)$, any vector $\vec{a}(\vec{r}, t)$ will be represented by its associated complex vector $\vec{A}$ such that $\vec{a}(\vec{r}, t) = \mathrm{Re}[\vec{A}(\vec{r})\ \exp(-i\omega t)]$. A similar definition applies to scalar fields. $\vec{E}$ and $\vec{H}$ denoting the complex vectors of electric and magnetic field, the time-harmonic Maxwell equations in linear optics take the form:

$$\mathrm{curl}\vec{E} = i\omega\mu\vec{H} \tag{I.8}$$

$$\mathrm{curl}\vec{H} = -i\omega\varepsilon\vec{E} \tag{I.9}$$

with ε and μ being the electric and magnetic permittivities. When projected on the coordinate axis, these equations lead to a set of six coupled partial differential equations:

$$\frac{\partial E_z}{\partial y} - \frac{\partial E_y}{\partial z} = i\omega\mu H_x \tag{I.10}$$

$$\frac{\partial E_x}{\partial z} - \frac{\partial E_z}{\partial x} = i\omega\mu H_y \tag{I.11}$$

$$\frac{\partial E_y}{\partial x} - \frac{\partial E_x}{\partial y} = i\omega\mu H_z \tag{I.12}$$

$$\frac{\partial H_z}{\partial y} - \frac{\partial H_y}{\partial z} = -i\omega\varepsilon E_x \tag{I.13}$$

$$\frac{\partial H_x}{\partial z} - \frac{\partial H_z}{\partial x} = -i\omega\varepsilon E_y \tag{I.14}$$

$$\frac{\partial H_y}{\partial x} - \frac{\partial H_x}{\partial y} = -i\omega\varepsilon E_z\,. \tag{I.15}$$

The next section demonstrates how to transform this set into two simpler sets.

I.2.2 Two fundamental polarizations

In many cases with special geometry, the set of Eqs. (I.10)-(I.15) can be simplified. The most common problem is the classical one-dimensional grating (Fig. I.1), with grooves parallel to the Oz axis and their shape given by the equation y = g(x) with period d and groove depth a. The grating surface separates the superstrate (with real refractive index ν_2) and the substrate (with real or complex refractive index ν_1). The magnetic permittivity is equal to μ in the entire space.

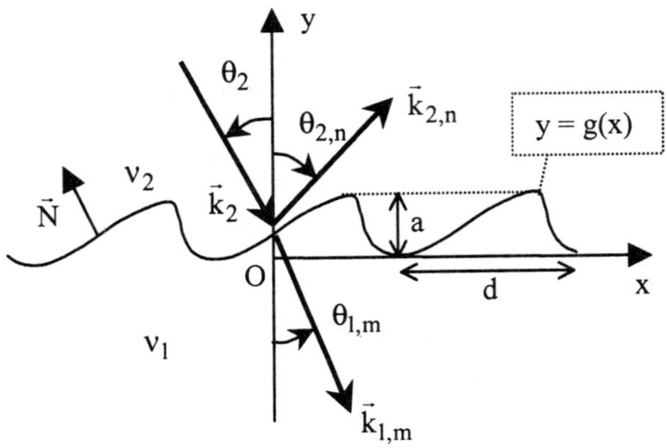

Fig. I.1. Schematic representation of a one-dimensional grating.

Usually gratings are used in in-plane mounts, i.e., with the incident direction lying in the plane perpendicular to the grooves. The diffraction orders are also lying in the same plane. The more general case of off-plane mount is called conical mounting, because the diffracted orders lie on a cone. The incident angle θ_2 is measured from the normal to the mean plane of the grating. The incident wavevector is expressed in the Oxyz coordinate system in the form $\vec{k}_2 = \nu_2 \frac{\omega}{c}(\sin\theta_2, -\cos\theta_2,\ 0)$, where c is the speed of light on vacuum. The incident electric field vector is:

$$\vec{E}^{(i)} = \vec{A}\exp(i\vec{k}_2.\vec{r}) = \vec{A}\exp[ik_2(x\sin\theta_2 - y\cos\theta_2)] \qquad \text{(I.16)}$$

where $k_2 = |\vec{k}_2|$ and $\vec{A}$ is the polarization vector. Since the incident plane wave and the geometry of the system are z-invariant, the set ((I.10)-(I.15) splits into two decoupled sets, Eqs. (I.10'), (I.11') and (I.15') with unknown functions H_x, H_y and E_z, and Eqs.(I.12'), (I.13') and (I.14') with unknown functions E_x, E_y and H_z:

$$\frac{\partial E_z}{\partial y} = i\omega\mu H_x \tag{I.10'}$$

$$-\frac{\partial E_z}{\partial x} = i\omega\mu H_y \tag{I.11'}$$

$$\frac{\partial E_y}{\partial x} - \frac{\partial E_x}{\partial y} = i\omega\mu H_z \tag{I.12'}$$

$$\frac{\partial H_z}{\partial y} = -i\omega\varepsilon E_x \tag{I.13'}$$

$$-\frac{\partial H_z}{\partial x} = -i\omega\varepsilon E_y \tag{I.14'}$$

$$\frac{\partial H_y}{\partial x} - \frac{\partial H_x}{\partial y} = -i\omega\varepsilon E_z\,. \tag{I.15'}$$

Thus the diffraction of an arbitrary polarized light can be studied by resolving two independent cases of polarization, called fundamental ones, denoted by TE (Eqs. (I.10'), (I.11') and (I.15')) and TM (Eqs. (I.12'), (I.13') and (I.14')). In TE polarization the polarization vectors of the incident wave $\vec{A}$ and of all the diffraction orders are parallel to the grooves, i.e., the z-axis. In TM polarization the same is valid for the magnetic field component. In addition, one of the equations in each of the sets can be eliminated by expressing the y-components of the field through the x and z-components. In TE polarization one arrives at a set of two equations and two unknown functions:

$$\frac{\partial \tilde{E}_z}{\partial y} = H_x \tag{I.17}$$

$$\frac{\partial H_x}{\partial y} = -k^2(x,y)\tilde{E}_z - \frac{\partial^2 \tilde{E}_z}{\partial x^2} \tag{I.18}$$

by using the simplified notations $k^2(x,y) = \omega^2\mu\varepsilon(x,y)$ and $\tilde{E}_z = E_z / i\omega\mu$. Substituting Eq. (I.17) into (I.18) leads to a single propagation equation:

$$\Delta E_z(x,y) + k^2(x,y)E_z(x,y) = 0\,. \tag{I.19}$$

This was the basic equation in the early works on the classical differential method [I.27], and it was believed that the TE case of diffraction reduces to a scalar problem [I.51]. However, when comparing Eqs. (I.17) and (I.18) with (I.19), it appears that during the integration of the latter one needs to determine the two components of the electromagnetic field, as far as the first derivative in y of E_z is proportional to H_x. This is why in later developments of the differential method the first-order set (I.17), (I.18) was sometimes used instead of (I.19).

TM polarization is treated in the same way to produce another set of two equations:

$$\frac{\partial H_z}{\partial y} = k^2(x,y)\tilde{E}_x \tag{I.20}$$

$$\frac{\partial \tilde{E}_x}{\partial y} = -H_z - \frac{\partial}{\partial x}\left(\frac{1}{k^2}\frac{\partial H_z}{\partial x}\right), \tag{I.21}$$

where $\tilde{E}_x = E_x / i\omega\mu$. Again, the problem can be reduced to a single second-order propagation equation:

$$\operatorname{div}\left[\frac{1}{k^2(x,y)}\operatorname{grad} H_z(x,y)\right] + H_z(x,y) = 0 \tag{I.22}$$

and similar comments apply as for TE polarization.

I.2.3 Pseudoperiodicity of the field

The existence and the uniqueness of the solution of the diffraction problem is a difficult mathematical problem that we do not intend to discuss here. The interested reader can refer to [I.11] and [I.52]. We shall simply assume by physical evidence that the solution exists and that it is unique. With this in hand and staying in linear optics, the grating can be associated to a linear operator $\Re$ that transforms the incident field $u^{(i)}$ into a total field u:

$$u(x,y) = \Re\, u^{(i)}(x,y) \tag{I.23}$$

And indeed, if u denotes any component of the set $(\vec{E},\vec{H})$, any solution u of Eqs. (I.10)-(I.15) when multiplied by a constant C is again a solution Thus, if the incident part of the total field is multiplied by C, the total field is multiplied by the same constant:

$$C\,u(x,y) = \Re\,[C\,u^{(i)}(x,y)\,]. \qquad \text{(I.23')}$$

On the other hand, the operator $\Re$ is invariant with respect to a translation with the grating period d along the x-axis, since the grating surface is periodic and infinite. Such translations (changing x into x+d) multiply the incident wave $u^{(i)}$ given by Eq. (I.16), by a phase factor $\exp(i\alpha_0 d)$, with $\alpha_0 = k_2 \sin\theta_2$:

$$u^{(i)}(x+d,y) = C\,u^{(i)}(x,y) \equiv \exp(i\alpha_0 d)\,u^{(i)}(x,y). \qquad \text{(I.24)}$$

A function satisfying Eq. (I.24) is said to be pseudoperiodic in x. The linearity of the grating operator, Eq. (I.23'), then implies:

$$C\ \ u(x,y) = \Re[u^{(i)}(x+d,y)]\ . \qquad \text{(I.25)}$$

On the other hand, Eq. (I.23) directly gives for the translated field:

$$u(x+d,y) = \Re\,u^{(i)}(x+d,y) \qquad \text{(I.26)}$$

wherefrom we can obtain the relation:

$$u(x+d,y) = C\,u(x,y) \equiv \exp(i\alpha_0 d)\,u(x,y). \qquad \text{(I.27)}$$

This shows that the total field is also pseudoperiodic in x with the grating period d.

I.2.4 Fourier expansion of the field

The pseudoperiodicity of the field, established in the previous section, ensures that the function

$$v(x,y) = \exp(-i\alpha_0 x)\,u(x,y)$$

is strictly periodic in x with a period d, and that it can be represented by its Fourier series with its Fourier components $u_n(y)$:

$$v(x,y) = \sum_{n=-\infty}^{\infty} u_n(y)\exp(inKx) \qquad \text{(I.28)}$$

where $K = 2\pi/d$ is the grating wavenumber. The last two equations enable us to derive an expression for the pseudoperiodic function u(x,y):

$$u(x,y) = \exp(i\alpha_0 x) \sum_{n=-\infty}^{\infty} u_n(y) \exp(inKx) \tag{I.29}$$

or

$$u(x,y) = \sum_{n=-\infty}^{\infty} u_n(y) \exp(i\alpha_n x) \tag{I.29'}$$

with

$$\alpha_n = \alpha_0 + nK \tag{I.29"}$$

The representation (I.29) is called a pseudo-Fourier series, corresponding to the Fourier series for the periodic functions. Further on, the qualifier "pseudo" will sometimes be omitted for simplicity.

The aim of any diffraction theory is to determine the functions $u_n(y)$, which are linked to the diffraction orders, propagating or evanescent in the superstrate (and in the substrate, if any). Fortunately, outside the modulated region, u_n can be represented in a simple form, using the so-called "Rayleigh expansion."

I.2.5 Rayleigh expansion

In any homogeneous region, k(x,y) is constant, proportional to the refractive index ν, $k \equiv k_j = \nu_j \omega / c$, j = 1, 2. Then the two propagation equations (I.19) and (I.22) are reduced to the Helmoltz equation:

$$\Delta u(x,y) + k_j^2 u(x,y) = 0 \tag{I.30}$$

where u(x,y) is the only nonzero z-component of the field, depending on the polarization. The grating diffraction problem in the majority of cases assumes that the cladding and the substrate are homogeneous. We can then distinguish three different regions in the space: the upper (y > a) and the lower (y < 0) homogeneous region and the modulated region. In the homogeneous regions, introducing the Fourier expansion (I.29) into Eq. (I.30) leads to

$$\exp(i\alpha_0 x) \sum_{n=-\infty}^{\infty} \left[\frac{d^2}{dy^2} + (k_j^2 - \alpha_n^2) \right] u_n(y) \exp(inKx) = 0 \tag{I.31}$$

which represents a Fourier expansion of a null function and implies that all the Fourier coefficients are null:

$$\left[\frac{d^2}{dy^2}+(k_j^2-\alpha_n^2)\right]u_n(y)=0 \quad . \tag{I.32}$$

This equation can easily be solved in terms of complex exponential functions as soon as the square root $\beta_n^{(j)}$ of $(k_j^2-\alpha_n^2)$ is defined. This requires a special attention when the refractive index is complex. Usually the cladding (j = 2) is lossless, i.e., for y > a, k_2 is real. Then, by definition,

$$\begin{aligned}\beta_n^{(2)}&=\sqrt{k_2^2-\alpha_n^2}, \text{ for } k_2^2-\alpha_n^2\geq 0,\\ \beta_n^{(2)}&=i\sqrt{\alpha_n^2-k_2^2}, \text{ for } k_2^2-\alpha_n^2<0,\end{aligned} \tag{I.33}$$

and the solution of Eq. (I.32) can be written as:

$$u_n(y)=A_n^{(2)}\exp(-i\beta_n^{(2)}y)+B_n^{(2)}\exp(i\beta_n^{(2)}y)$$

with $A_n^{(2)}$ and $B_n^{(2)}$ being the integration constants. The total field is then given by:

$$u(x,y)=\sum_{n=-\infty}^{+\infty}A_n^{(2)}\exp(i\alpha_n x-i\beta_n^{(2)}y)+\sum_{n=-\infty}^{+\infty}B_n^{(2)}\exp(i\alpha_n x+i\beta_n^{(2)}y) \tag{I.34}$$

which represents a series of propagating and evanescent plane waves. The two parts of u(x,y), denoted as

$$u^-=\sum_{n=-\infty}^{+\infty}A_n^{(2)}\exp(i\alpha_n x-i\beta_n^{(2)}y) \text{ and } u^+=\sum_{n=-\infty}^{+\infty}B_n^{(2)}\exp(i\alpha_n x+i\beta_n^{(2)}y)$$

represent two different sets of solutions. The first part u^- consists of a finite number of downgoing (incident) propagating plane waves (with $k_2^2-\alpha_n^2>0$) and an infinite number of evanescent waves $(k_2^2-\alpha_n^2<0)$ that diverge at $y\to+\infty$. The second part u^+ is a sum of a finite number of outgoing propagating plane waves and an infinite number of evanescent waves that tend to zero when $y\to+\infty$. In fact, these are the diffracted orders propagating or evanescent in the cladding. This means that once the incidence angle has been fixed, the Maxwell equations predict that the total field can be represented on the $\exp(i\alpha_n x)$ basis, i.e., as a sum of diffracted (propagating or evanescent) orders u^+ and incident (propagating or antievanescent) orders u^-. However, the diffraction problem will have an unique solution only if the incident field is completely specified. Excluding the unbounded solutions (diverging when $y \to +\infty$) and keeping a single incident wave with n = 0, Eq. (I.34) takes the form, known as a Rayleigh expansion:

$$u(x,y) = A_0^{(2)} \exp(i\alpha_0 x - i\beta_0^{(2)} y) + \sum_{n=-\infty}^{+\infty} B_n^{(2)} \exp(i\alpha_n x + i\beta_n^{(2)} y) \qquad (I.35)$$

with $B_n^{(2)}$ being the Rayleigh coefficients or diffracted order amplitudes. The propagating diffracted orders are sometimes called *spectral orders*.

Below the groove bottom ($y < 0$), the refractive index can be real or complex (losses, metals). If ν_1 is complex, the sign of the square root of the complex number $k_1^2 - \alpha_n^2$ is not uniquely defined. To avoid that, it is necessary to introduce a cut in the complex plane. The physical requirement is the diffracted orders to be bounded when $y \to +\infty$, which leads to

$$\beta_n^{(1)} = \sqrt{k_1^2 - \alpha_n^2} \quad \text{with} \quad \mathrm{Im}(\beta_n^{(1)}) \geq 0 . \qquad (I.36)$$

Sometimes, it is more convenient during the numerical calculations to use another cut, which does not coincide with the real β-axis, for example, defined by the equation $\mathrm{Re}(\beta_n^{(1)}) + \mathrm{Im}(\beta_n^{(1)}) \geq 0$. However, it is always necessary to have in mind which are the physical incident and diffracted waves, distinguished by the sign of $\mathrm{Im}(\beta_n^{(1)})$.

In the substrate the solution of Eq. (I.32) is:

$$u(x,y) = \sum_{n=-\infty}^{+\infty} A_n^{(1)} \exp(i\alpha_n x - i\beta_n^{(1)} y) + \sum_{n=-\infty}^{+\infty} B_n^{(1)} \exp(i\alpha_n x + i\beta_n^{(1)} y) \qquad (I.37)$$

with the second term containing waves propagating upward when $\beta_n^{(1)}$ is real. They correspond to the waves incident on the grating from the substrate side. If $\beta_n^{(1)}$ is complex, these waves diverge when $y \to -\infty$ and must be excluded from the solution.

Without incidence from the substrate, Eq. (I.37) reduces to the Rayleigh expansion of the transmitted field:

$$u(x,y) = \sum_{n=-\infty}^{+\infty} A_n^{(1)} \exp(i\alpha_n x - i\beta_n^{(1)} y) \quad . \qquad (I.38)$$

I.2.6 Fraunhofer equation

As soon as the x- and y-components α_n and $\beta_n^{(j)}$ (j = 1 or 2) of the diffracted order wavevectors are known, the directions of propagation are perfectly determined. This means that Eqs. (I.35) and (I.38) must contain the famous Fraunhofer equation, Eq. (I.5), for both reflected and transmitted orders. And indeed, if we consider at first the cladding, the x-component α_n of the n-th diffracted order wavevector is given by $k_{2,x,n} = k_2 \sin\theta_{2,n}$, where $\theta_{2,n}$ is the angle of diffraction between the y-axis and the direction of diffraction, Fig. I.1. On the other hand, according to Eq. (I.29"), $\alpha_n = \alpha_0 + nK$. If λ_0 and λ_j designate, respectively, the wavelength in vacuum, and in the j-th medium, the Fraunhofer equation in the cladding is straightly derived:

$$\sin\theta_{2,n} = \sin\theta_2 + n\frac{\lambda_2}{d} = \sin\theta_2 + n\frac{\lambda_0}{\nu_2 d}. \tag{I.39}$$

Concerning the transmission orders, the angles of diffractions $\theta_{1,n}$ are defined as the angles between the negative y-axis direction and the propagation direction (Fig. I.1, where $\theta_{1,n} > 0$). It is worth noting that this convention is consistent with the choice for the reflection orders, because if the angles of diffraction of the transmission orders are defined as the angles between the direction of propagation and the *positive* y-axis (as it is for the reflection orders), the transmission order angles will change into $\pi - \theta_{1,n}$, having the same sinus as $\theta_{1,n}$.

Similar calculations as for the reflected orders lead to:

$$k_{1,x,n} = k_1 \sin\theta_{1,n} = \alpha_n = \alpha_0 + nK = k_2 \sin\theta_2 + nK,$$

which leads to

$$\nu_1 \sin\theta_{1,n} = \nu_2 \sin\theta_2 + n\frac{\lambda_0}{d}. \tag{I.40}$$

For the zeroth order (n = 0), this equation is reduced to Snell's law of refraction.

I.2.7 Grating efficiency

In spectroscopic applications, the main interesting properties are the dispersion and the efficiencies of gratings. The dispersion is studied in all classical textbooks [I.47] and is derived by differentiating the Fraunhofer equation with respect to wavelength. The efficiency in the reflected or transmitted order is defined as the modulus of the ratio of the flux through the same surface parallel to the mean plane of the grating, of the Poynting vector of that order and of the Poynting vector of the incident wave. For defining the reflected efficiencies $e_n^{(r)}$, we consider a rectangle $BCC'B'$ located in a plane $y = a' > a$, whose side CB (see Fig. I.2) has length d and the side CC', parallel to the z-axis, has unit length.

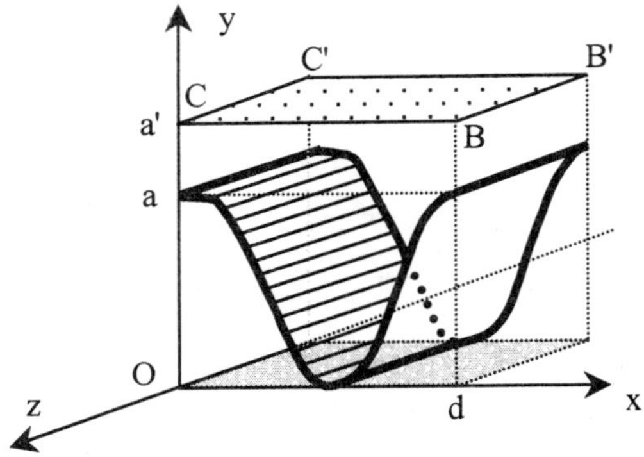

Fig. I.2. A period of the grating.

From Rayleigh expansion (I.35), the field corresponding to the nth order is a plane wave for which the simple relation $|\vec{H}| = |\vec{E}|/Z_2$, with Z_2 being the impedance of the superstrate, holds. As a result, the corresponding complex Poynting vector $\vec{P}$, equal to $\vec{E}\times\vec{H}^*/2$, where the star denotes the complex conjugate, is independent of x and y and simply reduces to:

$$|\vec{P}| = B_n^{(2)} B_n^{(2)*} \eta_2, \tag{I.41}$$

where $\eta_2 = 1/(2Z_2)$ for TE polarization and $\eta_1 = 2Z_2$ for TM case. The flux $\Phi_n^{(2)}$ of the Poynting vector is defined by $\Phi_n^{(2)} = \int_0^d \int_0^1 \vec{P}\cdot\hat{y}\,dx\,dy$ and is then given by:

$$\Phi_n^{(2)} = B_n^{(2)} B_n^{(2)*} \cos(\theta_{2,n})\eta_2 d.$$

For the incident wave the flux $\Phi^{(i)}$ can be obtained through a similar calculation, and we get:

$$\left|\Phi^{(i)}\right| = A_0^{(2)} A_0^{(2)*} \cos(\theta_2)\eta_2 d$$

As a result, $e_n^{(r)} = \dfrac{B_n^{(2)} B_n^{(2)*} \cos\theta_{2,n}}{A_0^{(2)} A_0^{(2)*} \cos\theta}$. If we remark that $\beta_n^{(2)} = k_2 \cos\theta_{2,n}$, this expression is equivalent to $e_n^{(r)} = \dfrac{B_n^{(1)} B_n^{(1)*} \beta_n^{(2)}}{A_0^{(2)} A_0^{(2)*} \beta_0^{(2)}}$. In linear optics, it is always possible to choose unit amplitude for the incident field: $A_0^{(2)} = 1\text{V/m}$ or 1A/m, depending on polarization. Thus, in many works or books [I.51], the latter expression is simplified into:

$$e_n^{(r)} = B_n^{(1)} B_n^{(1)*} \frac{\beta_n^{(2)}}{\beta_0^{(2)}}. \tag{I.42}$$

Similar calculations are done for the flux of the Poynting vector of the nth transmitted order through a rectangle with sides d and unity, located in the substrate ($y \leq 0$), whose impedance is Z_1. Defining $\eta_1 = 1/(2Z_1)$ in TE polarization and $\eta_1 = 2Z_1$ in TM polarization, we now obtain, for the flux $\Phi_n^{(1)}$ of the Poynting vector:

$$\left|\Phi_n^{(1)}\right| = A_n^{(1)} A_n^{(1)*} \cos(\theta_{1,n})\eta_1 d.$$

The corresponding transmitted efficiencies are:

$$e_n^{(t)} = \frac{A_n^{(1)} A_n^{(1)*} \cos(\theta_{1,n})\eta_1 d}{A_0^{(2)} A_0^{(2)} \cos(\theta_2)\eta_2 d}, \tag{I.43}$$

and we have to distinguish between the two cases of polarizations.

For TE polarization $\eta_2 = 2\nu_2 Z_0$, $\eta_1 = 2\nu_1 Z_0$, where Z_0 is the vacuum impedance, and Eq. (I.43) simplifies into

$$e_n^{(t)} = \frac{A_n^{(1)} A_n^{(1)*} \beta_n^{(1)}}{A_0^{(2)} A_0^{(2)} \beta_0^{(2)}} = A_n^{(1)} A_n^{(1)*} \frac{\beta_n^{(1)}}{\beta_0^{(2)}}, \tag{I.44}$$

similar to Eq. (I.42). For TM polarization $\eta_2 = 1/(2\nu_2 Z_0)$, $\eta_1 = 1/(2\nu_1 Z_0)$, and we get:

$$e_n^{(t)} = A_n^{(1)} A_n^{(1)*} \frac{\beta_n^{(1)}}{\beta_0^{(2)}} \left(\frac{\nu_2}{\nu_1} \right)^2 . \tag{I.45}$$

I.2.8 Energy-balance criterion

For lossless gratings, e.g., perfectly conducting metallic gratings and lossless dielectric gratings, the conservation of energy implies that the incident energy is totally distributed into the diffracted orders. For perfectly conducting gratings, if we denote by $U^{(2)}$ the set of values of subscript n for which the reflected orders are propagating, the result reads:

$$\sum_{n \in U^{(2)}} e_n^{(r)} = 1 \tag{I.46}$$

A mathematical demonstration of this intuitive result can be found in [I.51].

For lossless dielectric gratings, if $U^{(1)}$ is the set of values of n for which the orders are propagating in the substrate, we obtain:

$$\sum_{n \in U^{(2)}} e_n^{(r)} + \sum_{n \in U^{(1)}} e_n^{(t)} = 1. \tag{I.47}$$

For lossy materials like real metals, the difference between unity and the left-hand member in Eq. (I.47) gives the percentage of energy dissipated in the grating material.

I.2.9 Reciprocity

The concept of reciprocity is a basic law of electromagnetism, which applies from the elementary theory of networks in electrokinetics to antennas theory [I.53]. Roughly speaking, it states that the intensity of an electromagnetic signal is not affected by an interchange between the source and the detector. Applied to diffraction gratings, the theorem can be formulated in the following way.

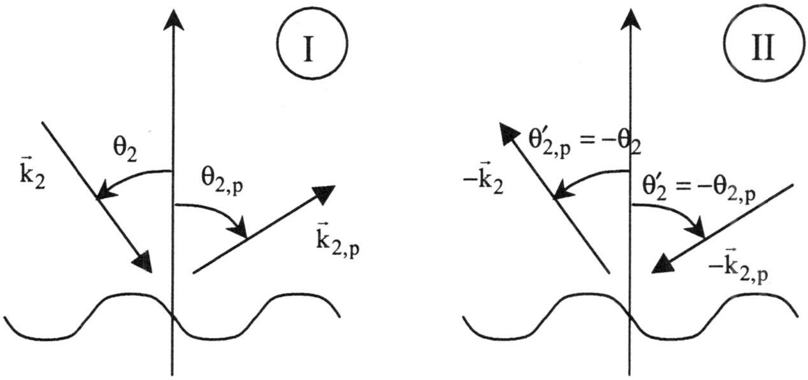

Fig. I.3. Illustration of the incidence conditions that lead to reciprocity relations.

Let us consider the two situations (I and II) illustrated in Fig. I.3. First, a linearly polarized incident plane wave falls on a grating under incidence θ_2 and with an incident wavevector $\vec{k}_2$. It generates several orders diffracted in reflection; the pth order propagates with wavevector $\vec{k}_{2,p}$ in a direction $\theta_{2,p}$ which is given by the Fraunhofer equation (I.39) $\sin\theta_{2,p} = \sin\theta_2 + p\lambda_2/d$ and has an efficiency $e_p^{(r)}$. In the second step (II), we reverse the sense of light propagation, which means that the incident field now has wavevector $-\vec{k}_{2,p}$ and the same polarization as in I; as a result, it falls under an incidence angle $\theta'_2 = -\theta_{2,p}$, due to the difference in orientation of angles of incidence and diffraction. It is easy to verify that the pth diffracted order propagates in a direction opposite to $\vec{k}_2$, denoted by angle $\theta'_{2,p}$ which is given by

$$\sin\theta'_{2,p} = \sin\theta'_2 + p\lambda_2/d = -\sin\theta_{2,p} + p\lambda_2/d .$$

Thus we obtain

$$\sin\theta'_{2,p} = -(\sin\theta_2 + p\lambda_2/d) + p\lambda_2/d = -\sin\theta_2 .$$

The reciprocity theorem states that the efficiency $e_p^{(r)\prime}$ of the pth order in situation II is equal to $e_p^{(r)}$ of situation I. Similar relations holds for gratings used in transmission [I.54, I.55]. Thus the reciprocity theorem clearly generalizes what it called in classical textbooks [I.47] the principle of reverse propagation of light. More general relations have been established on the Rayleigh coefficients of evanescent orders, for which the corresponding efficiencies vanish [I.56].

A simple and immediate consequence of the reciprocity theorem is that the efficiency in the 0th order is unchanged when the grating is rotated by 180° around its normal. This result, obvious for symmetrical grating profile, is not intuitive in the general case of an arbitrary groove shape.

References

I.1. D. Rittenhouse, "An optical problem proposed by F. Hopkinson and solved," J. Am. Phil. Soc. 201, 202-206, 1786

I.2. M. C. Hutley, *Diffraction Gratings,* (Academic Press, London, 1982)

I.3. E. G. Loewen and E. Popov, *Diffraction Gratings and Applications* (Marcel Dekker, Inc., New York, 1997)

I.4. M. Nevière, E. Popov, R. Reinisch, G. Vitrant, *Electromagnetic Resonances in Nonlinear Optics*, (Gordon and Breach Science Publishers, Amsterdam, 2000)

I.5. E. Yablonovitch, "Inhibited spontaneous emission in solid-state physics and electronics," Phys. Rev. Lett. 58, 2059-2062, 1987

"Photonic crystals," J. Modern Opt., 41,173-194, 1994.

I.6. T. Young, "On the theory of light and colors," Phil. Trans. II, 399-408, 1803

I.7. J. Fraunhofer, "Kurtzer Bericht von the Resultaten neuerer Versuche über die Gesetze des Lichtes, und die Theorie derselbem," Gilberts Ann. Phys. 74, 337-378, 1823

I.8. J. Fraunhofer, "Über die Brechbarkeit des electrishen Lichts," K. Acad. d. Wiss. zu München, April-June 1824, pp.61-62

I.9. R. Petit, ed., *Electomagnetic Theory of Gratings* (Springer-Verlag, Berlin, 1980)

I.10. O. M. Lord Rayleigh, "On the dynamical theory of gratings," Proc. Royal Soc. (London) A79, 399-416,1907

I.11. M. Cadilhac, "Some mathematical aspects of the grating theory," in *Electromagnetic Theory of Gratings*, R. Petit ed. (Springer-Verlag Berlin, 1980)

I.12. E. G. Loewen and E. Popov, *Diffraction Gratings and Applications* (Marcel Deker, Inc., New York, 1997), ch.10

I.13. M. Nevière and E. Popov, "Grating electromagnetic theory 'User guide'," J. Imaging Sci. Technol. 41, 315-323, 1997

I.14a. R. Petit, "Contribution à l'étude de la diffraction par un réseau métallique," Rev. Opt. 45, 249-276, 1966

I.14b. A. Wirgin, Thèse A.O. 1429, Faculté des Sciences d'Orsay, France, 1967

I.14c. A. Wirgin, "Selected papers from the URSI symp.," Alta Freq. 38, 327-338, 1969

I.14d. J. Pavajeau and J. Bousquet, "Diffraction par un réseau conducteur nouvelle méthode de résolution," Opt. Acta 17, 469-478, 1970

I.15. D. Maystre, "Sur la diffraction d'une onde plane par un réseau métallicque de conductivité finie," Opt. Commun. 6, 50-54, 1972

I.16. C. Botten, M. S. Craig, R. C. McPhedran, J. L. Adams, and J. R. Andrewartha, "The dielectric lamellar diffraction grating," Opt. Acta 28, 413-428, 1981.

"The finitely conducting lamellar diffraction grating," Opt. Acta 28, 1087-1102, 1981

I.17a. J. R. Andrewartha, G. H. Derrick, and R. C. McPhedran, "A general modal theory for reflection gratings," Opt. Acta 28, 1501-1516, 1981

I.17b. J. Y. Suratteau, M. Cadilhac, and R. Petit, "Sur la détermination numerique des efficacités de certain réseaux diélectriques profonds," J. Optics (Paris), 14, 273-282, 1983

I.18a. C. B. Burckhardt, "Diffraction of a plane wave at a sinusoidally stratified dielectric grating," J. Opt. Soc. Am. 56, 1502-1509, 1966

I.18b. S. T. Peng, T. Tamir, and H. L. Bertoni, "Theory of periodic dielectric waveguides," IEEE Trans. Microwave Theory and Techn. MTT-23, 123-133, 1975

I.19. E. I. Krupitsky and B. C. Chernov, "Rigorous analysis of arbitrary slanted volume holographic gratings," (in Russian), Proc. IX All-Union School of Holography, Leningrad, 1977, pp 84-95

I.20. M. G. Moharam and T. K. Gaylord, "Rigorous coupled-wave analysis of planar-grating diffraction," J. Opt. Soc. Am. 71, 811-818, 1981

"Rigorous coupled-wave analysis of dielectric surface-relief gratings," J. Opt. Soc. Am. 72, 1385-1392, 1982

I.21a. M. Nevière, M. Cadilhac, and R. Petit, "Applications of conformal mapping to the diffraction of electromagnetic waves by grating," IEEE Trans. Ant. Propag. AP-21, 37-46, 1973

I.21b. R. A. Depine and J. M. Simon, "Diffraction grating efficiencies. Conformal mapping method for a good real conductor," Optica Acta, 29, 1459, 1982

I.22a. J. Chandeson, D. Maystre, and G. Raoult, "A new theoretical method for diffraction gratings and its numerical application," J. Optics (Paris) 11, 235-241, 1980

I.22b. J. Chandeson, M. T. Dupuis, G. Gornet, and D. Maystre, "Multicoated gratings: a differential formalism applicable in the entire optical region," J. Opt. Soc. Am. 72, 839-846, 1982

I.23. R. Petit, "Diffraction d'une onde plane par une réseau métalique," Rev. Opt. 45, 353-370, 1966

I.24. G. Cerutti-Maori, R. Petit, and M. Cadilhac, "Etude Numérique du champ diffracté par un réseau," C. R. Ac. Sc. Paris 268, 1060-1063, 1969

I.25. M. A. Melkanoff, T. Sawada, and J. Raynal, "Nuclear optical model calculations," in *Methods in Computationnal Physics*, 1, 1-80, (Academic Press, New York, 1966)

I.26a. M. Nevière, R. Petit, and M. Cadilhac, "About the theory of optical grating coupler-waveguide systems," Opt. Commun. 8, 113-117, 1973

I.26b. M. Nevière, P. Vincent, R. Petit, and M. Cadilhac, "Systematic study of resonances of holographic thin film couplers," Opt. Commun. 9, 48-53, 1973

I.27a. M. Nevière, G. Cerutti-Maori, and M. Cadilhac, "Sur une nouvelle méthode de résolution du problème de la diffraction d'une onde plane par un réseau infiniment conducteur," Opt. Commun. 3, 48-52, 1971

I.27b. M. Nevière, P. Vincent, and R. Petit, "Sur la théorie du réseau conducteur et ses applications à l'optique," Nouv. Rev. Opt. 5, 65-77, 1974

I.27c. P. Vincent, "Differential methods," in *Electromagnetic Theory of Gratings*, R. Petit, ed. (Springer-Verlag Berlin, 1980), ch. 4

I.28a. M. Nevière and J. Flamand, "Electromagnetic theory as it applies to X-rays and XUV gratings," Nucl. Instrum. Methods 172, 273-279, 1980

I.28b. M. Nevière, J. Flamand, and J. M. Lerner, "Optimization of gratings for soft X-rays monochromators," Nucl. Instrum. Methods 195, 183-189, 1982

I.28c. A. J. F. den Boggende, P. A. J. de Korte, P. H. Videler, A. C. Brinkman, S. M. Kahn, W. W. Craig, C. J. Hailey, and M. Nevière, "Efficiencies of X-ray reflection gratings," in *X-rays Instruments, Multilayers and Sources*, J. F. Marshall, ed. Proc. SPIE 982, 283-289, 1988

I.29. P. Vincent, "New improvement of the differential formalism for high-modulated grating," in *Periodic Structures, Gratings, Moiré Patterns and Diffraction Phenomena*, C. H. Chi, E. G. Loewen, and C. L. O'Bryan III, eds. Proc. SPIE 240, 147-154, 1980

I.30. E. Popov, M. Nevière, B. Gralak, and G. Tayeb, "Staircase approximation validity for arbitrary shaped gratings," J. Opt. Soc. Am. A 19, 33-42, 2002

I.31. M. Nevière and P. Vincent, "Differential theory of gratings: answer to an objection on its validity for TM polarization," J. Opt. Soc. Am. A 5, 1522-1524, 1988

I.32. G. Tayeb, Thèse "Contribution à l'étude de la diffraction des ondes électromagnétiques par des réseaux. Reflexion sur les méthodes existantes et sur leur extension aux milieux anisotropes," Université Aix-Marseille III (1990)

I.33. D. M. Pai and K. A. Awada, "Analysis of dielectric gratings of arbitrary profiles and thicknesses," J. Opt. Soc. Am. A 8, 755-762, 1991

I.34. L. Li, "Multilayer modal method for diffraction gratings of arbitrary profile, depth, and permittivity," J. Opt. Soc. Am. A 10, 2581-2593, 1993

I.35. M. Nevière and F. Montiel, "Deep gratings: a combination of the differential theory and the multiple reflection series," Opt. Commun. 108, 1-7, 1994

I.36. F. Montiel and M. Nevière, "Differential theory of gratings: extention to deep gratings of arbitrary profile and permittivity through the R-matrix propagation algorithm," J. Opt. Soc. Am. A 11, 3241-3250, 1994

I.37. N. Chateau and J. P. Hugonin, "Algorithm for the rigorous coupled-wave analysis of grating diffraction," J. Opt. Soc. Am. A 11, 1321-1331, 1994

I.38. L. Li, "Formulation and comparison of two recursive matrix algorithms for modeling layered diffraction gratings," J. Opt. Soc. Am. A 13, 1024-1035, 1996

I.39. M. G. Moharam, D. A. Pommet, E. B. Grann, and T. K. Gaylord, "Stable implementation of the rigorous coupled-wave analysis for surface-relief gratings: enhanced transmission matrix approach," J. Opt. Soc. Am. A 12, 1077-1086, 1995

I.40. R. H. Morf, "Exponentially convergent and numerically efficient solution of Maxwell equations for lamellar gratings," J. Opt. Soc. Am. A 12, 1043-1056, 1995

I.41. F. Montiel, M. Nevière, and P. Peyrot, "Waveduide confinement of Cerenkov second-harmonic generation through a graded-index grating coupler: electromagnetic optimization," J. Mod. Opt. 45, 2169-2186, 1988

I.42. P. Lalanne and G. M. Morris, "Highly improved convergence of the coupled-wave method for TM polarization," J. Opt. Soc. Am. A 13, 779-784, 1996

I.43. G. Granet and B. Guizal, "Efficient implementation of the coupled-wave method for metallic gratings in TM polarization," J. Opt. Soc. Am. A 13, 1019-1023, 1996

I.44. L. Li, "Use of Fourier series in the analysis of discontinuous periodic structures," J. Opt. Soc. Am. A 13, 1870-1876, 1996

I.45. E. Popov and M. Nevière, "Grating theory: new equations in Fourier space leading to fast converging results for TM polarization," J. Opt. Soc. Am. A 17, 1773-1784, 2000

I.46. E. Popov and M. Nevière, "Maxwell equations in Fourier space: fast converging formulation for diffraction by arbitrary shaped, periodic, anisotropic media," J. Opt. Soc. Am. A 18, 2886-2894, 2001

I.47. M. Born and E. Wolf, *Principles of Optics*, (Pergamon Press, New York, 1959)

I.48. R. Wood, "On a remarkable case of uneven distribution of light in a diffraction grating spectrum," Phil. Mag. 4, 396-402, 1902

I.49. G. R. Harrison, "The production of diffraction gratings: I. Development of the ruling art; II. The desing of echelle gratings and spectrographs," J. Opt. Soc. Am. 39, 413-426, 522-528, 1949

I.50. R. P. Madden and J. Strong, *Concepts of Classical Optics*, (Freeman, San Francisco, 1958)

I.51. R. Petit, "A tutorial introduction," in *Electromagnetic Theory of Gratings*, R. Petit ed. (Springer-Verlag Berlin, 1980), ch. 1

I.52. C. H. Wilcox, "Scattering theory for the D'Alembert equation in exterior domains," in *Lecture Notes in Mathemetics*, vol. 442, (Springer, Berlin, 1975)

I.53. P. M. Morse and H. Feshbach, *Methods of Theoretical Physics*, (McGraw-Hill Company, New York, 1953)

I.54. M. Nevière and P. Vincent, "Sur une propriété de symétrie des réseaux diélectriques," Optica Acta 23, 557-568, 1976

I.55. P. Vincent and M. Nevière, "The reciprocity theorem for corrugated surfaces used in conical diffraction mountings," Optica Acta, 26, 889-898, 1979

I.56. L. C. Botten, N. A. Nicorovici, A. A. Asatryan, R. C. McPhedran, C. M. de Sterke, and P. A. Robinson, "Formulation for Electromagnetic Scattering and Propagation through grating stacks of metallic and dielectric cylinders for photonic crystal calculations. Part 2: Properties and implementation," J. Opt. Soc. Am. A17, 2177-2190, 2000

Chapter II
Basic Principles of the Differential Theory of Gratings

II.1 Classical differential theory of gratings

When Maxwell equations are applied to periodic media, they can be projected onto a Fourier basis with respect to one or more coordinates, as explained in detail in Chapter I. This procedure is quite useful, because it reduces the order of the differential equations. In the case of diffraction gratings used in in-plane mounts, the Fourier decomposition enables us to transform Maxwell equations into a set of ordinary differential equations, which can be integrated using well-developed numerical methods.

However, as already discussed, when using a truncated Fourier series, the truncation can cause numerical problems for highly conducting materials in TM polarization. Another numerical problem appears in both TE and TM cases, due to the exponentially growing terms during the integration. How to reformulate the differential theory in order to avoid these two problems will be explained in detail further on. However, for better understanding the fundamentals of the differential mehod, in this chapter we give its most simple presentation. It is worth noting, that this presentation is not of purely academic interest, because it enables treating some cases of great practical importance, in which the problems of Fourier series truncation and of the growing-term contamination play a minor role. Such are the diffraction gratings used in the X-ray domain and the usual gratings used in TE polarization.

II.1.1 TE polarization

II.1.1.1 The boundary-value problem

As shown in Section I.2.2, Maxwell equations applied to a TE polarized field can be reduced to a single second-order propagation equation (I.19) with E_z as an unknown function:

$$\Delta E_z + k^2(x,y)E_z(x,y) = 0.$$

This equation, valid in the entire space including on the grating profile, reduces to a Helmholtz equation (I.30) in any homogeneous media. As shown in Section I.2.5, the introduction of the pseudo-Fourier series of the field (I.29) into Eq. (I.30) allows expressing the field in terms of Rayleigh expansion, Eqs. (I.35) and (I.38). As previously stated in Section I.1.2, these expressions, in general, are not valid in the modulated region [I.11]. In the modulated region ($0 < y < a$) only expansions (I.29) can be used; it means that the unknown functions $u_n(y)$ cannot be expressed in terms of exponential functions. The only possibility is a numerical determination of these functions with a computer.

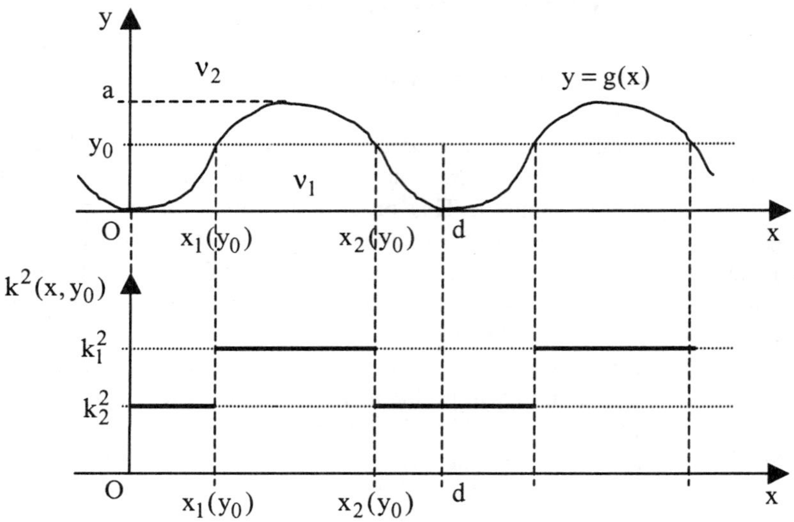

Fig. II.1. Function $k^2(x, y_0)$ for a dielectric grating.

To that end, we first remark that at a given ordinate y_0 inside the modulated region, $k^2(x,y_0)$ is a step function that is equal to k_1^2 inside the bumps and to k_2^2 inside the grooves, the width of the plateaus and the valleys depending on y_0 as illustrated in Fig. II.1. Thus $k^2(x,y)$ is a periodic function with respect to x, with period d, and can be expressed by its Fourier series:

$$k^2(x,y) = \sum_{n=-\infty}^{+\infty} (k^2)_n(y) \exp(inKx). \tag{II.1}$$

It is worth noting that the product $k^2(x,y)E_z(x,y)$ is pseudoperiodic with respect to x and thus can be expressed on the $\exp(i\alpha_n x)$ basis. Introducing Eqs. (I.29) and (II.1) into Eq. (I.19) we then obtain:

$$\forall n: \frac{d^2E_n(y)}{dy^2} + \sum_{m=-\infty}^{+\infty} (k^2)_{n-m}(y)E_m(y) - \alpha_n^2 E_n(y) = 0, \tag{II.2}$$

where $E_n(y)$ are the Fourier components of E_z. Of course, this set of coupled differential equations decouples into Eq. (I.32) in any homogeneous medium in which only the coefficient $[k^2(y)]_0$ is different from zero (and equal to k_2^2 if $y > a$ and k_1^2 if $y < 0$). Inside the modulated region, in view of finding a numerical solution, the infinite set of coupled differential equations must be truncated. To that end, we assume that the field is accurately described by 2N+1 Fourier components $E_n(y)$, with $n \in [-N,+N]$, where the value of N will have to be determined by convergence tests. We then introduce matrix notation to simplify what follows. Let us consider the column vector [E(y)] with 2N+1 components $E_n(y)$, and the (2N+1)x(2N+1) square matrix M^{TE} with elements $M_{nm}^{TE}(y) = -(k^2)_{n-m}(y) + \alpha_n^2 \delta_{nm}$. Eq. (II.2) simply becomes:

$$\frac{d^2\,[E(y)]}{dy^2} = M^{TE}(y)\ \ [E(y)]. \tag{II.2'}$$

Outside the modulated region, the solution [E(y)] must identify with the expressions resulting from Rayleigh expansions (I.35) and (I.38). Since the tangential component of $\vec{E}$ must be continuous, E_z must be continuous at $y = a$ and $y = 0$. This implies that:

$$E_n(a) = A_0^{(2)} \exp(-i\beta_0^{(2)} a)\delta_{n,0} + B_n^{(2)} \exp(i\beta_n^{(2)} a) \tag{II.3}$$

$$E_n(0) = A_n^{(1)} \exp(-i\beta_n^{(1)} 0) = A_n^{(1)}. \tag{II.4}$$

Another condition concerns the first derivative of $E_n(y)$ and can be derived from the continuity of the tangential component of $\vec{H}$. For TE polarization, Maxwell Eq. (I.8) implies:

$$\vec{H} = \frac{1}{i\omega\mu}\mathrm{curl}(\vec{E}) \equiv \frac{1}{i\omega\mu}\mathrm{curl}(E\vec{z}) = \frac{1}{i\omega\mu}(\mathrm{grad}E)\times\vec{z}\ ,$$

where $\vec{z}$ is the unit vector of the z-axis.

Thus, if $\vec{N}$ is the unit vector of the normal to the grating surface:

$$\vec{N}\times\vec{H} = \frac{1}{i\omega\mu}\vec{N}\times\left[(\mathrm{grad}E)\times\vec{z}\right] = -\frac{1}{i\omega\mu}(\vec{N}\cdot\mathrm{grad}E)\vec{z}\ ,$$

which shows that the tangential component of $\vec{H}$ is proportional to the normal derivative of E, with a *factor independent of the medium*. At the frontiers of the modulated region, $\vec{N} = \pm\vec{y}$, where $\vec{y}$ is the unit vector along the y-axis. Thus the normal derivative is $\pm\frac{dE}{dy}$ and the tangential component of $\vec{H}$ is H_x. Eq. (I.10') then also gives the link between them. Thus, on top of functions $E_n(y)$, $\frac{dE_n(y)}{dy}$ must be continuous at y = a and y = 0, and we must state:

$$\frac{dE_n(a)}{dy} = -i\beta_0^{(2)}A_0^{(2)}\exp(-i\beta_0^{(2)}a)\delta_{n,0} + i\beta_n^{(2)}B_n^{(2)}\exp(i\beta_n^{(2)}a) \qquad \text{(II.5)}$$

$$\frac{dE_n(0)}{dy} = -i\beta_n^{(1)}A_n^{(1)} . \qquad \text{(II.6)}$$

The diffraction problem is now reduced to the resolution of a boundary-value problem, which results from the differential equation (II.2') and the boundary conditions (II.3)-(II.6). This boundary-value problem is transformed into an initial-value problem by what is usually called a "shooting method" in mathematics.

II.1.1.2 The shooting method for TE polarization

Several numerical algorithms [II.1] can be used to find the numerical solution of a second-order differential equation (or set of equations) over a given range (0, a] of the y-variable, provided the initial values of the function and its first derivative are known at the first step of the integration process (i.e., at y = 0). This is not the case here, since coefficients $A_n^{(1)}$ are not known. We only have a link between the functions and their derivatives, resulting from Eqs. (II.4) and (II.6):

$$\left.\frac{dE_n(y)}{dy}\right|_{y=0} = -i\beta_n^{(1)}E_n(0) \qquad \text{(II.7)}$$

while Eqs. (II.3) and (II.5) would give a similar link at y = a. The shooting method overcomes the difficulty.

In view of finding the solution of the boundary-value problem, we first construct *2N+1 particular solutions* $[\hat{E}(y)]_p$ (p = –N, –N+1, ..., +N), each one corresponding to a transmitted diffracted field with unit amplitude ($A_n^{(1)} = \delta_{n,p}$ in Eq. (I.38)). Thus we choose

$$\hat{E}_{n,p}(0) = \delta_{n,p} \tag{II.8}$$

and

$$\left.\frac{d\hat{E}_{n,p}}{dy}\right|_{y=0} = -i\beta_p^{(1)}\delta_{n,p}. \tag{II.9}$$

It is worth noticing that the process implies performing 2N+1 consecutive numerical integrations. The first integration (p = –N) will start with the initial vector

$$\left[\hat{E}(0)\right]_{-N} = \begin{pmatrix} 1 \\ 0 \\ \vdots \\ \vdots \\ 0 \end{pmatrix} \text{ and } \left[\hat{E}'(0)\right]_{-N} = \begin{pmatrix} -i\beta_{-N}^{(1)} \\ 0 \\ \vdots \\ \vdots \\ 0 \end{pmatrix}.$$

A suitable numerical algorithm computes $[\hat{E}(y)]_{-N}$ and $[\hat{E}'(y)]_{-N}$, $\forall y \in (0, a]$. The second integration starts with

$$\left[\hat{E}(0)\right]_{-N+1} = \begin{pmatrix} 0 \\ 1 \\ 0 \\ \vdots \\ \vdots \\ 0 \end{pmatrix},$$

which is linearly independent of $[\hat{E}(0)]_{-N}$, and with

$$\left[\hat{E}'(0)\right]_{-N+1} = \begin{pmatrix} 0 \\ -i\beta_{-N+1}^{(1)} \\ 0 \\ \vdots \\ \vdots \\ 0 \end{pmatrix}$$

and provides $[\hat{E}(y)]_{-N+1}$ and $[\hat{E}'(y)]_{-N+1}$, and the process continues until $[\hat{E}(y)]_N$ and $[\hat{E}'(y)]_N$ are found $\forall y \in (0,a]$.

Of course, none of the particular solutions $[\hat{E}(y)]_p$ produce, at y = a, the total field verifying Eq. (I.35). They produce total fields given by Eq. (I.34) with wrong values of $A_n^{(2)}$ and $B_n^{(2)}$. We will take advantage of the linearity of the present problem to construct the exact total field $[E(y)]$.

Since we are concerned with linear electromagnetism, the basic propagation differential set of equations (II.2) is linear. It means that a linear operator $\mathbb{L}$ exists that relates any particular solution of (II.2) at the bottom of the grooves (y = 0) to the corresponding solution at the top:

$$\mathbb{L}\left([\hat{E}(0)]_p\right) = [\hat{E}(a)]_p \tag{II.10}$$

Of course the same relation applies to the exact total field $[E(y)]$:

$$\mathbb{L}\left([E(0)]\right) = [E(a)] \ . \tag{II.11}$$

Let us now consider the linear superposition

$$[U] = \sum_{p=-N}^{+N} A_p^{(1)} \left[\hat{E}(0)\right]_p \tag{II.12}$$

where $A_p^{(1)}$ are arbitrary constants.

From the definition of linear operators,

$$\begin{aligned}\mathbb{L}([U]) &= \mathbb{L}\left(\sum_{p=-N}^{+N} A_p^{(1)}\left[\hat{E}(0)\right]_p\right) = \sum_{p=-N}^{+N} A_p^{(1)}\mathbb{L}\left[\hat{E}(0)\right]_p \\ &= \sum_{p=-N}^{+N} A_p^{(1)}\left[\hat{E}(a)\right]_p \ .\end{aligned} \tag{II.13}$$

From the starting values of $[\hat{E}(y)]_p$ given in Eq. (II.8), we first calculate the nth component U_n of U:

$$U_n = \sum_{p=-N}^{+N} A_p^{(1)} \delta_{n,p} = A_n^{(1)} \equiv E_n(0).$$

Thus [U] identifies with [E(0)] provided the unknown coefficients $A_p^{(1)}$ have the correct values $A_n^{(1)}$ that we are looking for. From Eq. (II.13), we then derive:

$$\mathbb{L}\left[E(0)\right] = \sum_{p=-N}^{+N} A_p^{(1)} \left[\hat{E}(a)\right]_p , \tag{II.14}$$

and comparison with Eq. (II.11) gives:

$$\sum_{p=-N}^{+N} A_p^{(1)} \left[\hat{E}(a)\right]_p = \left[E(a)\right] . \tag{II.15}$$

More generally, $\forall y \in [0, a]$,

$$\left[E(y)\right] = \sum_{p=-N}^{+N} A_p^{(1)} \left[\hat{E}(y)\right]_p . \tag{II.16}$$

A similar reasoning made with

$$\left[\frac{dU}{dy}\right] = \sum_{p=-N}^{+N} A_p^{(1)} \left[\frac{d\hat{E}}{dy}(0)\right]_p$$

and making use of Eq. (II.7) will lead to:

$$\mathbb{L}\left[\frac{dE}{dy}(0)\right] = \sum_{p=-N}^{+N} A_p^{(1)} \left[\frac{d\hat{E}}{dy}(a)\right]_p = \left[\frac{dE}{dy}(a)\right]. \tag{II.17}$$

We only have to replace in Eq. (II.15) the value of [E(a)] given by Eq. (II.3) and to replace in Eq. (II.17) the value of $\frac{dE}{dy}(a)$ given by Eq. (II.5) to obtain:

$$\sum_{p=-N}^{+N} A_p^{(1)} \hat{E}_{np}(a) = A_0^{(2)} \exp(-i\beta_0^{(2)} a)\delta_{n,0} + B_n^{(2)} \exp(i\beta_n^{(2)} a) \tag{II.18}$$

$$\sum_{p=-N}^{+N} A_p^{(1)} \frac{d\hat{E}_{np}}{dy}(a) = -i\beta_0^{(1)} A_0^{(2)} \exp(-i\beta_0^{(2)} a)\delta_{n,0} + i\beta_n^{(2)} B_n^{(2)} \exp(i\beta_n^{(2)} a) \quad \text{(II.19)}$$

Eqs. (II.18)-(II.19) are a set of 2x(2N+1) linear algebraic equations with the same number of unknown coefficients. Elimination of the 2N+1 unknown coefficients $B_n^{(2)}$ leads to a set of 2N+1 equations, which can be written in matrix form:

$$T_A \left[A^{(1)} \right] = \left[V_A^{(2)} \right], \quad \text{(II.20)}$$

in which the vector $[A^{(1)}]$ has components $A_p^{(1)}$, T_A is a known square matrix (see later Eqs. (III.1,III.4,and III.9)), and $[V_A^{(2)}]$ is a vector depending on the incident wave with components equal to $V_{A,n}^{(2)} = i\delta_{n,0}(\beta_n^{(2)} + \beta_0^{(1)})\exp(-i\beta_0^{(2)} a)$. If the diffraction problem has a unique solution, the determinant of T_A is different from zero so that $\left[A^{(1)}\right] = T_A^{-1}\left[V_A^{(2)}\right]$. When $A_n^{(1)}$ are known, Eq. (II.18) provides $B_n^{(2)}$, and the field outside the modulated area is known. From the Rayleigh coefficients, the reflected and transmitted efficiencies can be calculated with Eqs. (I.42) and (I.44). They are the quantities of interest for the spectroscopists, as well as in many other uses of gratings. However, in some particular problems, the field inside the modulated area may be of interest too. It can be computed through Eqs. (II.16) and (I.29) as soon as the correct coefficients $A_p^{(1)}$ have been found, and provided the particular solutions $\hat{E}_{np}(y)$ have been stored in the computer memory $\forall y \in [0, a]$. Local field maps can then be constructed if necessary. If $\hat{E}_{np}(y)$ have not been stored, it suffices to start a last numerical integration with the exact starting values

$$E_n(0) = A_n^{(1)} \text{ and } \left.\frac{dE_n}{dy}\right|_{y=0} = -i\beta_n^{(1)} A_n^{(1)}$$

to obtain the correct values of $E_n(y)$, $\forall y \in (0, a]$, and thus, to have the total field through Eq. (I.29).

Let us say some words about the numerical algorithm used in the integration process. Having to integrate a second-order differential set without first derivative, the Numerov's algorithm [I.25] was chosen in the pioneering work [I.27]. It turned out to be simple and very fast, a quality of interest at the time of the advent of the first powerful computers. The speed of present computers has nowadays strongly reduced this advantage, in the sense that any standard algorithm existing in computer libraries can be used. In particular, the second-order differential set (II.2') can be transformed into a first-order differential one with twice as many equations, as it will be done in the following chapter for TM polarization, and integrated with the classical Runge-Kutta technique [II.1]. A change in the algorithm generally produces differences only on the 6th digit of the numerical results.

Numerical experiments show that in many grating problems, stable, converging results are obtained with a truncation index N between 5 and 15 (11 to 31 Fourier components) and a number K_N of steps in the integration process between 60 and 200. After convergence tests have been done on the values of N and K_N , the results must verify the energy-balance criterion (I.47) for lossless gratings, and the reciprocity theorem in reflection and/or transmission. But the best proof of their validity is obtained when they agree with results obtained through a completely different formalism or through measurements.

II.1.2 TM polarization

II.1.2.1 The boundary-value problem

We now start from the propagation equations (I.20) and (I.21). Defining H_n and $\tilde{E}_n$ as the pseudo-Fourier components of H_z and $\tilde{E}_x$ on the $\exp(i\alpha_n x)$ basis, Fourier analysis leads to:

$$\frac{dH_n}{dy} = \sum_{m=-\infty}^{+\infty} \left(k^2\right)_{n-m} \tilde{E}_m \tag{II.21}$$

$$\frac{d\tilde{E}_n}{dy} = \alpha_n \sum_{m=-\infty}^{+\infty} \alpha_m \left(\frac{1}{k^2}\right)_{n-m} H_m - H_n \tag{II.22}$$

If we use the same truncation process as for TE polarization, from –N to +N, we now obtain a first-order differential equation with 4N+2 equations and 4N+2 unknown functions $H_n(y)$ and $\tilde{E}_n(y)$. Introducing a vector [F] made of two blocks containing 2N+1 coefficients H_n and 2N+1 coefficients $\tilde{E}_n$, and a square matrix M^{TM} made of blocks M^{TM}_{ij}, i, j = 1 and 2, with (2N+1) x (2N+1) elements:

$$M^{TM}_{1,1,n,m} = 0 = M^{TM}_{2,2,n,m}$$

$$M^{TM}_{1,2,m,n} = \left(k^2\right)_{n-m}$$

$$M^{TM}_{2,1,m,n} = \alpha_n \left(\frac{1}{k^2}\right)_{n-m} \alpha_m - \delta_{n,m} \ ,$$

the basic equation of the TM boundary-value problem reads:

$$\frac{d[F(y)]}{dy} = M^{TM}(y) \ \ [F(y)] \ . \tag{II.23}$$

Moreover Eqs. (I.20) and (I.21) which are equivalent to Eq. (I.22) reduce to a Helmholtz equation in each homogeneous region, so that Rayleigh expansions (I.35) and (I.38) apply to the TM field $u(x, y) \equiv H_z$. It means that

$$H_n(a) = A_0^{(2)} \exp(-i\beta_0^{(2)} a)\delta_{n,0} + B_n^{(2)} \exp(i\beta_n^{(2)} a) \tag{II.24}$$

and

$$H_n(0) = A_n^{(1)}. \tag{II.25}$$

Since we have to integrate a first-order differential set (II.23), only the initial values of the components of [F(0)] have to be given. Eq. (II.25) provides half of the answer, while $\widetilde{E}_n(0)$ can be derived from Eq. (I.20), which leads to:

$$\tilde{E}_x = \frac{1}{k^2(x,y)} \frac{\partial H_z}{\partial y}. \tag{II.26}$$

Here $\dfrac{\partial H_z}{\partial y}$ will be derived from the Rayleigh expansion of H_z at y = 0. Fourier analysis leads to:

$$\widetilde{E}_n(y) = \sum_{m=-N}^{+N} \left(\frac{1}{k^2}\right)_{n-m} \frac{dH_m}{dy}. \tag{II.27}$$

Since at y = 0, k^2 (x,y) reduces to k_1^2, we thus obtain:

$$\widetilde{E}_n(0) = \frac{1}{k_1^2} \frac{dH_n}{dy}(0) = \frac{1}{k_1^2}(-i\beta_n^{(1)})A_n^{(1)}. \tag{II.28}$$

We have to remark that a similar relation can be written at y = a, where $k^2(x,y) = k_2^2$:

$$\tilde{E}_n(a) = \frac{1}{k_2^2}\frac{dH_n}{dy}(a) = \frac{1}{k_2^2}\left[-i\beta_0^{(2)}A_0^{(2)}\exp(-i\beta_0^{(2)}a)\delta_{n,0} + i\beta_n^{(2)}B_n^{(2)}\exp(i\beta_n^{(2)}a)\right]. \quad \text{(II.29)}$$

Eqs. (II.23)-(II.25) and (II.28)-(II.29) constitute the TM boundary-value problem.

II.1.2.2 The shooting method for TM polarization

The numerical solution of the boundary-value problem is conducted along similar lines as for TE polarization. We first construct 2N+1 particular solutions $\left[\hat{F}(y)\right]_p$ corresponding to outgoing or evanescent transmitted waves with unit amplitudes $A_n^{(1)} = \delta_{n,p}$. It means that we initiate the 2N+1 vectorial integration process with vectors $\left[\hat{F}(0)\right]_p$ with elements $\hat{H}_n(0) = \delta_{n,p}$ and $\hat{\tilde{E}}(0) = \frac{1}{k_1^2}(-i\beta_p^{(1)})\delta_{n,p}$. With the initial values of vector $\left[\hat{F}(0)\right]_p$, a suitable algorithm (e.g., Runge-Kutta technique) computes the corresponding particular solutions of Eq. (II.23). The same reasoning on the linearity of the problem as for TE polarization applies, so that the linear superposition $[F(y)] = \sum_{p=-N}^{+N} A_p^{(1)}\left[\hat{F}(y)\right]_p$ matches the corresponding term in the Rayleigh expansion on the top of the grooves, Eqs. (II.24) and (II.29):

$$\sum_{p=-N}^{+N} A_p^{(1)}\hat{H}_{np}(a) = A_0^{(2)}\exp(-i\beta_0^{(2)}a)\delta_{n,0} + B_n^{(2)}\exp(i\beta_n^{(2)}a) \quad \text{(II.30)}$$

$$\sum_{p=-N}^{+N} A_p^{(1)}\hat{\tilde{E}}_{n,p}(a) = \frac{1}{k_2^2}\left[-i\beta_0^{(2)}A_0^{(2)}\exp(-i\beta_0^{(2)}a)\delta_{n,0} + i\beta_n^{(2)}B_n^{(2)}\exp(i\beta_n^{(2)}a)\right]. \quad \text{(II.31)}$$

Elimination of $B_n^{(2)}$ between Eqs. (II.30), (II.31) again leads to a linear algebraic set whose unknowns are $A_p^{(1)}$ (see also Appendix IV). Matrix inversion provides the correct transmitted amplitudes $A_n^{(1)}$ and Eq. (II.30) then gives the reflected ones $B_n^{(2)}$.

When applied on dielectric gratings the present method gives stable and converging results with about the same number of Fourier components as for TE polarization, although often a little higher. On the other hand, when applied to metallic gratings, poor convergence with respect to N was observed. On deep gratings, the energy balance may be violated. These numerical problems, pointed out in 1974 [I.27], took about 20 years to be cleared, and the efforts of the scientific community have been surveyed in Section I.1.4. We will show with a simple example how the numerical contamination appears.

II.2 Numerical contamination

II.2.1 A simple example of sensitivity to initial conditions

Modern science has pointed out several domains in which an extremely small error in an initial value of a solution of an equation produces unpredictable results. Well-known examples are meteorology, or the theory of deterministic chaos. We choose to illustrate the problem on a simple, single differential equation.

Let us considerer the following initial-value problem:
We want to solve the equation

$$\frac{d^2u(y)}{dy^2} = C^2u(y), \tag{II.32}$$

over interval (0,a] (with $a > 0$ and C a real number). We search the solution that satisfies the initial conditions

$$u(0) = 1 \tag{II.33}$$

$$\frac{du}{dy}(0) = -C. \tag{II.34}$$

As is well known, Eq. (II.32) has two particular solutions $u_1 = \exp(Cy)$ and $u_2 = \exp(-Cy)$, so that the general solution is

$$u(y) = \tau_1 \exp(Cy) + \tau_2 \exp(-Cy), \tag{II.35}$$

with τ_1 and τ_2 being integration constants; solution u_2 satisfies the initial conditions (II.33), (II.34). It means that in the general solution (II.35), we must take $\tau_1 = 0$ and $\tau_2 = 1$ in order to compute the desired solution through a numerical integration process. However, all computers work with a finite number of digits, typically 16 in the usual "double precision." It means that the initial values (II.33) and (II.34) are only satisfied with errors $\hat{\delta}_1$ and $\hat{\delta}_2$ equal to $\pm 10^{-16}$. The result is that instead of computing the solution given by Eq. (II.35) with $\tau_1 = 0$ and $\tau_2 = 1$, we compute another solution which satisfies:

$$1 + \hat{\delta}_1 = \tau_1 + \tau_2$$

$$-C + \hat{\delta}_2 = C\tau_1 - C\tau_2.$$

As a result, τ_2 is slightly different from unity, and τ_1 is equal to $\frac{1}{2}\left(\hat{\delta}_1 + \frac{\hat{\delta}_2}{C}\right)$, which is of the order of magnitude of $\hat{\delta}_1$. It means that our numerical solution includes at the first step of the integration process a small amount of the undesired solution $u_1(y)$, which is a growing exponential function. If the integration process is continued far enough, the undesired solution may become greater than the desired one, and the results have no sense. If we tolerate a 10^{-6} relative error, we can calculate the range y_0 of y beyond which the contamination appears:

$$|\tau_1| \exp(Cy_0) \geq 10^{-6} \tau_2 \exp(-Cy_0),$$

which leads to:

$$\exp(2Cy_0) \geq \frac{10^{-6}}{10^{-16}} C = 10^{10} C,$$

or

$$y_0 = \frac{\ln(10^{10} C)}{2C}. \tag{II.36}$$

The higher C, the smaller is the range limit y_0.

II.2.2 Application to the grating problem

How does the growing exponential contamination concern our grating problem? Indeed, it is somehow difficult to demonstrate this in the most general case, for which the solution in the modulated region cannot be expressed by a superposition of plane and evanescent waves, but can only be determined numerically. However, it has been established that for sinusoidal profiles [I.11] Rayleigh expansions (I.34) and (I.37) are valid to represent the field in the modulated region as far as the modulation a/d is less than 0.143. Then, inside the modulated region, the pseudo-Fourier components $E_n(y)$ of E_z (in TE polarization case) are given by:

$$\begin{aligned} E_n(y) &= A_n^{(2)} \exp(-i\beta_n^{(2)} y) + B_n^{(2)} \exp(i\beta_n^{(2)} y) \quad \text{if} \quad y > g(x) \\ &= A_n^{(1)} \exp(-i\beta_n^{(1)} y) + B_n^{(1)} \exp(i\beta_n^{(1)} y) \quad \text{if} \quad y < g(x). \end{aligned} \tag{II.37}$$

From these expressions, it is clear that in this particular case, the numerical solution of the coupled set of differential equations (II.2') will have exponential behavior for the evanescent orders. As established in the previous section, the higher $\left|\beta_n^{(2)}\right|$ or $\left|\beta_n^{(1)}\right|$, the smaller the safe range of integration y_0 will be. Since $\left|\beta_n^{(1)}\right|$ increases with n and with $\left|\nu_1\right|$, instabilities are expected sooner on highly conducting metals (good reflectors) and for large truncation limits 2N+1. This is exactly what numerical experiments show, both on sinusoidal gratings and on other grating profiles. It means that even when Eqs. (II.37) cannot be used to represent the field in the modulated region, the numerical solution keeps a growing behavior similar to the exponential one that rules the homogeneous regions, so that the same contamination occurs.

This assertion is fully justified for lamellar (rectangular) profiles. In that case, as will be explained in Chapter VII, inside the grooves the field can be represented by the propagating and evanescent modes of a rectangular cavity; hence the y-exponential growing. For arbitrary groove shape, the Yasuura Improved Point-Matching Method [I.11] shows that although Rayleigh expansions are not rigorously valid, they may be used to find the best least-square approximant of the field. This confirms that, indeed, the field keeps an exponential-like growth even when the groove shape is arbitrary.

When Eqs. (II.37) apply, and more generally anytime the exponentially-dependence of $E_n(y)$ is known, there is a simple way to avoid the contamination by the growing exponential functions $A_n^{(j)} \exp(-i\beta_n^{(j)} y)$ for which $\beta_n^{(j)}$ are complex numbers with positive imaginary part. It consists of rewriting the Rayleigh expansion (II.37) in the form:

$$E_n(y) = \tilde{A}_n^{(j)} \exp\left[-i\beta_n^{(j)}(y-a)\right] + B_n^{(j)} \exp(i\beta_n^{(j)} y), \tag{II.38}$$

with $j = 2$ if $y > g(x)$, and $j = 1$ if $y < g(x)$.

During the integration process, (y-a) remains negative; the growing exponential functions remain bounded by unity. Thus they cannot contaminate the numerical solution (the decaying exponential functions).

This simple way of avoiding the contamination, however, cannot be used for an arbitrary profile for which the y-dependence of E_n (y) is not explicitly known. We would have no mean to deduce the original Rayleigh coefficients $A_n^{(j)}$ when $\tilde{A}_n^{(j)}$ are determined, while previously, we simply had:

$$A_n^{(j)} = \tilde{A}_n^{(j)} \exp(i\beta_n^{(j)} a) \quad \text{or} \quad \tilde{A}_n^{(j)} = A_n^{(j)} \exp(-i\beta_n^{(j)} a). \tag{II.39}$$

However, such changes in Rayleigh coefficients may be useful to avoid a second numerical problem, i.e., the overflows and underflows that may occur during the computation, and stop the work.

Presently, best way to eliminate the contamination by growing exponential functions consists of dividing the deep grating into different M-2 slices, which are thin enough so that no contamination occurs. A suitable matrix algorithm will then allow resolving the diffraction problem of the whole grating without producing numerical contamination. To that end, it is necessary to be able to analyze a stack of gratings instead of a single one.

Reference

II.1. P. Henrici, *Discrete Variable Methods in Ordinary Differential Equations*, (John Willey, New York, 1962)

Chapter III

Stacks of Gratings

Correctly modeling multilayered structures of plane or corrugated interfaces is important for practical grating design. While many dielectric gratings and corrugated dielectric waveguides are naturally made of several layers, even the bare metallic gratings, which could be considered as a single-grating interface, are normally covered with some oxide or protecting layer that must be taken into account in some cases, for example in the near UV region.

Moreover, as explained in Chapter II, the numerical modeling of a single-interface grating requires dividing the modulated region into several layers to avoid the contamination due to exponentially growing terms.

III.1 Reflection and transmission matrices of a grating

Before analyzing stacks of gratings, we first introduce the reflection and transmission matrix, which generalizes the classical reflection and transmission Fresnel coefficient to a grating. Instead of only having an incident plane wave, we now assume that we have (2N+1) incident and "anti-evanescent" waves of the form $A_n^{(2)} \exp(-i\beta_n^{(2)} y) \exp(i\alpha_n x)$ allowed by Rayleigh expansion (I.34) truncated to (2N+1) terms. We then define three vectors with 2N+1 components, which represent the incident, reflected and transmitted field at the top or bottom of the grooves as:

$$V_A^{(2)} = \begin{pmatrix} \vdots \\ \vdots \\ A_n^{(2)} \exp(-i\beta_n^{(2)} a) \\ \vdots \\ \vdots \end{pmatrix}, \qquad V_B^{(2)} = \begin{pmatrix} \vdots \\ \vdots \\ B_n^{(2)} \exp(i\beta_n^{(2)} a) \\ \vdots \\ \vdots \end{pmatrix}, \qquad V_A^{(1)} = \begin{pmatrix} \vdots \\ \vdots \\ A_n^{(1)} \\ \vdots \\ \vdots \end{pmatrix}.$$

As soon as the fields are represented on a finite basis, namely on the 2N+1 $\exp(i\alpha_n x)$ functions, the linear operator $\mathbb{L}$ defined in Eq. (II.10) or Eq. (II.11) can be represented by a matrix.

Other convenient matrices representing linear operators can be introduced in our grating problem. Let us define matrices T_A and T_B by:

$$V_A^{(2)} = T_A V_A^{(1)} \tag{III.1}$$

$$V_B^{(2)} = T_B V_A^{(1)}. \tag{III.2}$$

As is well known in mathematics, a linear application, which acts on a finite dimensional space, is entirely determined when the image of a basis is known. Thus, if for $V_A^{(1)}$ we successively choose (2N+1) linearly independent vectors $V_{A,p}^{(1)}$ ($p \in [-N,+N]$), with elements

$$V_{A,n,p}^{(1)} = \delta_{n,p}, \tag{III.3}$$

Eq. (III.1) shows that *matrix T_A will be identified with the set of (2N+1) vectors* $V_{A,p}^{(2)}$. If we remark that Eq. (III.3) identifies with Eq. (II.8), we see that *the shooting method directly provides matrix T_A and T_B at the end of the integration process*. It suffices to extract $V_{A,n,p}^{(2)}$ and $V_{B,n,p}^{(2)}$ from the values of $\hat{E}_{n,p}(a)$ and $\frac{d\hat{E}_{n,p}}{dy}(a)$ for TE polarization. From Eqs. (II.3) and (II.5), we obtain:

$$T_{A,n,p} = V_{A,n,p}^{(2)} = A_{n,p}^{(2)} \exp(-i\beta_n^{(2)} a) = \frac{1}{2}\left[\hat{E}_{n,p}(a) - \frac{1}{i\beta_n^{(2)}} \frac{d\hat{E}_{n,p}}{dy}(a)\right] \tag{III.4}$$

$$T_{B,n,p} = V_{B,n,p}^{(2)} = B_{n,p}^{(2)} \exp(i\beta_n^{(2)} a) = \frac{1}{2}\left[\hat{E}_{n,p}(a) + \frac{1}{i\beta_n^{(2)}} \frac{d\hat{E}_{n,p}}{dy}(a)\right]. \tag{III.5}$$

We note that in these equations the Rayleigh coefficients have now two subscripts; $A_{n,p}^{(2)}$ is the trial value of the nth Rayleigh coefficient of a downgoing wave obtained with the pth starting value of Eq. (III.3).

When the matrices T_A and T_B are known, it is straightforward to relate the reflected field to the incident one; this relation defines the reflection matrix r:

$$V_B^{(2)} = r\, V_A^{(2)}. \tag{III.6}$$

We immediately obtain, from Eqs. (III.1) and (III.2):

$$r = T_B (T_A)^{-1} \tag{III.7}$$

The transmission matrix (t) defined by

$$V_A^{(1)} = t V_A^{(2)} \tag{III.8}$$

is simply given by

$$t = (T_A)^{-1}. \tag{III.9}$$

It is important to remark that as soon as matrices r and t have been determined, the reflected and transmitted fields $V_B^{(2)}$ and $V_A^{(1)}$ corresponding to any incident field of the form $\sum_{n=-N}^{+\infty} A_n^{(2)} \exp(i\alpha_n x - i\beta_n^{(2)} y)$ can be determined by simple matrix products, without performing any new numerical integration. When only one incident plane wave falls on the grating ($A_n^{(2)} = \delta_{n,0}$), the reflected field $V_B^{(2)}$ reduces to the central column of matrix r:

$$V_{B,n}^{(2)} = r_{n,0} \exp(-i\beta_0^{(2)} a). \tag{III.10}$$

In a similar way,

$$V_{A,n}^{(1)} = t_{n,0} \exp(-i\beta_0^{(2)} a). \tag{III.11}$$

III.2 The transmission-matrix algorithm

When a grating profile is included into a stack of a given number of modulated or homogeneous interfaces as illustrated in Fig. III.1 and Fig. III.2, all profiles except the extreme ones receive incident and outgoing waves both from top and bottom. The grating region is divided into a number of different horizontal layers, which not only depend on the number of interfaces, but on their geometry, as well. The numbering starts from the substrate, which has number 1 and is considered as a layer with infinite thickness. Each layer can contain either a homogeneous or a modulated region, and its interfaces are chosen to intersect no grating surface. For example, a grating consisting of two corrugated interfaces should be divided into different numbers of layers, depending on the mutual position of the interfaces. If the profiles interpenetrate, there are two layers, including the substrate (Fig. III.1a); when the profiles are touching each other, it is necessary to consider three layers (Fig. III.1b); they become four when there is a distance between the grating surfaces (Fig. III.1c). As shown in Fig. III.2, where the notations are defined, the superstrate has now a number M, while the substrate has the number 1.

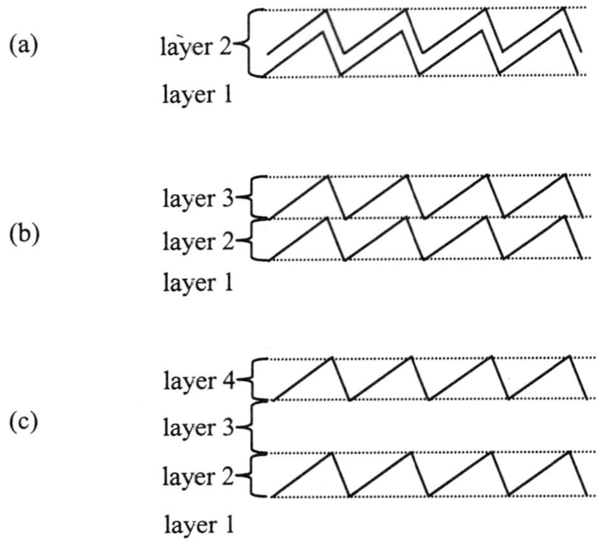

Fig. III.1 Three cases with a stack of two corrugated interfaces. The substrate is layer 1 and has an infinite thickness.

We can then generalize the concept of transmission matrix, now called T-matrix, in the following way, a concept that will also concern homogeneous layers. At each ordinate y_j, $j \in [1, M-1]$, which separates modulated and/or homogeneous regions, we first introduce infinitely thin layers of *the same medium as the superstrate*, in which Rayleigh expansion is valid; at ordinate y_j, we thus write

$$E(x,y) = \sum_{n=-N}^{+N} \left[A_n'^{(j)} \exp(-i\beta_n^{(M)} y_j) + B_n'^{(j)} \exp(i\beta_n^{(M)} y_j) \right] \exp(i\alpha_n x)$$

where $\beta_n^{(M)}$ are defined by Eq. (I.33). It is worth noting that $A_n'^{(1)} = A_n^{(1)}$ and $B_n'^{(1)} = B_n^{(1)}$, due to the fact that $y_1 = 0$. We also have $A_n'^{(M-1)} = A_n^{(M)}$ and $B_n'^{(M-1)} = B_n^{(M)}$ because M is the superstrate number. We then consider vectors $V^{(j)}$ with 2(2N+1) components made of two blocks of Fourier components of waves downgoing and upgoing in the infinitely thin layer at ordinate y_j:

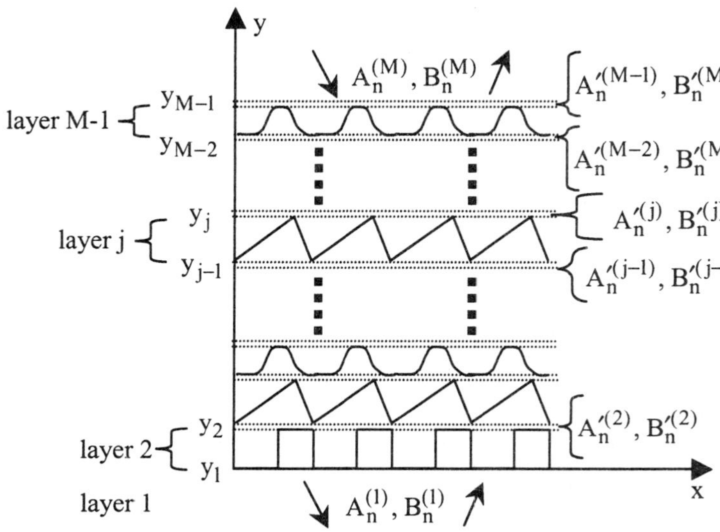

Fig. III.2 Stack of modulated and homogeneous regions and the notations used for the Rayleigh coefficients.

$$V^{(j)} = \begin{pmatrix} \vdots \\ A_n'^{(j)} \exp(-i\beta_n^{(M)} y_j) \\ \vdots \\ \hline \vdots \\ B_n'^{(j)} \exp(i\beta_n^{(M)} y_j) \\ \vdots \end{pmatrix}. \tag{III.12}$$

It is worth noticing that this column matrix consists of two blocks, derived from $V_A^{(2)}$ and $V_B^{(2)}$ (Section III.1) by replacing a by y_j, $A_n^{(2)}$ by $A_n'^{(j)}$, and $B_n^{(2)}$ by $B_n'^{(j)}$.

We define a new transmission matrix $T^{(j)}$ of the jth slice by

$$V^{(j)} = T^{(j)} V^{(j-1)} \tag{III.13}$$

Its main interest lies in the fact that the T matrix of a stack of M-2 horizontal layers, containing modulated and/or homogeneous slices can be simply obtained through the product of the M-2 $T^{(j)}$ matrices. Indeed:

$$\begin{pmatrix} \vdots \\ A_n^{(M)} \exp(-i\beta_n^{(M)} y_{M-1}) \\ \vdots \\ \hline \vdots \\ B_n^{(M)} \exp(i\beta_n^{(M)} y_{M-1}) \\ \vdots \end{pmatrix} = T^{(M-1)} \ldots\ldots T^{(j)} \ldots\ldots T^{(3)} T^{(2)} \begin{pmatrix} \vdots \\ A_n^{(1)} \\ \vdots \\ \hline \vdots \\ B_n^{(1)} \\ \vdots \end{pmatrix} . \tag{III.14}$$

Thus

$$T = T^{(M-1)} \ldots\ldots T^{(j)} \ldots\ldots T^{(3)} T^{(2)}. \tag{III.15}$$

When the T matrix of a stack is known, the outgoing-wave condition below the lowest profile, which states that $B_n^{(1)} = 0$ $(\forall n)$ allows finding the Rayleigh coefficients. Dividing matrix T into four blocks $T_{i,j}$, $i, j = 1,2$, we have:

$$\begin{pmatrix} \vdots \\ A_n^{(M)} \exp(-i\beta_n^{(M)} y_{M-1}) \\ \vdots \\ \hline \vdots \\ B_n^{(M)} \exp(i\beta_n^{(M)} y_{M-1}) \\ \vdots \end{pmatrix} = \begin{pmatrix} T_{11} & T_{12} \\ T_{21} & T_{22} \end{pmatrix} \begin{pmatrix} \vdots \\ A_n^{(1)} \\ \vdots \\ \hline \vdots \\ 0 \\ \vdots \end{pmatrix} . \tag{III.16}$$

Eq. (III.16) gives:

$$A_n^{(M)} \exp(-i\beta_n^{(M)} y_{M-1}) = \sum_{m=1}^{2N+1} T_{1,1,n,m} A_m^{(1)} \tag{III.17}$$

and

$$B_n^{(M)} \exp(i\beta_n^{(M)} y_{M-1}) = \sum_{m=1}^{2N+1} T_{2,1,n,m} A_m^{(1)}. \tag{III.17'}$$

Inversion of Eq. (III.17) and comparison with Eq. (III.8) immediately shows that for the stack of gratings

$$t = (T_{1,1})^{-1}, \tag{III.18}$$

while Eqs. (III.6), (III.17) and (III.17') lead to:

$$r = T_{2,1}(T_{1,1})^{-1}. \tag{III.19}$$

When these matrices are known, the Rayleigh coefficients are derived through Eqs. (III.10) and (III.11):

$$B_n^{(M)} = r_{n,0} \exp\left[-i(\beta_n^{(M)} + \beta_0^{(M)}) y_{M-1} \right] \tag{III.20}$$

$$A_n^{(1)} = t_{n,0} \exp\left(-i\beta_0^{(M)} y_{M-1} \right) . \tag{III.21}$$

III.3 Determination of the T matrix of a grating

The computation of the T matrix of a grating follows a shooting method similar to the one used to determine matrices T_A and T_B in Section III.1. Indeed, comparison of Eqs. (III.1), (III.2) and (III.16) shows that $T_A \equiv T_{1,1}$ and $T_B \equiv T_{2,1}$. The difference is that on top of T_A and T_B, we also have T_{12} and T_{22} to determine; it means than twice more independent starting vectors will be necessary, namely 2x(2N+1) vectors. Indeed, since the field below the jth profile now includes 2N+1 upgoing and 2N+1 downgoing waves, we will compute the field and its y-derivative at the top of the jth profile corresponding to each of these above-mentioned waves with unit amplitude. If y_{j-1} is the ordinate of the bottom of the grooves and y_j the ordinate of the top, we thus initiate 2N+1 numerical integrations with starting vectors $V_p^{(j-1)}$ such that:

$$\begin{aligned} V_{n,p}^{(j-1)} &\equiv A_n'^{(j-1)} \exp(-i\beta_n^{(M)} y_{j-1}) = \delta_{n,p} \\ V_{n+2N+1,p}^{(j-1)} &\equiv B_n'^{(j-1)} \exp(i\beta_n^{(M)} y_{j-1}) = 0 \end{aligned}$$

$n, p \in [-N, +N]$, with derivatives given by:

$$\begin{aligned} \frac{dV_{n,p}^{(j-1)}}{dy} &= -i\beta_n^{(M)} \delta_{n,p} \\ \frac{dV_{n+2N+1,p}^{(j-1)}}{dy} &= 0. \end{aligned}$$

With these starting values, we obtain (2N+1) vectors $V_p^{(j)}$ at the end of the integration process thanks to Eqs. (III.4) and (III.5) in which a is replaced by y_j.

In a second step, we again perform (2N+1) integrations with starting vectors verifying:

$$V_{n,p}^{(j-1)} \equiv A_n'^{(j-1)} \exp(-i\beta_n^{(M)} y_{j-1}) = 0$$
$$V_{n+2N+1,p}^{(j-1)} \equiv B_n'^{(j-1)} \exp(i\beta_n^{(M)} y_{j-1}) = \delta_{n,p}$$

with derivatives:

$$\frac{dV_{n,p}^{(j-1)}}{dy} = 0$$
$$\frac{dV_{n+2N+1,p}^{(j-1)}}{dy} = i\beta_n^{(M)} \delta_{n,p}$$

and compute the 2N+1 corresponding $V_p^{(j)}$ vectors. From all these starting values, it is clear that $V_{n,p}^{(j-1)} = \delta_{n,p}$, $\forall n, p \in [-N, 3N+1]$. As a result, from Eq. (III.13), matrix $T^{(j)}$ identifies with the collection of 4N+2 vectors $V_p^{(j)}$, i.e., $T_{n,p}^{(j)} = V_{n,p}^{(j)}$.

III.4 T matrix of a homogeneous layer

When a slice reduces to a homogeneous layer, no numerical integration is necessary since the field inside it can be expressed by a Rayleigh expansion. The relation between $V^{(j)}$ and $V^{(j-1)}$ can be derived by simply writing the continuity conditions at the two interfaces. If the homogeneous slice lies between ordinates y_{k-1} and y_k, and is made with a medium with refractive index ν_k, we first define constant $\beta_n^{(k)}$ by:

$$\beta_n^{(k)} = \sqrt{\left(\frac{\omega}{c}\nu_k\right)^2 - \alpha_n^2}$$

with

$$\text{Im } \beta_n^{(k)} \geq 0,$$

i.e., in the same way as stated in Eq. (I.36). We then first remark that *inside this homogeneous medium*, when y is increased by δ, the upgoing and downgoing waves are simply multiplied by $\exp(\pm i\beta_n^{(k)}\delta)$:

$$\begin{pmatrix} \vdots \\ A_n^{(k)} \exp[-i\beta_n^{(k)}(y+\delta)] \\ \vdots \\ \hline \vdots \\ B_n^{(k)} \exp[i\beta_n^{(k)}(y+\delta)] \\ \vdots \end{pmatrix} = \begin{pmatrix} e^{-i\beta_n^{(k)}\delta} & 0 \\ 0 & e^{i\beta_n^{(k)}\delta} \end{pmatrix} \begin{pmatrix} \vdots \\ A_n^{(k)} \exp(-i\beta_n^{(k)} y) \\ \vdots \\ \hline \vdots \\ B_n^{(k)} \exp(i\beta_n^{(k)} y) \\ \vdots \end{pmatrix}. \tag{III.22}$$

In the same manner as is done for the modulated layers, we can introduce at both sides of the homogeneous layer two infinitely thin layers having refractive index ν_M. When crossing the interface with y_k ordinate, the continuity of the field and its y-derivative in TE polarization introduces another matrix $C^{(k)}$ such that:

$$V^{(k)} = C^{(k)} \begin{pmatrix} \vdots \\ A_n^{(k)} \exp(-i\beta_n^{(k)} y_k) \\ \vdots \\ \hline \vdots \\ B_n^{(k)} \exp(i\beta_n^{(k)} y_k) \\ \vdots \end{pmatrix}, \tag{III.23}$$

where $C^{(k)}$ is made of four diagonal blocks:

$$C^{(k)} = \left(\begin{array}{ccc|ccc} \ddots & \cdots & 0 & \ddots & \cdots & 0 \\ \vdots & \dfrac{\beta_n^{(k)} + \beta_n^{(M)}}{2\beta_n^{(k)}} & \vdots & \vdots & \dfrac{\beta_n^{(k)} - \beta_n^{(M)}}{2\beta_n^{(k)}} & \vdots \\ 0 & \cdots & \ddots & 0 & \cdots & \ddots \\ \hline \ddots & \cdots & 0 & \ddots & \cdots & 0 \\ \vdots & \dfrac{\beta_n^{(k)} - \beta_n^{(M)}}{2\beta_n^{(k)}} & \vdots & \vdots & \dfrac{\beta_n^{(k)} + \beta_n^{(M)}}{2\beta_n^{(k)}} & \vdots \\ 0 & \cdots & \ddots & 0 & \cdots & \ddots \end{array}\right). \tag{III.24}$$

Similarly, at $y = y_{k-1}$, we obtain:

$$\begin{pmatrix} \vdots \\ A_n^{(k)} \exp(-i\beta_n^{(k)} y_{k-1}) \\ \vdots \\ \hline \vdots \\ B_n^{(k)} \exp(i\beta_n^{(k)} y_{k-1}) \\ \vdots \end{pmatrix} = C'^{(k)} V^{(k-1)} \tag{III.25}$$

where $C'^{(k)}$ is given by:

$$C'^{(k)} = \left(\begin{array}{ccc|ccc} \ddots & \cdots & 0 & \ddots & \cdots & 0 \\ \vdots & \dfrac{\beta_n^{(M)} + \beta_n^{(k)}}{2\beta_n^{(M)}} & \vdots & \vdots & \dfrac{\beta_n^{(M)} - \beta_n^{(k)}}{2\beta_n^{(M)}} & \vdots \\ 0 & \cdots & \ddots & 0 & \cdots & \ddots \\ \hline \ddots & \cdots & 0 & \ddots & \cdots & 0 \\ \vdots & \dfrac{\beta_n^{(M)} - \beta_n^{(k)}}{2\beta_n^{(M)}} & \vdots & \vdots & \dfrac{\beta_n^{(M)} + \beta_n^{(k)}}{2\beta_n^{(M)}} & \vdots \\ 0 & \cdots & \ddots & 0 & \cdots & \ddots \end{array}\right). \tag{III.26}$$

From Eqs. (III.22), (III.23), and (III.25), it follows that the T matrix of the homogeneous slice is given by:

$$T^{(k)} = C^{(k)} \begin{pmatrix} e^{-i\beta_n^{(k)}(y_k - y_{k-1})} & 0 \\ 0 & e^{i\beta_n^{(k)}(y_k - y_{k-1})} \end{pmatrix} C'^{(k)}. \tag{III.27}$$

For TM polarization, matrices $C^{(k)}$ and $C'^{(k)}$ are derived from Eqs. (III.24) and (III.26) by replacing $\beta_n^{(k)}$ by $\beta_n^{(k)} / (\nu_k)^2$.

In fact, the transmission matrix of a homogeneous layer can be found directly without introducing intermediate infinitely thin layers with index ν_M, which will reduce the number of the matrix operations, but we have made this choice in order to preserve the same presentation for the grating and the homogeneous layer. Anyway, the choice makes no importance, because the T-matrix approach is not suitable for programming due to reasons explained in the next section.

III.5 Limitation of the T-matrix propagation algorithm

The T-matrix propagation algorithm has been applied with success to study Bragg multilayer gratings [III.1], [III.2] used in X-ray domain. It has also been applied with success in integrated optics [III.3] to study corrugated waveguides with double corrugation, on top and bottom interfaces. However, when applied to deep gratings with high conductivity, the method which consist in cutting the grating into M-2 slices, computing the various $T^{(j)}$ matrices, obtaining the T matrix through their product and finding the reflection and transmission matrices r and t through Eqs. (III.18) and (III.19) does not work. The use of the T-matrix propagation algorithm does not enlarge the domain of application of the differential method; the numerical contamination appears as soon as when we perform a unique numerical integration from bottom to top of the deep grating.

Indeed, when looking at Eqs. (III.14) and (III.15), we see that the (n,m) entry of the $T_{1,1}$ block matrix links $A_n^{(M)} \exp(-i\beta_n^{(M)} y_{M-1})$ to $A_m^{(1)}$, i.e., when $\beta_n^{(M)}$ is purely imaginary, it links two "antievanescent" waves at the two extreme ordinates y_{M-1} and 0 of the modulated area. Thus these entries are expected to be very large when n is high, for which $\left|\beta_n^{(M)}\right| \approx |nK|$. If we keep in mind that the transmitted efficiencies are numbers less than unity, that will be derived through Eqs. (III.21) and (III.18), we understand that their determination will be done through summations and differences of extremely high numbers, as required by the inversion process of $T_{1,1}$. It is then well known that digits will be lost and that the results may be completely wrong.

III.6 The S-matrix propagation algorithm

The S-matrix propagation algorithm avoids the loss of digits. In the T-matrix propagation algorithm, the final $T_{1,1}$ block matrix includes the product of the $T_{1,1}^{(j)}$ block matrices of the M-2 different slices, so that no compensation occurs between the growing exponential functions. The S-matrix propagation algorithm combines the various block matrices so that the exponentially growing functions never appears in the algorithm. Also, each block matrix of S has a physical interpretation that states it to be bounded.

The $S^{(q)}$ matrix of a stack of q slices (not to be confused with the S matrix of the qth slice!) is defined by:

$$\begin{pmatrix} \vdots \\ B_n'^{(q)} e^{i\beta_n^{(M)} y_q} \\ \vdots \\ A_n^{(1)} e^{-i\beta_n^{(1)} y_1} \\ \vdots \end{pmatrix} = S^{(q)} \begin{pmatrix} \vdots \\ B_n^{(1)} e^{i\beta_n^{(1)} y_1} \\ \vdots \\ A_n'^{(q)} e^{-i\beta_n^{(M)} y_q} \\ \vdots \end{pmatrix}. \tag{III.28}$$

It means that it gives the values of the outgoing or decaying waves from the values of the incoming or increasing ones, for the whole stack below the layer q+1. As a result, its four blocks $S_{ij}^{(q)}$ have simple interpretation meaning [I.38] as reflection or transmission matrices. For example, $S_{11}^{(q)}$ and $S_{12}^{(q)}$ are the transmission matrix and reflection matrix that give the upward wave amplitudes in the qth infinitely thin layer respectively resulting from the transmission of the upward incident waves in medium 1 (substrate) and from the reflection by the stack of the incident downward waves incident in the qth layer $(S_{12}^{(q)} \equiv r)$. As a result, the elements of these blocks are bounded. If we are given unit incident amplitudes at ordinate y_1 $(y_1 = 0)$, $S_{1,1,n,m}^{(q)}$ are bounded to unity if n and m correspond to propagating waves, and by decaying exponential functions for the other terms. Thus no large exponential functions are expected to appear in the $S^{(q)}$-matrix elements.

This assertion can be proved on the simple case of a single homogeneous layer with thickness t_1 included below a modulated stack. In that case, Eqs. (III.22)-(III.27) give a transmission matrix $T^{(2)}$ in close form. If we define

$$c_n^{(k)} = (\beta_n^{(k+1)} + \beta_n^{(k)})/(2\beta_n^{(k+1)})$$

$$d_n^{(k)} = (\beta_n^{(k+1)} - \beta_n^{(k)})/(2\beta_n^{(k+1)})$$

$$e_n^- = \exp(-i\beta_n^{(2)} t_1)$$

$$e_n^+ = 1/e_n^-,$$

we simply obtain for the four blocks of the $T^{(2)}$ matrix:

$$\begin{aligned}
T^{(2)}_{1,1,n,m} &= (c^{(1)}_n e^-_n c^{(2)}_n + d^{(1)}_n e^+_n d^{(2)}_n)\delta_{n,m} \\
T^{(2)}_{1,2,n,m} &= (c^{(1)}_n e^-_n d^{(2)}_n + d^{(1)}_n e^+_n c^{(2)}_n)\delta_{n,m} \\
T^{(2)}_{2,1,n,m} &= (d^{(1)}_n e^-_n c^{(2)}_n + c^{(1)}_n e^+_n d^{(2)}_n)\delta_{n,m} \\
T^{(2)}_{2,2,n,m} &= (d^{(1)}_n e^-_n d^{(2)}_n + c^{(1)}_n e^+_n c^{(2)}_n)\delta_{n,m}.
\end{aligned} \tag{III.29}$$

These equations contain growing and decaying exponential functions, which means that when n is large enough, the terms e^-_n will produce overflows and/or numerical contamination during the inversion process.

Let us now derive the $S^{(2)}$ matrix of the layer. Comparison of the definitions of the $S^{(q)}$ matrix, Eq. (III.28), and of the $T^{(q)}$ matrix, Eqs. (III.14) and (III.16), immediately shows that:

$$S^{(2)}_{22} = \left(T^{(2)}_{11}\right)^{-1}. \tag{III.30}$$

Moreover, we also get:

$$S^{(2)}_{12} = T^{(2)}_{21}\left(T^{(2)}_{11}\right)^{-1}. \tag{III.31}$$

As a result, since matrix $T^{(2)}$ of the slice is diagonal,

$$S^{(2)}_{2,2,n,m} = \delta_{n,m} / (c^{(1)}_n e^-_n c^{(2)}_n + d^{(1)}_n e^+_n d^{(2)}_n)$$

$$S^{(2)}_{1,2,n,m} = \frac{d^{(1)}_n e^-_n c^{(2)}_n + c^{(1)}_n e^+_n d^{(2)}_n}{c^{(1)}_n e^-_n c^{(2)}_n + d^{(1)}_n e^+_n d^{(2)}_n}\delta_{nm}.$$

When $e^-_n \to \infty$ while $e^+_n \to 0$, we obtain that $S^{(2)}_{2,2,n,m} \to 0$ and $S^{(2)}_{1,2,n,m} \to \frac{d^{(1)}_n}{c^{(1)}_n}$. Thus all elements of $S^{(2)}_{1,2}$ and $S^{(2)}_{2,2}$ remain bounded. As will be shown later, they are the only blocks that are used in the S-matrix propagation algorithm. Hence the stability of the algorithm is expected, at least for homogeneous layers.

The S-matrix propagation algorithm is constructed in writing Eq. (III.28) at levels q and q+1. If we define column vectors $\tilde{A}^{(q)}$ and $\tilde{B}^{(q)}$ with components

$$\left(\tilde{B}^{(q)}\right)_n = \tilde{B}_n^{(q)} = B_n'^{(q)} \exp(i\beta_n^{(M)} y_q) \tag{III.32}$$

and

$$\left(\tilde{A}^{(q)}\right)_n = \tilde{A}_n^{(q)} = A_n'^{(q)} \exp(-i\beta_n^{(M)} y_q), \tag{III.33}$$

we obtain the block matrix relations:

$$\tilde{B}^{(q+1)} = S_{11}^{(q+1)}\tilde{B}^{(1)} + S_{12}^{(q+1)}\tilde{A}^{(q+1)} \tag{III.34}$$

$$\tilde{A}^{(1)} = S_{21}^{(q+1)}\tilde{B}^{(1)} + S_{22}^{(q+1)}\tilde{A}^{(q+1)} \tag{III.35}$$

and

$$\tilde{B}^{(q)} = S_{11}^{(q)}\tilde{B}^{(1)} + S_{12}^{(q)}\tilde{A}^{(q)} \tag{III.36}$$

$$\tilde{A}^{(1)} = S_{21}^{(q)}\tilde{B}^{(1)} + S_{22}^{(q)}\tilde{A}^{(q)} \tag{III.37}$$

We then relate the downgoing and upgoing waves through the T matrix, Eq. (III.13):

$$\tilde{A}^{(q+1)} = T_{11}^{(q+1)}\tilde{A}^{(q)} + T_{12}^{(q+1)}\tilde{B}^{(q)} \tag{III.38}$$

$$\tilde{B}^{(q+1)} = T_{21}^{(q+1)}\tilde{A}^{(q)} + T_{22}^{(q+1)}\tilde{B}^{(q)}. \tag{III.39}$$

Putting $\tilde{B}^{(q)}$ given by Eq. (III.36) into Eq. (III.39) gives:

$$\tilde{B}^{(q+1)} = \left(T_{21}^{(q+1)} + T_{22}^{(q+1)}S_{12}^{(q)}\right)\tilde{A}^{(q)} + T_{22}^{(q+1)}S_{11}^{(q)}\tilde{B}^{(1)} \tag{III.40}$$

Eq. (III.37) gives:

$$\tilde{A}^{(q)} = \left(S_{22}^{(q)}\right)^{-1}\left(\tilde{A}^{(1)} - S_{21}^{(q)}\tilde{B}^{(1)}\right), \tag{III.41}$$

which, when put into Eq. (III.40), gives:

$$\tilde{B}^{(q+1)} = \left(T_{21}^{(q+1)} + T_{22}^{(q+1)} S_{12}^{(q)}\right) \left(S_{22}^{(q)}\right)^{-1} \tilde{A}^{(1)}$$

$$+ \left[T_{22}^{(q+1)} S_{11}^{(q)} - T_{21}^{(q+1)} + T_{22}^{(q+1)} S_{12}^{(q)} \left(S_{22}^{(q)}\right)^{-1} S_{21}^{(q)} \right] \tilde{B}^{(1)}$$

Replacing $\tilde{A}^{(1)}$ by the right-hand member of Eq. (III.35) leads to:

$$\begin{aligned} &\tilde{B}^{(q+1)} = \\ &\Big[\left(T_{21}^{(q+1)} + T_{22}^{(q+1)} S_{12}^{(q)}\right)\left(S_{22}^{(q)}\right)^{-1} S_{21}^{(q+1)} \\ &+ T_{22}^{(q+1)} S_{11}^{(q)} - \left(T_{21}^{(q+1)} + T_{22}^{(q+1)} S_{12}^{(q)}\right)\left(S_{22}^{(q)}\right)^{-1} S_{21}^{(q)} \Big] \tilde{B}^{(1)} \\ &+ \left(T_{21}^{(q+1)} + T_{22}^{(q+1)} S_{12}^{(q)}\right)\left(S_{22}^{(q)}\right)^{-1} S_{22}^{q+1)} \tilde{A}^{(q+1)} . \end{aligned} \tag{III.42}$$

In a similar way, putting $\tilde{A}^{(q)}$ given by Eq. (III.41) into Eq. (III.38) gives:

$$\tilde{A}^{(q+1)} = T_{11}^{(q+1)} \left(S_{22}^{(q)}\right)^{-1} \tilde{A}^{(1)} - T_{11}^{(q+1)} \left(S_{22}^{(q)}\right)^{-1} S_{21}^{(q)} \tilde{B}^{(1)} + T_{12}^{(q+1)} \tilde{B}^{(q)}. \tag{III.43}$$

Replacing $\tilde{B}^{(q)}$ by its value given by Eq. (III.36) gives:

$$\tilde{A}^{(q+1)} = T_{11}^{(q+1)} \left(S_{22}^{(q)}\right)^{-1} \tilde{A}^{(1)} - T_{11}^{(q+1)} \left(S_{22}^{(q)}\right)^{-1} S_{21}^{(q)} \tilde{B}^{(1)} + T_{12}^{(q+1)} S_{11}^{(q)} \tilde{B}^{(1)}$$

$$+ T_{12}^{(q+1)} S_{12}^{(q)} \left(S_{22}^{(q)}\right)^{-1} \left(\tilde{A}^{(1)} - S_{21}^{(q)} \tilde{B}^{(1)}\right)$$

or

$$\tilde{A}^{(q+1)} = \left(T_{11}^{(q+1)} + T_{12}^{(q+1)} S_{12}^{(q)}\right) \left(S_{22}^{(q)}\right)^{-1} \tilde{A}^{(1)}$$

$$- \left(T_{11}^{(q+1)} + T_{12}^{(q+1)} S_{12}^{(q)}\right)\left(S_{22}^{(q)}\right)^{-1} S_{21}^{(q)} \tilde{B}^{(1)} + T_{12}^{(q+1)} S_{11}^{(q)} \tilde{B}^{(1)} .$$

Introducing a matrix $Z^{(q)}$ defined by:

$$Z^{(q)} = \left(T_{11}^{(q+1)} + T_{12}^{(q+1)} S_{12}^{(q)}\right)^{-1}, \tag{III.44}$$

the previous equation reads:

$$S_{22}^{(q)} Z^{(q)} \tilde{A}^{(q+1)} = \tilde{A}^{(1)} - S_{21}^{(q)} B^{(1)} + S_{21}^{(q)} Z^{(q)} T_{12}^{(q)} S_{11}^{(q)} \tilde{B}^{(1)},$$

from which we derive:

$$\tilde{A}^{(1)} = S_{21}^{(q)} \left(\mathbb{1} - Z^{(q)} T_{12}^{(q+1)} S_{11}^{(q)} \right) \tilde{B}^{(1)} + S_{22}^{(q)} Z^{(q)} \tilde{A}^{(q+1)} \quad . \tag{III.45}$$

Comparison with Eq. (III.35) leads to:

$$S_{22}^{(q+1)} = S_{22}^{(q)} Z^{(q)}. \tag{III.46}$$

With that result in mind, comparison of Eqs. ((III.34) and (III.42) leads to:

$$S_{12}^{(q+1)} = \left(T_{21}^{(q+1)} + T_{22}^{(q+1)} S_{12}^{(q)} \right) \; Z^{(q)} \tag{III.47}$$

Of course, $S_{11}^{(q)}$ and $S_{21}^{(q)}$ could be expressed in a similar way; we could find:

$$S_{21}^{(q+1)} = S_{21}^{(q)} - S_{22}^{(q+1)} T_{12}^{(q+1)} S_{11}^{(q)}$$

and

$$S_{11}^{(q+1)} = -S_{12}^{(q+1)} T_{12}^{(q+1)} S_{11}^{(q)} + T_{22}^{(q+1)} S_{11}^{(q)}$$

but, generally, the last two equations are not used in diffraction theory, as it will be shown in what follows. The algorithm is started by taking $S_{12}^{(1)} = 0$ and $S_{22}^{(1)} = \mathbb{1}$, which means that when there is no slice, no reflection occurs and the transmission is unity for all incoming waves. As a result,

$$Z^{(1)} = \left(T_{11}^{(2)} \right)^{-1}$$

$$S_{12}^{(2)} = T_{21}^{(2)} \left(T_{11}^{(2)} \right)^{-1}$$

$$S_{22}^{(2)} = \left(T_{11}^{(2)} \right)^{-1}.$$

These equations were already derived from direct comparison of the definitions of matrices $T^{(2)}$ and $S^{(2)}$ (see Eqs. (III.30) and (III.31)). They may be used to start the algorithm. From the values of $S_{12}^{(2)}$ and $S_{22}^{(2)}$, the values of $S_{12}^{(M)}$ and $S_{22}^{(M)}$ for the entire stack are found through the recursive process specified by

Eqs. (III.46), (III.47) and (III.44). The four blocks of the $T^{(q+1)}$ matrix are computed using the shooting method described in Section III.3, which requires 2(2N+1) starting vectors. It is worth noticing that Eqs. (III.46) and (III.47) do not make use of blocks $S_{11}^{(q)}$ and $S_{21}^{(q)}$. However, some applications require the knowledge of the four blocks to the S-matrix; for example, in the calculation of the forbidden gaps of photonic crystal (see Chapter X). Following Eq. (III.28) and taking into account that $B_n'^{(M-1)} = B_n^{(M)}$ and $A_n'^{(M-1)} = A_n^{(M)}$, the entire S-matrix of the stack will have a block form:

$$\begin{pmatrix} \vdots \\ B_n^{(M)} e^{i\beta_n^{(M)} y_{M-1}} \\ \vdots \\ \hline \vdots \\ A_n^{(1)} \\ \vdots \end{pmatrix} = \begin{pmatrix} S_{11}^{(M-1)} & S_{12}^{(M-1)} \\ S_{21}^{(M-1)} & S_{22}^{(M-1)} \end{pmatrix} \begin{pmatrix} \vdots \\ B_n^{(1)} \\ \vdots \\ \hline \vdots \\ A_n^{(M)} e^{-i\beta_n^{(M)} y_{M-1}} \\ \vdots \end{pmatrix}. \quad \text{(III.48)}$$

The outgoing-wave condition that specifies that $B_n^{(1)} = 0 \ \forall n$ allows finding the Rayleigh coefficients:

$$B_n^{(M)} \exp(i\beta_n^{(M)} y_{M-1}) = \sum_{m=-N}^{+N} S_{1,2,n,m}^{(M-1)} A_m^{(M)} \exp(-i\beta_m^{(M)} y_{M-1}) \quad \text{(III.49)}$$

$$A_n^{(1)} = \sum_{m=-N}^{+N} S_{2,2,n,m}^{(M-1)} A_m^{(M)} \exp(-i\beta_m^{(M)} y_{M-1}). \quad \text{(III.50)}$$

Equations (III.49)-(III.50) clearly show that only half of the S-matrix is useful to resolve the diffraction problem. Moreover, since $A_m^{(M)} = \delta_{m,0}$, we simply obtain:

$$B_n^{(M)} = S_{1,2,n,0}^{(M-1)} \exp\left[-i(\beta_n^{(M)} + \beta_0^{(M)}) y_{M-1}\right] \quad \text{(III.51)}$$

$$A_n^{(1)} = S_{2,2,n,0}^{(M-1)} \exp(-i\beta_0^{(M)} y_{M-1}). \quad \text{(III.52)}$$

Thus the Rayleigh coefficients are readily derived from the central column of the right-half of the S-matrix. This is a strong advantage over the previously used R-matrix propagation algorithm [I.36, I.38], which, on top of computing the four blocks of matrix R, required inversion of a (4N+2)x(4N+2) matrix to get the Rayleigh coefficients. Some special applications (e.g., in nonlinear optics) require the knowledge of the field inside the stack of slices. How to determine it is explained at the end of Chapter XV. Studies of photonic crystals need to calculate the eigenvalues of the T matrix. How to obtain them when the S-matrix is known is shown in Chapter X.

Eqs. (III.51) and (III.52) are fully consistent with eqs. (III.20) and (III.21). And indeed, using the definitions of the four blocks of the S-matrix, Eq. (III.48), it follows that $S_{1,2}^{(M-1)} \equiv r$ and that $S_{2,2}^{(M-1)} \equiv t$. The major advantage of Eqs. (III.51) and (III.52) is that the two blocks $S_{1,2}^{(M-1)}$ and $S_{2,2}^{(M-1)}$ are obtained without numerical contamination.

How can the stability of the S-matrix propagation algorithm be explained in the general case of an arbitrary profile? Although arguments can be found in the simple case of homogeneous layers and for lamellar profiles for which the y-dependence of the field is known, as far as we know, nothing convincing has been written in the general case. A physical interpretation [III.4]* of the S-matrix propagation algorithm has been done by comparing it with the Bremmer series method [III.5]. However, as pointed out in [III.4], the two methods are not identical. The study done in [I.35] shows that Bremmer series method fails in TM polarization and metallic gratings, while the S-matrix propagation method works. The only argument that we can present is that the T-matrix propagation method requires inversion of the global T_{11} block, which, as explained in Section III.5, has elements that may overflow. On the other hand, the S-matrix propagation algorithm only makes use of two blocks of matrices $S^{(q)}$, which were explained to be bounded, and compute the inverse $Z^{(q)}$ of the linear combination: $T_{11}^{(q+1)} + T_{12}^{(q+1)} S_{12}^{(q)}$. As soon as the number M-1 of slices has been chosen large enough so that no numerical contamination occurs in the computation of the $T^{(q)}$ matrices, the previous inversion is made without any difficulty. Indeed, the elements of the matrices $T^{(q)}$ relate "anti-evanescent" waves at the two ordinates y_q and y_{q-1}, which are close enough to avoid too large numbers, contrary to what occurred in computing T_{11}. This time, the algorithm does not make the products of $T_{ij}^{(q)}$ elements; ratios of such elements appear in Eq. (III.47), as they appeared on the simple example of a homogeneous slice.

Concerning the determination of the number of slices that must be chosen to ensure a noncontaminated determination of the $T^{(j)}$ matrices, we must ensure that the highest exponential functions in Rayleigh expansion are less than 10^{16} or 10^{15}, otherwise digits begin to be lost. This implies

---

* In this paper, the S-matrix was unfortunately called R-matrix.

$$\left|\exp\left[(\operatorname{Im}\beta_{\pm N}^{(M)})\, y_{M-1}/(M-2)\right]\right| < 10^{15},$$

which is satisfied if

$$\left[\left|\operatorname{Im}\beta_{\pm N}^{(M)}\right| y_{M-1}/(M-2)\right] < 15\times 2.3.$$

This equation gives a value of M that can be increased at the demand without significant increase of the computation time.

It is interesting to remark the similarities and differences between the S-matrix propagation algorithm and the improvement of the differential method proposed by Tayeb [III.6]. In the differential theory, the numerical solution [F(y)] determined by the shooting method is a linear superposition of particular vectors $[\hat{F}(y)]_p$ (see for example Eq. (II.16)). At the starting step of the integration process (at $y = 0$), the different vectors $[\hat{F}(0)]_p$ are orthonormalized. But due to the coupling between the various Fourier components inherent to the set of equations that one integrates, the vectors do not remain orthogonal during the integration process. Tayeb suspected that the numerical contamination could be linked with this loss of orthogonality. He proposed a method to estimate the "orthogonality degree." Then, when this criterion becomes too small while y is increased, the integration process is stopped, say, at ordinate y_j . A new orthogonal set of vectors is derived from the computed vectors $[\hat{F}(y_j)]_p$ via the Schmidt orthogonalization process; then a new integration process starts with these new orthogonal vectors. The process has shown good improvement in the calculation of the reflected efficiencies of metallic gratings [III.6]. No numerical result can be found in [III.6] concerning the transmitted efficiencies of thick dielectric gratings. It is worth noting that in the S-matrix propagation algorithm, we also start each integration at the bottom y_j of each slice by a set of orthonormalized vectors $[\hat{F}(y_j)]_p$.

Another possibility exists for formulating the S-matrix algorithm [I.38, I.4], which is easier to program, but less attractive from a pedagogical point of view. It consists in dealing with the x and z-components of the field, instead of using the up- and downgoing wave components in the definition of the S matrix, Eq. (III.28). For example, in TE polarization it takes the form

$$\begin{pmatrix} \vdots \\ E_n(y_q) \\ \vdots \\ H_n(0) \\ \vdots \end{pmatrix} = S_F^{(q)} \begin{pmatrix} \vdots \\ E_n(0) \\ \vdots \\ H_n(y_q) \\ \vdots \end{pmatrix} \qquad \text{(III.53)}$$

where the subscript F of S indicates that this is the S-matrix of the field components. Working with S_F instead of S eliminates the use of artificially introduced infinite thin layers with index ν_M and thus reduces the number of matrix operations. In addition, this approach has the same advantage as the approach presented above in eliminating the numerical instabilities due to the growing terms. However, it does not bring insight as to why the S-matrix algorithm eliminates numerical contamination.

III.7 Remaining limitation of the differential method

The S-matrix propagation algorithm is the simplest, fastest and most-elegant way to analyze thick modulated stacks. It has blown up the numerical contamination that plagued the differential theory during a quarter of century. Applied to thick phase gratings and multilayer structures with arbitrary refractive index profile [III.7], numerical integrations have been carried out over ranges of hundred thousand wavelengths, requiring hundred thousand slices in the algorithm, without significant numerical contamination.

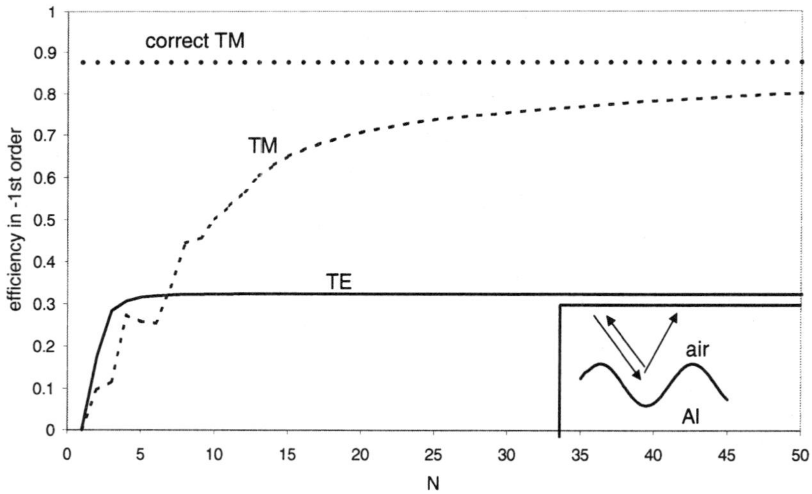

Fig. III.3. Convergence rates for TE and TM polarization as a function of the truncation parameter N. Efficiency in -1st order for a sinusoidal aluminum grating with period d = 0.5 μm and groove depth 0.2 μm illuminated at 30° angle of incidence with 0.6328 μm wavelength. The "correct TM" value is obtained by the integral method.

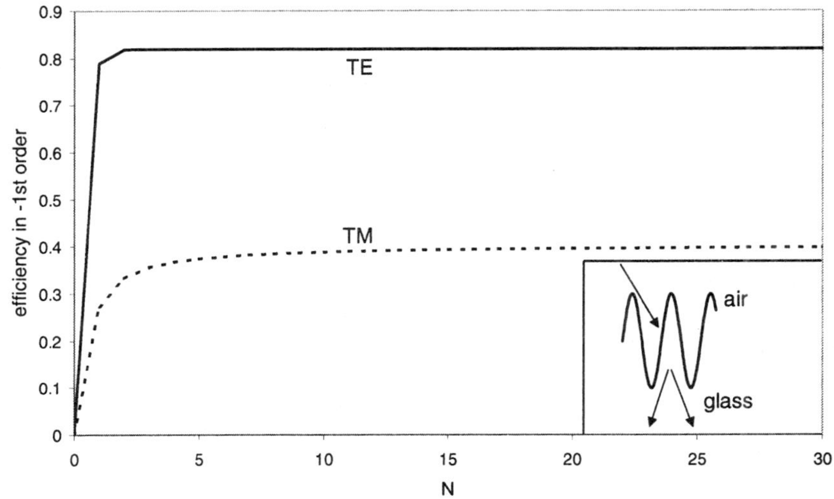

Fig. III.4. Same as in Fig. III.3, except for deep sinusoidal glass grating, working in transmission. Period d = 0.5 μm and groove depth 1.0 μm illuminated at 30° angle of incidence with 0.6328 μm wavelength. ν_1 = 1.5.

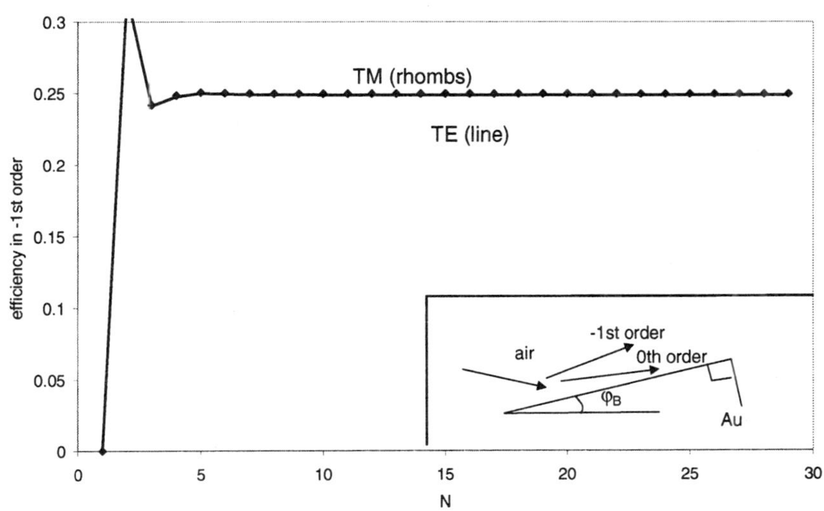

Fig. III.5. Same as Fig. III.3 except for gold echelette grating (blaze angle $\varphi_B = 0.75^\circ$) in grazing incidence (angle of incidence 87.5°) in X-ray domain, $\lambda = 1.334$ nm.

In X-ray domain, Bragg-Fresnel linear zone plates [III.8], which are nonperiodic objects, have been analyzed, which required integrating 200 coupled differential equations; similar numerical problems were encountered for analyzing multilayer coated echelles for X-rays [III.9]. They were resolved with success. When the S-matrix propagation algorithm was applied to analyze deep metallic gratings used in TM polarization, no overflow or numerical contamination was observed. However, a poor convergence was obtained on the efficiencies when increasing the number (2N+1) of the Fourier components of the field. A typical example is presented in Fig. III.3, where the convergence rates with respect to N are compared in TE and TM polarization for a sinusoidal metallic grating, widely used in laser applications. As can be observed, less than 11 Fourier components (N = 5) are sufficient to correctly represent the field in TE polarization, whereas in TM case even 101 components are not sufficient to reduce the error to less than 5%. Two counter-examples are given in Figs. III.4 and III.5. Fig. III.4 shows that for a dielectric material (refractive index 1.5) the classical differential method can be used with success in both TE and TM polarization, even with a depth-to-period ratio exceeding 200%. Fig. III.5 shows that gratings made of good conducting metals can also be analyzed in some particular spectral domains. This is the case in the X-ray domain, where the refractive indices of most of the materials do not significantly differ from unity, and gratings usually have low modulation.

It looked like as if the Fourier basis was not adequate to represent the field near good conductors; this was in contradiction with the results obtained with the conformal mapping technique [I.21], which gave a fast convergent representation of the field near perfectly conducting gratings in terms of Fourier series. As previously explained in Section I.1.4, the remarkable work of Li [I.44] prompted us to suspect the basic propagation equations written in a truncated Fourier space. We then developed what we now call the fast Fourier factorization (FFF) method, which produces much faster converging results when studying diffraction by highly conducting devices.

References

III.1. M. Nevière, "Bragg-Fresnel multilayer gratings: electromagnetic theory," J. Opt. Soc. Am., A11, 1835-1845, 1994

III.2. B. Vidal, P. Vincent, P. Dhez, and M. Nevière, "Thin films and gratings: theories used to optimize the high reflectivity of mirrors and gratings for X-ray optics," in *Applications of thin film multilayered structures to figured X-ray optics*, G. F. Marshall, ed., Proc. Soc. Photo-Opt. Instrum. Eng. 563, 142-149, 1985

III.3. J. C. Brazas, L. Li, and A. McKeon, "High efficiency input coupling into optical waveguide using grating with double surface corrugation," Appl. Opt. 34, 604-609, 1995

III.4. L. Li, "Bremmer series, R-matrix propagation algorithm, and numerical modeling of diffraction gratings," J. Opt. Soc. Am., A11, 2829-2836, 1994

III.5. H. Bremmer, "The W. K. B. approximation as the first term of a geometrical-optical series," Commun. Pure. Appl. Math. 4, 105-115, 1951

III.6. G. Tayeb, "Contribution à l'étude de la diffraction des ondes électromagnétiques par des réseaux. Reflections sur les méthodes existantes et sur leur extention aux milieux anisotropes," thesis presented on 5 December 1990 at the Université de Droit, d'Economie et des Sciences d'Aix-Marseille III, Faculté des Sciences et Techniques de St Jérôme, Marseille, France

III.7. B. Bozkhov, M. Nevière, and E. Popov, "Differential method as applied to multilayered structures with arbitrary refractive index profiles," Bulg. J. Physics, to be published

III.8. F. Montiel and M. Nevière, "Elecromagnetic theory of Bragg-Fresnel linear zone plates," J. Opt. Soc. Am., A12, 2672-2678, 1995

III.9. M. Nevière and F. Montiel, "Soft X-ray multilayer-coated echelle gratings: electromagnetic and phenomenological study," J. Opt. Soc. Am., A13, 811-818, 1996

Chapter IV

Fast Fourier Factorization (FFF) Method

The S-matrix propagation algorithm, presented in detail in Chapter III, completely eliminates the numerical problems linked with the integration process and matrix inversion in the classical differential method. However, as explained in the first two chapters, the Fourier series of the field for highly conducting gratings used in TM polarization remained slowly converging. This problem remained a mystery for quite long time. Of course, one could try to convince himself of the basic difference in nature between TE and TM polarization problems in saying that in TE polarization, we only represent into Fourier series a continuous function $E_z(x,y)$, while in TM polarization, on top of $H_z(x,y)$, we also represent into Fourier series a discontinuous function E_x whose coefficients $E_n(y)$ only decrease as $1/n$, instead of $1/n^2$. Thus the desire of working with continuous functions. Today, it is clear that the difficulties of the differential method for TM case lie in the correct presentation of the Fourier series of different products of discontinuous functions, when truncated to a finite number of Fourier components.

In fact, the main difference between the basic propagation equations (II.20), (II.21) and (II.17), (II.18) for TM and TE polarization lies in the fact that in TE polarization, we will have to represent by a truncated Fourier series the product $\varepsilon(x,y)E_z$, which is the product of a *discontinuous function* $\varepsilon(x,y)$ by a *continuous function* E_z, while in TM polarization, two products (εE_x and $\frac{1}{\varepsilon}\frac{\partial H_z}{\partial x}$) will have to be developed, which are both products of *two discontinuous functions*. The question is whether Eqs. (II.21) and (II.22), when truncated to order N, will give the best Fourier representation of these products and will converge to the exact equations when $N \to \infty$.

This question is not treated in the textbooks on Fourier analysis, because, when working with infinite series, the answer is obvious and does not depend on whether the functions are continuous or not. The trivial answer to how to factorize the Fourier series of a product is given by the Laurent's rule:

$$(fg)_n = \sum_{m=-\infty}^{+\infty} f_{n-m}g_m = \sum_{m=-\infty}^{+\infty} g_{n-m}f_m. \tag{IV.1}$$

If we designate by $[\![f]\!]$ the Toeplitz matrix whose (n, m) entry is the $(n-m)^{th}$ Fourier component of f, i.e., f_{n-m}, and by [g] the vector constructed with the Fourier components g_n of function g, Eq. (IV.1) can be written in a matrix form:

$$[fg]=[\![f]\!][g]=[\![g]\!][f].$$

IV.1 Truncated products: a simple example

Let us consider a very simple example in order to illustrate the importance of the truncation of the Fourier series of products when dealing with periodic discontinuous functions. Both functions f(x) and g(x) are defined in the interval $[-d/2,+d/2]$ and are discontinuous at x = 0, but their product is continuous and constant, Fig. IV.1. The functions are defined as:

$$\left.\begin{matrix} f=0.5 \\ g=2 \end{matrix}\right\} x\in\left(-\frac{d}{2},0\right) \quad \text{and} \quad \left.\begin{matrix} f=2 \\ g=0.5 \end{matrix}\right\} x\in\left(0,+\frac{d}{2}\right). \qquad \text{(IV.2)}$$

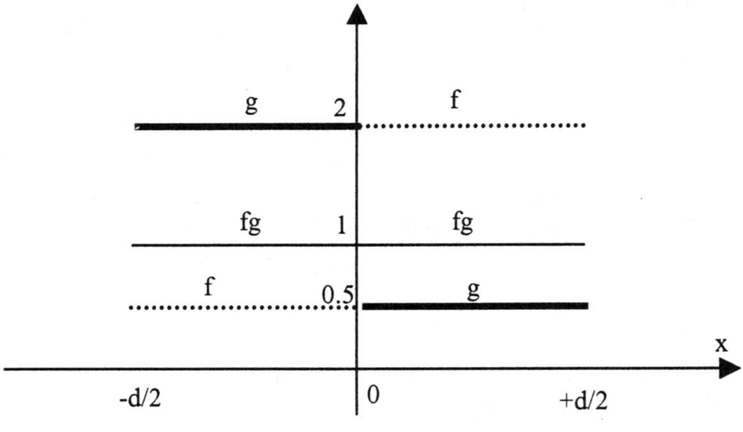

Fig. IV.1 Graphical representations of two discontinuous functions f and g with continuous product fg over one period.

Their product fg = 1 is constant, so that its only nonvanishing Fourier component is the zeroth one, $(fg)_m=\delta_{m,0}$. Let us try to use the Laurent's rule in order to factorize the Fourier series of the product fg, by truncating to the extreme the Fourier series of f(x) and g(x), i.e., by considering only their zeroth Fourier component (mean value):

$$[f]_0 = \frac{1}{d}\int_{-d/2}^{+d/2} f(x)dx = 1.25 \text{ and } [g]_0 = \frac{1}{d}\int_{-d/2}^{+d/2} g(x)dx = 1.25 .$$

Applying the extremely truncated Laurent's rule, the mean value of fg is given by: $[fg]_0 = [f]_0 [g]_0 = 1.5625$, which presents an error greater than 56%. However, the correct result (fg = 1) can be obtained by replacing $[f]_0$ by $[1/f]_0^{-1}$ (assuming that $[1/f]_0 \neq 0$):

$$[fg]_0 = \left[\frac{1}{f}\right]_0^{-1} [g]_0 = \left(\frac{0.5}{2} + \frac{0.5}{0.5}\right)^{-1} 1.25 = \frac{1}{1.25} 1.25 = 1.$$

This factorization will be called "inverse rule" in the next section.

IV.2 Exact propagation equations in a truncated Fourier basis for TM polarization

With the notations introduced in the previous section, Eqs. (II.21) and (II.22) can be written in the form:

$$\left[\frac{\partial H_z}{\partial y}\right] = -i\omega[\![\varepsilon]\!][E_x] \tag{IV.3}$$

$$\left[\frac{\partial E_x}{\partial y}\right] = \frac{i}{\omega}\alpha\left[\!\!\left[\frac{1}{\varepsilon}\right]\!\!\right]\alpha[H_z] - i\omega\mu[H_z] \tag{IV.4}$$

where α is a diagonal matrix with elements $\alpha_n = \alpha_0 + n\,2\pi/d$.

Although these equations are rigorous as far as they are not truncated, following the simple example in the previous section, one can suspect that in view of truncating to order N it could be better to replace them by

$$\left[\frac{\partial H_z}{\partial y}\right] = -i\omega\left[\!\!\left[\frac{1}{\varepsilon}\right]\!\!\right]^{-1}[E_x] \tag{IV.5}$$

$$\left[\frac{\partial E_x}{\partial y}\right] = \frac{i}{\omega}\alpha [\![\varepsilon]\!]^{-1}\alpha[H_z] - i\omega\mu[H_z]. \tag{IV.6}$$

In 1995, Eqs. (IV.5) and (IV.6) were applied [I.42] and [I.43] to lamellar gratings to give much faster convergence than Eqs. (IV.3) and (IV.4). That empirical discovery received a solid mathematical foundation from Li [I.44]. Using three theorems of Fourier factorization, Li established the extremely important following conclusion. Although the propagation equations of the original differential theory were exact, both those written in spatial variables and those in discrete Fourier space, the latter ones become ill-suited as soon as they are truncated by any numerical treatment. The truncated equations no longer preserve the continuity of the appropriate field components across the discontinuities of the permittivity function. Li proposed appropriate factorization rules to preserve that continuity.

It is worth pointing out that when in 1975 one of the authors (Nevière) tried to use the same set of equations (IV.5) and (IV.6), as well as any other possible combination, set (IV.3) and (IV.6) or set (IV.4) and (IV.5), applied for sinusoidal profiles, this didn't work to ameliorate the convergence. The explanation of why this happened and why Eqs. (IV.5) and (IV.6) are the best choice for lamellar profiles is our next aim. After that, we use the factorization rules to establish new propagation equations for the differential method that can be truncated without any numerical problem. These equations are established for arbitrary profiles, of which equations (IV.5) and (IV.6) used in [I.42] and [I.43] appear as particular cases. Their advantages will be shown through numerical results in various selected examples.

IV.2.1 Factorization rules

Eqs. (II.21) and (II.22) which are the foundation of the classical differential method are rigorous in so far as the summation over subscript m runs over an infinite number of terms. But for making computations, these series have to be truncated from –N to +N limits, and all authors who have worked on the differential theory have assumed that such a process is valid provided that N is large enough. The great merit of the work of Li was to study the legitimacy of the truncation process and to show that the equations derived from a truncation of Eqs. (II.21) and (II.22) are exact almost everywhere, but not everywhere, a fact that makes a world of difference in the computation.

Indeed, Eqs. (II.21) and (II.22) are derived from Eqs. (I.20) and (I.21) assuming Laurent's rule [IV.1], which states that the Fourier components h_n of the product $h(x)$ of two arbitrary periodic functions $f(x)$ and $g(x)$ is simply given by:

$$h_n = \sum_{m=-\infty}^{+\infty} f_{n-m} g_m, \tag{IV.7}$$

or in matrix notation:

$$[h] = [\![f]\!][g] \tag{IV.8}$$

Here both the vectors and the Toeplitz matrix have infinite size. The problem that we address is whether Eq. (IV.8) still holds when the vectors and the Toeplitz matrix are truncated to a finite order N. The difficulty comes from the fact that, depending on the kind of discontinuities of functions f(x) and g(x), the truncation of Eq. (IV.8) will, or will not, introduce errors on the reconstructed function h(x) obtained through its Fourier harmonics. Li clearly derived three important conclusions [I.44]:

1. A product of two piecewise-smooth, bounded, periodic functions that have no concurrent jump discontinuities can be factorized by Laurent's rule:

$$h_n^{\{N\}} = \sum_{m=-N}^{N} f_{n-m} g_m. \tag{IV.9}$$

"No concurrent jump discontinuities" means that functions f and g are not discontinuous at the same abscissa x. This is of course satisfied if g is continuous while f is discontinuous. In that case, Eq. (IV.9) gives a better approximation $h_n^{\{N\}}$ of h_n than $\sum_{m=-N}^{N} g_{n-m} f_m$, because it truncates more severely the Fourier components of the continuous function which decrease as $1/m^2$, while f_m only decrease as $1/m$.

The validity of Eq. (IV.9) becomes obvious if g(x) can be well represented by 2N+1 Fourier components (e.g., a constant function is represented by 1 component, sine and cosine, by 3 components). In that case, the truncation of the infinite sum, Eq. (IV.7), to a finite one, Eq. (IV.9) is rigorous and does not introduce errors other than the round-off ones.

For a continuous function, the Fourier components decrease at least as $1/m^2$, so that if N in Eq. (IV.9) is large enough, all the Fourier components of g such that $|m| > N$ can be replaced by zeros with an insignificant error (which decreases as $1/N^2$). Then the truncation of the Fourier series of the discontinuous function f does not introduce any additional error, since the neglected coefficients of f would have been multiplied by zeros in the nontruncated series. Thus, for any continuous function g, Eq. (IV.9) can be used to approximate Eq. (IV.7) to any degree of accuracy.

This is not the case when g(x) is a discontinuous function. Then, any truncation of the series will introduce a Gibbs phenomenon and the error when applying Eq. (IV.9) becomes large. A simple example was given in Section IV.1, where a truncation to N = 0 was applied and the possibility of amelioration of the truncated series factorization was proposed. This leads us to the second rule.

2. A product of two piecewise-smooth, bounded, periodic functions that have only pairwise-complementary jump discontinuities cannot be factorized by Laurent's rule, but in most cases, it can be factorized by the inverse rule:

$$h_n^{\{N\}} = \sum_{m=-N}^{N} \left(\left[\!\left[\frac{1}{f} \right]\!\right]^{\{N\}} \right)_{n,m}^{-1} g_m . \qquad \text{(IV.10)}$$

"Functions that have only pairwise-complementary jump discontinuities" mean that both f and g are discontinuous at the same abscissa, and that the values of their jump are such that the product $h = fg$ is continuous. It is thus obvious to derive the "inverse rule" by applying the Laurent's rule at function $g = \frac{1}{f} h$:

$$g_n^{\{N\}} = \sum_{m=-N}^{+N} \left[\!\left[\frac{1}{f} \right]\!\right]_{n,m}^{\{N\}} h_m ,$$

which, after inversion, leads to Eq. (IV.10).

3. A product of two piecewise-smooth, bounded, periodic functions that have concurrent but not complementary jump discontinuities can be Fourier factorized neither by Laurent's rule, nor by the inverse rule.

The use of the second conclusion allowed Li to derive on a mathematical basis the equations previously proposed in [I.42] and [I.43] in the case of lamellar profiles. It can easily be verified that, for this simple profile, the product εE_x in Eq. (I.20) is continuous, since it is equal to the component of $\vec{D}$ normal to the profile. Moreover, $\frac{1}{\varepsilon} \frac{\partial H_z}{\partial x}$ in Eq. (I.21) is also continuous, since, as shown by Eq. (I.14'), it is proportional to E_y, which, in that case, is purely tangential. Thus the inverse rule (Eq. (IV.10)) applies for both equations.

Unfortunately, for an arbitrary groove shape, the two above-mentioned products are no longer continuous, and conclusion **3** applies. Our aim is to reformulate the propagation equations in such a way that conclusions **1** and **2** could be used.

IV.2.2 Derivation of the equations for TM polarization

Since E_x and E_y are, in general, discontinuous functions at the boundary of an arbitrary grating profile, we express these components in terms of the tangential (E_T) and normal (E_N) component. To that end, let us introduce the unit vector $\vec{T}$ of the tangent at the grating profile at a point (x_1, y) as the one illustrated in Fig. IV.2. Its components are simply given by:

$$T_x = \cos\theta;\ T_y = \sin\theta\ ,$$

where θ is defined as

$$\tan\theta = \frac{dg}{dx}\ ,\ \theta \in \left[-\frac{\pi}{2}, +\frac{\pi}{2}\right],$$

$$\theta = -\pi/2 \ \text{if}\ \frac{dg}{dx} \to -\infty,$$

$$\theta = \pi/2 \ \text{if}\ \frac{dg}{dx} \to +\infty.$$

Thus E_T and E_N are written as:

$$E_T = E_x \cos\theta + E_y \sin\theta \tag{IV.11}$$

$$E_N = -E_x \sin\theta + E_y \cos\theta. \tag{IV.12}$$

In what follows, we use the following abbreviations: $c = \cos\theta,\ s = \sin\theta$.

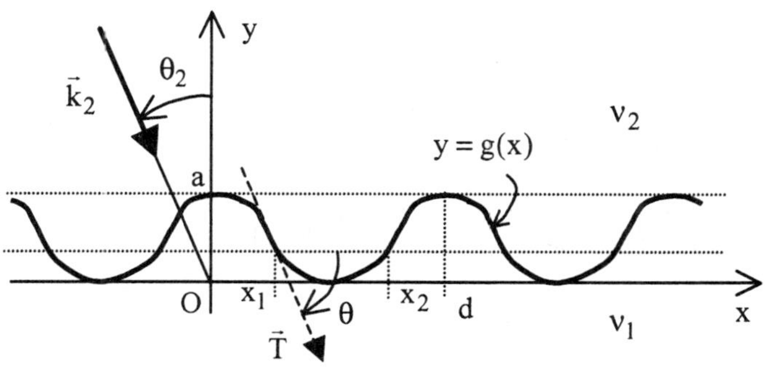

Fig. IV.2. Grating geometry and notations.

Inverting Eqs. (IV.11) and (IV.12) leads to :

$$E_x = cE_T - sE_N \tag{IV.13}$$

$$E_y = sE_T + cE_N. \tag{IV.14}$$

At a given ordinate y, Eqs. (IV.11) and (IV.12) are only valid for two values (x_2 and x_1) of x. However, we should define functions $E_T(x,y)$ and $E_N(x,y)$ through a suitable continuation of functions c(x) and s(x) explained in what follows. This extension makes the present theory valid not only for bare profiles, but also for dielectric coated gratings and multilayer coated gratings.

Eq. (IV.13) will first be used to calculate the nth Fourier component of $k^2\tilde{E}_x$ in Eq. (I.20) :

$$\varepsilon E_x = \varepsilon cE_T - \varepsilon sE_N. \tag{IV.15}$$

Let us study the continuity of the different terms in Eq. (IV.15). To that end, we extrapolate functions c(x) and s(x) outside points x_1 and x_2 by continuous functions of x (in fact, it suffices that c and s are continuous at the points where E_x, E_y and ε are discontinuous), in such a way that $\forall x$,

$$c^2(x) + s^2(x) = 1 \tag{IV.16}$$

so that $\vec{T}$ remains a unit vector everywhere.

The first term in the right-hand side of Eq. (IV.15) includes the product of the discontinuous function ε by a continuous function cE_T. Following conclusion **1** of Li, the Fourier component of the product can be found through Laurent's rule. Concerning the second term, the product εsE_N is continuous, while ε and sE_N are discontinuous; thus the calculation of the Fourier components of the product requires the use of conclusion **2**, i.e., the use of the inverse rule (IV.10). Using the Toeplitz matrix notation we then get:

$$[\varepsilon E_x] = [\![\varepsilon]\!][cE_T] - [\![1/\varepsilon]\!]^{-1}[sE_N].$$

Using Eqs. (IV.11) and (IV.12), we get

$$[\varepsilon E_x] = [\![\varepsilon]\!]\left[c^2E_x + csE_y\right] + [\![1/\varepsilon]\!]^{-1}\left[s^2E_x - csE_y\right],$$

and the use of Laurent's rule leads to

$$[\varepsilon E_x] = \left([\![\varepsilon]\!][\![c^2]\!] + [\![1/\varepsilon]\!]^{-1}[\![s^2]\!]\right)[E_x] + \left([\![\varepsilon]\!] - [\![1/\varepsilon]\!]^{-1}\right)[\![cs]\!]\left[E_y\right].$$

Let us introduce matrices Δ, A and B given by

$$\Delta = \left([\![\varepsilon]\!] - [\![1/\varepsilon]\!]^{-1} \right),\ \mathrm{A} = \Delta [\![c^2]\!],\ \mathrm{B} = \Delta [\![cs]\!].$$

Using relation: $[\![c^2]\!] + [\![s^2]\!] = \mathbb{1}$ (unit matrix), which can be readily derived from Eq. (IV.16), we finally obtain:

$$[\varepsilon E_x] = \left(\mathrm{A} + [\![1/\varepsilon]\!]^{-1} \right)[E_x] + \mathrm{B}[E_y]. \tag{IV.17}$$

Then Maxwell equation (I.14') leads to

$$[\varepsilon E_x] = \left(\mathrm{A} + \left[\!\!\left[\frac{1}{\varepsilon}\right]\!\!\right]^{-1} \right)[E_x] + \mathrm{B}\left[\frac{1}{i\omega\varepsilon}\frac{\partial H_z}{\partial x}\right]. \tag{IV.18}$$

It may seem a bit surprising when looking at Eq. (IV.17) that the calculation of the Fourier components of εE_x requires those of E_y. Indeed this comes from the truncation. If all vectors and matrices were infinite, we would have

$$[\![1/\varepsilon]\!]^{-1} = [\![\varepsilon]\!],\ \Delta = 0,\ \mathrm{A} = \mathrm{B} = 0.$$

Thus Eq. (IV.17) would reduce to :

$$[\varepsilon E_x] = [\![\varepsilon]\!][E_x]$$

which is merely the Laurent's rule used in the classical version of the theory. However, when truncated, the equivalence $[\![1/\varepsilon]\!]^{-1} = [\![\varepsilon]\!]$ does not hold anymore, in particular at the ends of the main diagonal of the matrices (ref. [I.44], p. 1876), which plays a significant role in the calculations. The more complicated form of Eq. (IV.17) comes from the fact that when working in a truncated basis, the Laurent's rule cannot be applied to εE_x; it can be applied to εE_T, and E_T depends on both E_x and E_y.

Here we arrive at the second key step of the theory. The product $\frac{1}{\varepsilon}\frac{\partial H_z}{\partial x}$ in the last braces of Eq. (IV.18) is discontinuous for any arbitrary grating profile, since it is equal to E_y. Moreover, both $\frac{1}{\varepsilon}$ and $\frac{\partial H_z}{\partial x}$ are discontinuous at x_1 and x_2. Thus the Fourier components of the product can be obtained neither through Laurent's rule, nor through the inverse rule. In order to derive them, we calculate the Fourier components of εE_y along the same lines as for εE_x. From Eq. (IV.14), we have:

$$\varepsilon E_y = \varepsilon s E_T + \varepsilon c E_N,$$

and the study of the continuity of the different products gives :

$$\left[\varepsilon E_y\right] = [\![\varepsilon]\!][sE_T] + \left[\!\!\left[\frac{1}{\varepsilon}\right]\!\!\right]^{-1}[cE_N].$$

Introducing Eqs. (IV.11) and (IV.12), we get:

$$\left[\varepsilon E_y\right] = [\![\varepsilon]\!]\left[csE_x + s^2E_y\right] + \left[\!\!\left[\frac{1}{\varepsilon}\right]\!\!\right]^{-1}\left[-csE_x + c^2E_y\right]$$

Thus:

$$\left[\varepsilon E_y\right] = [\![\varepsilon]\!]\left([\![cs]\!][E_x] + [\![s^2]\!][E_y] \right) + \left[\!\!\left[\frac{1}{\varepsilon}\right]\!\!\right]^{-1}\left(-[\![cs]\!][E_x] + [\![c^2]\!][E_y]\right),$$

i.e.,

$$\left[\varepsilon E_y\right] = B[E_x] + \left([\![\varepsilon]\!] - A\right)\left[E_y\right].$$

From this equation, we deduce that:

$$\left[E_y\right] = \left([\![\varepsilon]\!] - A\right)^{-1}\left(\left[\varepsilon E_y\right] - B[E_x]\right), \tag{IV.19}$$

or:

$$\left[\frac{1}{\varepsilon}\frac{\partial H_z}{\partial x}\right] = \left([\![\varepsilon]\!] - A\right)^{-1}\left(i\alpha[H_z] - i\omega B[E_x]\right), \tag{IV.20}$$

in which α is a matrix previously defined after Eq. (IV.4).

The use of Eqs. (I.20), (IV.18) and (IV.20) leads to the first basic equation of the differential theory, and Eqs. (I.21) and (IV.20) lead to the second one:

$$\left[\frac{\partial H_z}{\partial y}\right] = -i\omega\left(A + [\![1/\varepsilon]\!]^{-1} - B\left([\![\varepsilon]\!] - A\right)^{-1}B\right)[E_x]$$
$$- B\left([\![\varepsilon]\!] - A\right)^{-1} i\alpha[H_z], \tag{IV.21}$$

$$\left[\frac{\partial E_x}{\partial y}\right] = \frac{i}{\omega}\alpha\left([\![\varepsilon]\!] - A\right)^{-1}\alpha[H_z] - i\omega\mu[H_z] \\ -i\alpha\left([\![\varepsilon]\!] - A\right)^{-1}B[E_x]. \tag{IV.22}$$

Eqs. (IV.21) and (IV.22) are Maxwell equations written in a truncated Fourier space as they apply to the TM polarized field diffracted by a classical grating. The method for obtaining them, as explained in this section, has received the name of fast Fourier factorization (FFF) method.

The FFF method is not limited to this particular problem. It is applicable to any theory that requires the Fourier analysis of continuous products of discontinuous periodic functions. Such is D_N in electromagnetism, equal to εE_N, but similar examples could be found in acoustics, etc. The main idea consists in extending the concepts of tangential and normal component of a vector outside the discontinuity surface where they are defined, through a suitable continuation. The Cartesian components of the field can then be expressed in terms of these tangential and normal components everywhere, so that the discontinuous products of discontinuous functions can then be reduced to products of type **1** and **2** of Li, for which Laurent's rule or the inverse rule can be used.

IV.2.3 Discussion of the new equations

Despite their apparent complexity, Eqs. (IV.21) and (IV.22) allow us to discuss and explain several important points.

First, if we do not truncate the Fourier series of the field ($N \to \infty$), matrix Δ is null and so are matrices A and B. Most of the terms in Eqs. (IV.21) and (IV.22) vanish, and the equations reduce to Eqs. (II.21) and (II.22). We then conclude that the original untruncated equations written in the Fourier space were rigorous. The problems that occurred were due to truncation only.

Second, if the modulation ratio a/d of the grating tends toward zero, $\theta \to 0$. Thus $c \to 1$, $s \to 0$, which leads to $A = \Delta$ and $B = 0$. Eqs. (IV.21) and (IV.22) reduce to:

$$\left[\frac{\partial H_z}{\partial y}\right] = -i\omega[\![\varepsilon]\!][E_x],$$

$$\left[\frac{\partial E_x}{\partial y}\right] = \frac{i}{\omega}\alpha\left[\!\!\left[\frac{1}{\varepsilon}\right]\!\!\right]\alpha[H_z] - i\omega\mu[H_z].$$

These equations are identical with Eqs. (IV.3) and (IV.4) which are the truncated forms of Eqs. (II.21) and (II.22). This shows that the original truncated equations in the Fourier space are only exact at the limit $a/d \to 0$. It explains why, even on highly reflecting metals, a good convergence was obtained provided the modulation was low enough. We always observed that the higher the modulus of the permittivity of the metal, the smaller the groove depth we could analyze numerically with the original differential method. This comes from the fact that the elements of A and B increase with $|\varepsilon|$, where ε is the permittivity of the grating material. Thus the additional terms in Eqs. (IV.21) and (IV.22) that were neglected in Eqs. (IV.3) and (IV.4) play an increasing role, when $s \neq 0$.

Third, if $\varepsilon \to 1 + i0$, as occurs for all materials in X-ray region, the discontinuity of ε across the grating surface tends to zero. Thus all discontinuity problems vanish and all Fourier components of products can be factorized by Laurent's rule (IV.9). Indeed it is easy to verify that, in that case, $A \to 0$, $B \to 0$, since $[\![1/\varepsilon]\!]^{-1} \to [\![\varepsilon]\!]$. We understand why Eqs. (IV.3) and (IV.4), which have been used to study X-ray gratings for many years, turned out to be a reliable tool in that region. The convergence of the Fourier series of the field in TM polarization was about as good as for the TE case, even under circumstances in which results strongly depart from scalar theory predictions, as occurs for gratings used under high or grazing incidence.

Fourth, what happens to a lamellar profile? In that case, for any ordinate y during the integration process, $\theta = \pi/2$, thus c = 0, s = 1, A = 0, B = 0. Eqs. (IV.21) and (IV.22) reduce to Eqs. (IV.5) and (IV.6), which are identical to Eqs. (IV.3) and (IV.4) except that $[\![\varepsilon]\!]$ is replaced by $[\![1/\varepsilon]\!]^{-1}$ and $[\![1/\varepsilon]\!]$ by $[\![\varepsilon]\!]^{-1}$. These are exactly the transformations that were proposed by the authors of [I.42] and [I.43], which resulted in a high improvement of the convergence of the Fourier series of the field for a lamellar grating. We are now in a position to understand why when we tried to use Eqs. (IV.21) and (IV.22) to improve the convergence of the results on a sinusoidal profile 22 years ago, we obtained no improvement. Indeed, for an arbitrary profile, Eqs. (IV.5) and (IV.6) are as far from the exact Eqs. (IV.21) and (IV.22) as Eqs. (IV.3) and (IV.4) are. Moreover, we explain why Lalanne found that, in the small depth limit [IV.2], the conventional formulation of the differential theory is better conditioned than his new formulation, since at the limit the first one is rigorous.

It has to be pointed out that the advantage of the FFF method is due to the fact that it enables us working with continuous functions (εE_N and E_T) even in the TM case, by using the correct factorization rules of the product of functions, which are continuous or discontinuous on the interface between different media. Let us demonstrate that in the case of lamellar grating. Inside the modulated region, ε does not depend on y and Eq. (I.12') can be rewritten in the form:

$$\frac{\partial(\varepsilon E_x)}{\partial y} = \varepsilon \frac{\partial E_y}{\partial x} - i\omega\mu\varepsilon H_z .$$

The components $\varepsilon E_x \equiv D_N$, $E_y \equiv E_T$, and $H_z \equiv H_T$ are continuous functions along the vertical sides of the lamellae, so that applying the Laurent's rule and Eq. (I.20), the Fourier transform results in

$$\left[\frac{\partial(\varepsilon E_x)}{\partial y}\right] = i[\![\varepsilon]\!]\alpha\left[E_y\right] - i\omega\mu[\![\varepsilon]\!][H_z].$$

The second set of equations can be easily determined by a Fourier transform of Eq. (I.13') so that

$$\left[\frac{\partial H_z}{\partial y}\right] = -i\omega[\varepsilon E_x].$$

This choice allows integrating only continuous functions ($D_x = \varepsilon E_x$ and H_z) in the TM case in a manner similar to the TE case. However, the question remains how to determine the Fourier components of E_y through the components of H_z. Eq. (I.14') gives the link between E_y and H_z, $E_y = \frac{1}{i\omega\varepsilon}\frac{dH_z}{dy}$. The classical formulation of the differential theory (and of RCW method, in particular) used the Laurent's rule:

$$\left[E_y\right] = \frac{1}{\omega}\left[\!\!\left[\frac{1}{\varepsilon}\right]\!\!\right]\alpha\,[H_z].$$

However, this formulation suffers from the same numerical problems as the classical formulation, Eq. (IV.4), because E_y, which is a continuous function of x, is a product of two discontinuous functions. This fact requires applying the inverse rule, i.e.,

$$\left[E_y\right] = \frac{1}{\omega}[\![\varepsilon]\!]^{-1}\alpha\,[H_z].$$

Using this equation, the convergence rate was found to be the same as when the correctly truncated Eq. (IV.6) is used.

The last point, which is also clarified concerns the fast convergence of the Fourier series in TM polarization for perfectly conducting gratings observed when analyzed using the conformal mapping technique [I.21], contrary to the slow convergence of the classical differential method when applied to highly conducting metals. The explication lies in the fact that the conformal mapping technique uses functions continuous with respect to the abscissa of the transformed coordinate system, so that there is no convergence problem due to the incorrect factorization of Fourier series of discontinuous functions. Now we know that the poor convergence of the classical differential method was due to the bad way of making such factorization in a truncated Fourier space, rather than to essential physical reasons.

IV.2.4 Examples of convergence of the numerical results

The method has been implemented numerically and tested on various profiles, for both dielectric and metallic gratings. In order to appreciate the convergence of the results, we compare them with those obtained through the integral formalism [IV.3], which was used with a number of discretisation points (and terms in the kernel of the integral equations) high enough to have results stabilized to better than the fifth digit. What is called the error in the following is calculated by comparison with that reference.

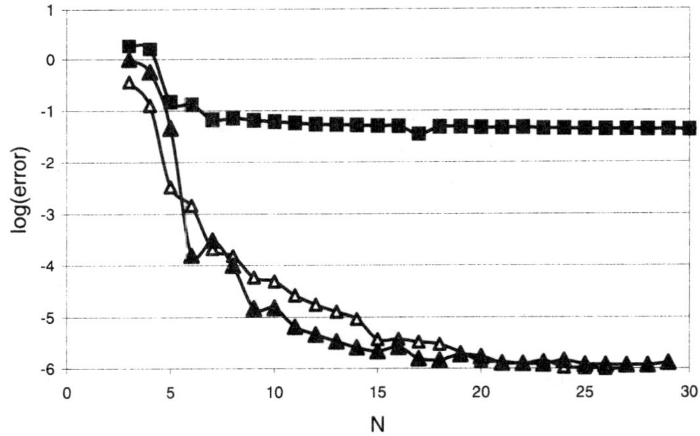

Fig. IV.3. Convergence of the previous and new version of the differential theory for the case of a dielectric sinusoidal grating with high contrast. Squares, old version of the differential theory for TM polarization; open triangles, new version, TM polarization; solid triangles, TE polarization (after [I.45]).

Fig. IV.3 shows the results for a sinusoidal dielectric grating with high contrast; the modulation ratio a/d is equal to unity, and the refractive index is 2.5. The period d is $1\ \mu m$. The grating is lighted under 30°-incidence angle with a $0.6328\ \mu m$ wavelength. Fig. IV.3 plots the logarithm of the relative mean square error as a function of the parameter N that defines the truncation. The error is defined as :

$$\frac{1}{N_O}\sum_{n=1}^{N_O}\left(\frac{e_n^{(r)}-\tilde{e}_n^{(r)}}{e_n^{(r)}}\right)^2,$$

with N_O the number of propagating orders, $e_n^{(r)}$ the exact efficiency in the nth propagating order computed with a code based on the integral theory[IV.3], and $\tilde{e}_n^{(r)}$ the corresponding efficiency computed through the methods for which we test the convergence.

The squares are obtained using Eqs. (IV.3) and (IV.4), and the open triangles are obtained by using Eqs. (IV.21) and (IV.22). A strong improvement is observed, an error of 10^{-4} being reached with $N \approx 8$, i.e., 17 Fourier components of the field; the convergence obtained with the new method is about the same as the convergence obtained in TE polarization, shown by solid triangles.

Many other convincing examples can be found in [I.46] on both metallic and dielectric gratings, as well as on various kinds of groove shapes, including rod gratings for which the profile is not given by an equation of the type $y = g(x)$.

IV.2.5 Calculation of Toeplitz matrices $[\![c^2]\!]$ *and* $[\![cs]\!]$

We show here how to calculate the Toeplitz matrices $[\![c^2]\!]$ and $[\![cs]\!]$ for some common grating profiles. The simplest case of lamellar gratings simply reduces to Eqs. (IV.5) and (IV.6) and is not addressed here.

For all other profiles, the quantities c and s are only defined at points x_1 and x_2 shown in Fig.IV.2, while Eqs. (IV.13) and (IV.14) have to be valid at any abscissa x. We thus have to define function c(x) and s(x) such that $c(x_1) = c$ and $s(x_1) = s$. Moreover, c(x) and s(x) must be everywhere equal to the components of the unit vector $\vec{T}(x)$ which defines E_T and E_N through Eqs. (IV.11) and (IV.12). We thus have $c^2(x) + s^2(x) = 1$, $\forall x$, otherwise we will obtain E_T and E_N components such that, $E_T^2 + E_N^2 \neq E^2$ and a metric tensor would appear in Eqs. (IV.11) - (IV.14).

The continuation of functions c(x) and s(x) must give functions that are continuous at the discontinuities x_1 and x_2 of k^2, E_x, and E_y in order to apply the factorization rules described in Section IV.2.1. We will stick to these rules for the various profiles, but when possible $c^2(x)$ and $c(x)s(x)$ will be chosen to be continuous functions $\forall x$ in order to ensure faster convergence of the Fourier series. Our choices, of course, are not the only possible ones.

We describe the method on the general case of a continuous profile defined by y = g(x). From the definition of θ, we have

$$\tan\theta = dg/dx \tag{IV.23}$$

Since $\cos\theta = (1+\tan^2\theta)^{-1/2}$, we define function c(x) by:

$$c(x) = \left[1+\left(\frac{dg}{dx}\right)^2\right]^{-1/2} \tag{IV.24}$$

Moreover, since $\sin\theta = \tan\theta/(1+\tan^2\theta)^{1/2}$, we define function c(x)s(x) by:

$$c(x)s(x) = \frac{dg}{dx}\Big/\left[1+\left(\frac{dg}{dx}\right)^2\right]. \tag{IV.25}$$

Such continuation ensures that $c^2(x)+s^2(x)=1,\ \forall x$, and has the advantage that their Fourier series are calculated only once for the entire integration process, since $c^2(x)$ and c(x)s(x) are independent of y. The geometrical interpretation of Eqs. (IV.24) and (IV.25) is the following one. Let us consider a point M with coordinates x and y inside the modulated area. At point M(x,y), Eqs. (IV.24) and (IV.25) define a unit vector $\vec{T}(x,y)$ which is parallel to the unit vector $\vec{T}(x)$ tangent to the profile at point $M_0(x,g(x))$.

A strong advantage of that continuation will then be found when analyzing dielectric coated gratings, in which the profile of the coating layers are deduced from the grating profile by translation along the y-axis. Since Eqs. (IV.24) and (IV.25) only make use of dg/dx instead of g(x), they apply to the overcoatings as well as to the grating surface, giving the correct values of c and s at each interface; this is made obvious by the previous geometrical interpretation.

For example, in the case of sinusoidal profile defined by

$$g(x) \equiv (a/2)(1+\cos Kx),\ K = 2\pi/d,$$

we obtain

$$c^2(x) = 1\Big/\left(1+\frac{\pi^2 a^2}{d^2}\sin^2(Kx)\right)$$

$$c(x)s(x) = \frac{-\pi a}{d}\sin(Kx)\Big/\left(1+\frac{\pi^2 a^2}{d^2}\sin^2 Kx\right).$$

The Fourier coefficients of c^2 and cs can then be determined by the standard fast Fourier transform method.

For a triangular profile, $\frac{dg}{dx}$ will take two discrete constant values on each slope. Functions $c^2(x)$ and $c(x)s(x)$ then can be defined as periodic step functions whose jumps must occur at the abscissa of the bottom and top of the grooves. Fourier coefficients can be derived analytically. Trapezoidal profiles lead to similar conclusions. A more complicated case that uses parametric equations of the profile is described in [I.46].

IV.3 Extension to conical diffraction

When the incident wave vector $\vec{k}_2$ is out of the cross-section plane xOy of the grating, the diffracted orders lie on a half cone whose axis is the direction $\vec{O}z$ of the rulings, and whose top half angle is the angle between $\vec{k}_2$ and $\overrightarrow{Oz}$; hence the name of "conical diffraction" [IV.4]. In this mounting, the incident wave u^i depends on x, y, and z coordinate, so that the diffraction problem is no longer z-independent.

For gratings with finite conductivity, the basic set of Eqs. (I.10)-(I.15) cannot be split into two independent sets. It means that the efficiencies cannot be determined by resolving separately the TE and the TM case of polarization for the total field. As will be shown, four first-order-coupled differential equations must be resolved simultaneously, even when the incident magnetic or electric field vector is perpendicular to the z-axis.

The incident field $\vec{E}_i$ now reads:

$$\vec{E}^{(i)} = \vec{A}\exp\left[i(\alpha_0 x - \beta_0 y + \gamma_0 z)\right],$$

where γ_0 is the z-component of the incident wavevector $\vec{k}_2$. Since the diffracting device is z-independent, the diffracted field must have the same z dependence as the incident one in order to satisfy the boundary conditions at the grating surface (Appendix I). Hence the total field has an $\exp(i\gamma_0 z)$ dependence, and the z-derivatives in Maxwell equations reduces to multiplications by $(i\gamma_0)$. In the Fourier space, we then get:

$$\frac{\partial[E_z]}{\partial y} - i\gamma_0\left[E_y\right] = i\omega\mu[H_x] \tag{IV.26}$$

$$i\gamma_0[E_x] - i\alpha[E_z] = i\omega\mu\left[H_y\right] \tag{IV.27}$$

$$i\alpha\left[E_y\right] - \frac{\partial[E_x]}{\partial y} = i\omega\mu[H_z] \tag{IV.28}$$

$$\frac{\partial[H_z]}{\partial y} - i\gamma_0\left[H_y\right] = -i\omega[D_x] \tag{IV.29}$$

$$i\gamma_0[H_x] - i\alpha[H_z] = -i\omega\left[D_y\right] \tag{IV.30}$$

$$i\alpha\left[H_y\right] - \frac{\partial[H_x]}{\partial y} = -i\omega[D_z]. \tag{IV.31}$$

The Fourier components of D_x, D_y and D_z will be calculated with the fast Fourier factorization (FFF) method. We denote by $\left[\vec{D}\right]$ a vector made of 3 blocks $[D_x]$, $\left[D_y\right]$ and $[D_z]$, each of these blocks having 2N+1 Fourier components. Introducing the unit vector $\vec{N}$ on the grating surface y = g(x), whose components are $N_x = -\frac{dg}{dx}\Big/\left[1+\left(\frac{dg}{dx}\right)^2\right]^{1/2}$, $N_y = 1\Big/\left[1+\left(\frac{dg}{dx}\right)^2\right]^{1/2}$ and $N_z = 0$, we can define the normal and tangential components of the field:

$$\vec{E}_N = \vec{N}\left(\vec{N}.\vec{E}\right)$$
$$\vec{E}_T = \vec{E} - \vec{E}_N.$$

Starting from:

$$\left[\vec{D}\right] = \left[\varepsilon\vec{E}\right] = \left[\varepsilon\vec{E}_T + \varepsilon\vec{E}_N\right],$$

we first remark that $\varepsilon\vec{E}_T$ is discontinuous, while $\varepsilon\vec{E}_N$ is continuous. The factorization rules give:

$$\begin{aligned}\left[\vec{D}\right] &= [\![\varepsilon]\!]\left[\vec{E}_T\right] + \left[\!\!\left[\frac{1}{\varepsilon}\right]\!\!\right]^{-1}\left[\vec{E}_N\right] \\ &= [\![\varepsilon]\!]\left[\vec{E} - \vec{N}\left(\vec{N}.\vec{E}\right)\right] + \left[\!\!\left[\frac{1}{\varepsilon}\right]\!\!\right]^{-1}\left[\vec{N}\left(\vec{N}.\vec{E}\right)\right] \\ &= [\![\varepsilon]\!]\left[\vec{E}\right] - \left([\![\varepsilon]\!] - \left[\!\!\left[\frac{1}{\varepsilon}\right]\!\!\right]^{-1}\right)\left[\vec{N}\left(\vec{N}.\vec{E}\right)\right].\end{aligned}$$

Introducing a square matrix denoted $\left(\vec{N}\vec{N}\right)$ the elements of which are given by $\left(\vec{N}\vec{N}\right)_{ij} = N_i N_j$, we then obtain:

$$\left[\vec{D}\right] = [\![\varepsilon]\!]\left[\vec{E}\right] - \left([\![\varepsilon]\!] - \left[\!\!\left[\frac{1}{\varepsilon}\right]\!\!\right]^{-1}\right)[\![\vec{N}\vec{N}]\!]\left[\vec{E}\right]. \qquad \text{(IV.32)}$$

This equation can be represented in the form:

$$\left[\vec{D}\right] = Q_\varepsilon\left[\vec{E}\right], \qquad \text{(IV.32')}$$

which defines a 3x3-block matrix Q_ε given by:

$$Q_\varepsilon = \begin{pmatrix} [\![\varepsilon]\!][\![N_y^2]\!] + \left[\!\!\left[\frac{1}{\varepsilon}\right]\!\!\right]^{-1}[\![N_x^2]\!] & -\left([\![\varepsilon]\!] - \left[\!\!\left[\frac{1}{\varepsilon}\right]\!\!\right]^{-1}\right)[\![N_x N_y]\!] & 0 \\ -\left([\![\varepsilon]\!] - \left[\!\!\left[\frac{1}{\varepsilon}\right]\!\!\right]^{-1}\right)[\![N_x N_y]\!] & [\![\varepsilon]\!][\![N_x^2]\!] + \left[\!\!\left[\frac{1}{\varepsilon}\right]\!\!\right]^{-1}[\![N_y^2]\!] & 0 \\ 0 & 0 & [\![\varepsilon]\!] \end{pmatrix}. \qquad \text{(IV.33)}$$

Eqs. (IV.32') and (IV.33) allow expressing $\left[E_y\right] = Q_{\varepsilon,yy}^{-1}\left\{\left[D_y\right] - Q_{\varepsilon,yx}\left[E_x\right]\right\}$ and Eq. (IV.30) leads to:

$$\left[E_y\right] = Q_{\varepsilon,yy}^{-1}\left\{\frac{\left[\alpha[H_z] - \gamma_0[H_x]\right]}{\omega} - Q_{\varepsilon,yx}\left[E_x\right]\right\}. \qquad \text{(IV.33')}$$

This expression is then put into Eqs. (IV.26) and (IV.28) for eliminating $\left[E_y\right]$. In a simpler way, Eq. (IV.27) gives:

$$\left[H_y\right] = \frac{\gamma_0\left[E_x\right] - \alpha\left[E_z\right]}{\omega\mu},$$

and this expression is put into Eqs. (IV.29) and (IV.31). We then get a set of 4 blocks of first-order differential equations of the form:

$$\frac{d}{dy}\begin{pmatrix}[E_x]\\ [E_z]\\ [H_x]\\ [H_z]\end{pmatrix} = M\begin{pmatrix}[E_x]\\ [E_z]\\ [H_x]\\ [H_z]\end{pmatrix}, \tag{IV.34}$$

where

$$M = i\begin{pmatrix} -\alpha Q_{\varepsilon,yy}^{-1}Q_{\varepsilon,yx} & 0 & -\alpha Q_{\varepsilon,yy}^{-1}\frac{\gamma_0}{\omega} & \frac{\alpha}{\omega}Q_{\varepsilon,yy}^{-1}\alpha-\omega\mu \\ -\gamma_0 Q_{\varepsilon,yy}^{-1}Q_{\varepsilon,yx} & 0 & \omega\mu-\frac{\gamma_0^2}{\omega}Q_{\varepsilon yy}^{-1} & \gamma_0 Q_{\varepsilon,yy}^{-1}\frac{\alpha}{\omega} \\ \frac{\alpha\gamma_0}{\omega\mu} & \omega Q_{\varepsilon zz}-\frac{\alpha^2}{\omega\mu} & 0 & 0 \\ \omega\left(Q_{\varepsilon,xy}Q_{\varepsilon,yy}^{-1}Q_{\varepsilon yx}-Q_{\varepsilon,xx}\right)+\frac{\gamma_0^2}{\omega\mu} & -\frac{\gamma_0\alpha}{\omega\mu} & Q_{\varepsilon,xy}Q_{\varepsilon,yy}^{-1}\gamma_0 & -Q_{\varepsilon,xy}Q_{\varepsilon,yy}^{-1}\alpha \end{pmatrix} \tag{IV.35}$$

Eqs. (IV.34) and (IV.35) are the basic propagation equations for conical diffraction. When $\vec{k}_2$ lies in the cross-section plane of the grating, $\gamma_0 = 0$. Then $M_{13} = 0 = M_{21} = M_{24} = M_{31} = M_{42} = M_{43}$, and Eq. (IV.34) decouples into TE and TM polarization cases. By expressing the components of $\vec{N}$ $\left(N_x = -s(x);\ N_y = c(x)\right)$ one can find that for TE polarization Eqs. (IV.34) and (IV.35) reduce to Eqs. (I.17) and (I.18), written in Fourier space, and to Eqs. (IV.21) and (IV.22) in TM polarization.

The boundary-value problem obtained for conical diffraction is resolved via a shooting method similar to the one used for TM polarization, except that since the number of unknown coefficients is twice more, the number of starting vectors will be 4N+2 instead of 2N+1. Also, the Fourier coefficients of the transverse components E_x and H_x will have to be derived from the chosen values of E_z and H_z at the starting integration point. Classical textbooks [IV.5] and [IV.6] establish in the frame of the waveguide theory that the transverse components of the field can be expressed in terms of E_z and H_z only. This demonstration is reproduced in Appendix III. Eqs. (A.III.12) and (A.III.13) thus give:

$$E_x = \frac{1}{k^2-\gamma_0^2}\left(i\omega\mu\frac{\partial H_z}{\partial y} + i\gamma_0\frac{\partial E_z}{\partial x}\right) \tag{IV.36}$$

$$H_x = \frac{1}{k^2 - \gamma_0^2}\left(-i\omega\varepsilon \frac{\partial E_z}{\partial y} + i\gamma_0 \frac{\partial H_z}{\partial x} \right). \tag{IV.37}$$

Inside the substrate, k^2 reduces to the constant k_I^2, so that there is no convergence problem when calculating the Fourier components of the products which appear in Eqs. (IV.36) and (IV.37). Moreover the y and x derivatives of function E_z and H_z are obtained through the Rayleigh expansions of the field. We thus obtain:

$$\begin{aligned} E_{x,n} &= \frac{1}{k_I^2 - \gamma_0^2}\left[\omega\mu\beta_n^{(1)} H_{z,n} - \gamma_0 \alpha_n E_{z,n} \right] \\ H_{x,n} &= \frac{1}{k_I^2 - \gamma_0^2}\left[-\omega\varepsilon_1 \beta_n^{(1)} E_{z,n} - \gamma_0 \alpha_n H_{z,n} \right]. \end{aligned} \tag{IV.38}$$

We start a first series of (2N+1) integrations $(p \in [-N, N])$ by choosing,

$$\begin{aligned} E_{z,n,p} &= \delta_{n,p} \\ H_{z,n,p} &= 0, \end{aligned} \tag{IV.39}$$

from which we deduce:

$$\begin{aligned} E_{x,n,p} &= -\gamma_0 \alpha_p \big/ \left(k_I^2 - \gamma_0^2\right) \\ H_{x,n,p} &= -\omega\varepsilon_1 \beta_p^{(1)} \big/ \left(k_I^2 - \gamma_0^2\right). \end{aligned}$$

With these starting values, Eq. (IV.34) is integrated through a standard algorithm as far as y = a (top of the grooves). The second series of (2N+1) integrations with $p \in [N+1, 3N+1]$ starts with:

$$\begin{aligned} E_{z,n,p} &= 0 \\ H_{z,n,p} &= \delta_{n,p} \end{aligned} \tag{IV.40}$$

from which we get:

$$\begin{aligned} E_{x,n,p} &= \omega\mu\beta_p^{(1)} \big/ \left(k_I^2 - \gamma_0^2\right) \\ H_{x,n,p} &= -\gamma_0 \alpha_p \big/ \left(k_I^2 - \gamma_0^2\right). \end{aligned}$$

The Fourier components $E_{x,n}(a)$, $E_{z,n}(a)$, $H_{x,n}(a)$, and $H_{z,n}(a)$ are then matched with the various Rayleigh expansions at the top of the grooves in a similar way as for TM polarization (Section II.1.2.2). The 4(2N+1) Rayleigh coefficients of E_z and H_z components of the reflected and transmitted field are obtained after inversion of a (8N+4)x(8N+4) matrix. The Rayleigh coefficients of E_x and H_x are then derived through Eqs. (IV.36) and (IV.37).

In fact, it is possible to eliminate half of the unknown amplitudes in order to reduce the dimension of the matrix to be inverted to (4N+2) instead of (8N+4). Details are given in Appendix IV.

References

IV.1. A. Zygmund, *Trigonometric Series*, (Cambridge U. Press, Cambridge, 1977), Vol.1, Ch. 4, Sec. 8, p. 159

IV.2. P. Lalanne, "Convergence performance of the coupled-wave and the differential method for thin gratings," J. Opt. Soc. Am. A14, 1583-1591, 1997

IV.3. D. Maystre, "Integral methods," in *Electromagnetic Theory of Gratings*, Vol. 22 of Topics in Current Physics, R. Petit ed., Springer-Verlag, Berlin, 1980, p. 63-100

IV.4. M. Nevière, D. Maystre, and R. W. Hunter, "On the use of classical and conical diffraction mountings for XUV gratings," J. Opt. Soc. Am. 68, 1106 - 1113, 1978

IV.5. R. Petit, "Ondes electromagnétiques en radioélectricité et en optique," (Masson, Paris, 1989), p. 173

IV.6. R, E. Collin, *Field Theory of Guided Waves*, 2nd ed., (IEEE, New York, 1991)

Chapter V

Maxwell Equations in Truncated Fourier Space

The fast Fourier factorization (FFF) method, which has resolved the problems of slow convergence of the Fourier series of the field diffracted by a grating used in TM polarization and in conical mounting, can be used [V.1] to tackle the most general situation in which the grating device can be 1-, 2-, or 3-dimensionally periodic, can be done of anisotropic or magnetooptic material, etc. This chapter presents the most general form of the theory.

V.1 General, fast converging propagation equations in Fourier space

We consider a periodic surface described by its Cartesian equation f(x,y,z) = 0. This surface is assumed to have a normal vector $\vec{N}$ everywhere, and it separates two media characterized by their tensors $\bar{\bar{\varepsilon}}$ and $\bar{\bar{\mu}}$, which degenerate into scalars for isotropic media. Harmonic Maxwell equations:

$$\mathrm{curl}\vec{E} = i\omega\vec{B} \tag{V.1}$$

$$\mathrm{curl}\vec{H} = -i\omega\vec{D} \tag{V.2}$$

lead to partial derivative equations for the Cartesian coordinates of $\vec{E}$ or $\vec{H}$. These components are discontinuous across the $\bar{\bar{\varepsilon}}$ and $\bar{\bar{\mu}}$ discontinuity surface, and we have to use continuous components in view of Fourier analysis. In the plane tangent to the periodic surface at any point M, we introduce two orthogonal unit vectors $\vec{T}_1$ and $\vec{T}_2$ such that $\vec{T}_2 \times \vec{T}_1 = \vec{N}$, $\vec{T}_1$ being arbitrary, but conveniently chosen parallel to the xOy plane. The tangential components of the electrical field are thus given by:

$$E_{T_1} = \vec{T}_1 \bullet \vec{E} \tag{V.3}$$

$$E_{T_2} = \vec{T}_2 \bullet \vec{E}\,. \tag{V.4}$$

Here the dot product is the well-known scalar product of two vectors, but we use a special notation (large dot) to introduce the more general dot product between matrices representing tensors and vectors of different ranks. The generalized dot product means contraction of one of the indices, i.e., a summation with respect to one index. It may then denote the product of a row vector by a square matrix, as well as the product of a square matrix by a column vector. With that notation, the normal component of $\vec{D}$ is given in the form:

$$D_N = \vec{N} \bullet \vec{D} = \vec{N} \bullet \bar{\bar{\varepsilon}} \bullet \vec{E} \tag{V.5}$$

From the laws of electromagnetism, vector $\vec{F}_\varepsilon = (E_{T_1}, D_N, E_{T_2})$ is continuous accross the periodic surface. The first step is to relate it to $\vec{E}$, in order to link the Fourier components of $\vec{D}$ to those of $\vec{E}$.

V.1.1 Relation between $\left[\vec{D}\right]$ *and* $\left[\vec{E}\right]$

Following the notation introduced in Section IV.3, $\left[\vec{D}\right]$ will denote a vector made with the three blocks $[D_x]$, $\left[D_y\right]$ and $[D_z]$. However, now each of these blocks will have 2N+1 Fourier components for a one-dimensional periodic medium, $(2N+1)^2$ components for a bidimensional one, having two periods of the same order of magnitude, etc. Eqs. (V.3) and (V.5) lead to:

$$E_{T_i} = T_{ix} E_x + T_{iy} E_y + T_{iz} E_z, \qquad i = 1,2 \tag{V.6}$$

$$\begin{aligned} D_N = {} & N_x \left(\varepsilon_{xx} E_x + \varepsilon_{xy} E_y + \varepsilon_{xz} E_z\right) + N_y \left(\varepsilon_{yx} E_x + \varepsilon_{yy} E_y + \varepsilon_{yz} E_z\right) \\ & + N_z \left(\varepsilon_{zx} E_x + \varepsilon_{zy} E_y + \varepsilon_{zz} E_z\right) \end{aligned} \tag{V.7}$$

If we introduce a matrix A_ε defined by

$$\vec{F}_\varepsilon = A_\varepsilon \bullet \vec{E} \equiv A_\varepsilon \vec{E} \tag{V.8}$$

Eqs. (V.6) and (V.7) immediately show that:

$$A_\varepsilon =$$

$$\begin{pmatrix} T_{1x} & T_{1y} & T_{1z} \\ N_x \varepsilon_{xx} + N_y \varepsilon_{yx} + N_z \varepsilon_{zx} & N_x \varepsilon_{xy} + N_y \varepsilon_{yy} + N_z \varepsilon_{zy} & N_x \varepsilon_{xz} + N_y \varepsilon_{yz} + N_z \varepsilon_{zz} \\ T_{2x} & T_{2y} & T_{2z} \end{pmatrix}$$

Since E_{T_i} and D_N are only defined on the surface of discontinuity of $\bar{\bar{\varepsilon}}$ and $\bar{\bar{\mu}}$, Eqs. (V.6)-(V.8) are only valid on this surface, defined by the equation f(x,y,z) = 0. However, as was previously done in Sections IV.2.2 and IV.2.5, we extend the validity of Eqs. (V.6)-(V.8) $\forall x, y, z$ through a suitable continuation of $\vec{T}_i$ and $\vec{N}$ outside the surface. The continuation must ensure that these vectors are continuous on the surface; they may be discontinuous at points where $\bar{\bar{\varepsilon}}$ and $\bar{\bar{\mu}}$ are continuous. They must also be unit vectors everywhere in order to ensure that for $\forall(x, y, z)$ $E_{T_1}^2 + E_{T_2}^2 + E_N^2 = E^2$.

Let us thus introduce a vector noted $\left(\vec{N} \bullet \bar{\bar{\varepsilon}}\right)$, where the dot product stands for a summation over one of the indices of $\bar{\bar{\varepsilon}}$, as already discussed. Then $D_N = \left(\vec{N} \bullet \bar{\bar{\varepsilon}}\right) \bullet \vec{E}$ and the matrix A_ε can simply be expressed as:

$$A_\varepsilon = \begin{pmatrix} T_{1x} & T_{1y} & T_{1z} \\ \left(\vec{N} \bullet \bar{\bar{\varepsilon}}\right)_x & \left(\vec{N} \bullet \bar{\bar{\varepsilon}}\right)_y & \left(\vec{N} \bullet \bar{\bar{\varepsilon}}\right)_z \\ T_{2x} & T_{2y} & T_{2z} \end{pmatrix}. \tag{V.9}$$

Since $\bar{\bar{\varepsilon}}$ never vanishes, the determinant of A_ε is a quadratic non-null form; thus A_ε has an inverse C_ε, so that $\vec{E} = C_\varepsilon \vec{F}_\varepsilon$. As a result, equation $\vec{D} = \bar{\bar{\varepsilon}} \bullet \vec{E}$ can be written as $\vec{D} = \bar{\bar{\varepsilon}} \bullet C_\varepsilon \vec{F}_\varepsilon = \bar{\bar{\varepsilon}} \bullet C_\varepsilon A_\varepsilon \vec{E}$. In that way, $\vec{D}$ is represented as a product of a discontinuous quantity $\bar{\bar{\varepsilon}} \bullet C_\varepsilon$ by a continuous one $\vec{F}_\varepsilon$ (or $A_\varepsilon \vec{E}$). Its Fourier components can then be obtained through Laurent's rule: $\left[\vec{D}\right] = \left[\!\left[\bar{\bar{\varepsilon}} \bullet C_\varepsilon\right]\!\right] \left[\vec{F}_\varepsilon\right]$. Of course, since both $\left[\vec{D}\right]$ and $\left[\vec{F}_\varepsilon\right]$ are vectors formed by blocks, the Toeplitz matrix which relates them *will be made of blocks that are the Toeplitz matrices of the different elements of matrix* $\bar{\bar{\varepsilon}} \bullet C_\varepsilon$. Using the inverse rule to find the Fourier components of the continuous product $A_\varepsilon \vec{E}$, we obtain

$$\left[\vec{D}\right] = \left[\!\left[\bar{\bar{\varepsilon}} \bullet C_\varepsilon\right]\!\right] \left[\!\left[C_\varepsilon\right]\!\right]^{-1} \left[\vec{E}\right]. \tag{V.10}$$

This is a key point of the theory. Tedious, but elementary calculations show that the inverse of A_ε is given by

$$C_\varepsilon = \frac{1}{\xi_\varepsilon}\begin{pmatrix} \left(\left(\vec{N}\bullet\bar{\bar{\varepsilon}}\right)\times\vec{T}_2\right)_x & N_x & -\left(\left(\vec{N}\bullet\bar{\bar{\varepsilon}}\right)\times\vec{T}_1\right)_x \\ \left(\left(\vec{N}\bullet\bar{\bar{\varepsilon}}\right)\times\vec{T}_2\right)_y & N_y & -\left(\left(\vec{N}\bullet\bar{\bar{\varepsilon}}\right)\times\vec{T}_1\right)_y \\ \left(\left(\vec{N}\bullet\bar{\bar{\varepsilon}}\right)\times\vec{T}_2\right)_z & N_z & -\left(\left(\vec{N}\bullet\bar{\bar{\varepsilon}}\right)\times\vec{T}_1\right)_z \end{pmatrix} \quad \text{(V.11)}$$

where ξ_ε is the determinant of A_ε and is equal to $\vec{N}\bullet\bar{\bar{\varepsilon}}\bullet\vec{N}$. Thus Eq. (V.10) can be written in a compact form:

$$\left[\vec{D}\right] = Q_\varepsilon\left[\vec{E}\right] \quad \text{(V.12)}$$

where $Q_\varepsilon = \left[\!\left[\bar{\bar{\varepsilon}}\bullet C_\varepsilon\right]\!\right]\left[\!\left[C_\varepsilon\right]\!\right]^{-1}$ is a known matrix.

It is worth noting that the expression for the matrix C_ε in Eq. (V.11) is not specific to Cartesian coordinates. It has a similar expression in any orthogonal coordinate system (e.g., cylindrical or spherical coordinates). Its columns are constructed with the components of the three vectors $(\vec{N}\bullet\bar{\bar{\varepsilon}})\times\vec{T}_1$, $\vec{N}$ and $-(\vec{N}\bullet\bar{\bar{\varepsilon}})\times\vec{T}_2$, expressed in the specific coordinate system. This property is used in Chapter XIV to establish Maxwell equations in a truncated Fourier space using cylindrical coordinates.

V.1.2 Relation between $\left[\vec{B}\right]$ *and* $\left[\vec{H}\right]$

For the most general problem, the situation is completely symmetric concerning the relation between $\left[\vec{B}\right]$ and $\left[\vec{H}\right]$, and $\left[\vec{D}\right]$ and $\left[\vec{E}\right]$. We thus introduce a vector $\vec{F}_\mu$ given by: $\vec{F}_\mu = \left(H_{T_1}, B_N, H_{T_2}\right)$, and a matrix A_μ defined by $\vec{F}_\mu = A_\mu\vec{H}$. The expressions of A_μ and its inverse C_μ are respectively derived from Eqs. (V.9) and (V.11) by replacing $\bar{\bar{\varepsilon}}$ by $\bar{\bar{\mu}}$. As a result, we obtain:

$$\left[\vec{B}\right] = Q_\mu\left[\vec{H}\right] \quad \text{(V.13)}$$

where $Q_\mu = \left[\!\left[\bar{\bar{\mu}}\bullet C_\mu\right]\!\right]\left[\!\left[C_\mu\right]\!\right]^{-1}$.

V.1.3 The set of differential equations in Fourier space

After projected on the coordinate axis, time-harmonic Maxwell equations governing the electromagnetic field in an arbitrary media are:

$$\frac{\partial E_z}{\partial y} - \frac{\partial E_y}{\partial z} = i\omega B_x \tag{V.14}$$

$$\frac{\partial E_x}{\partial z} - \frac{\partial E_z}{\partial x} = i\omega B_y \tag{V.15}$$

$$\frac{\partial E_y}{\partial x} - \frac{\partial E_x}{\partial y} = i\omega B_z \tag{V.16}$$

$$\frac{\partial H_z}{\partial y} - \frac{\partial H_y}{\partial z} = -i\omega D_x \tag{V.17}$$

$$\frac{\partial H_x}{\partial z} - \frac{\partial H_z}{\partial x} = -i\omega D_y \tag{V.18}$$

$$\frac{\partial H_y}{\partial x} - \frac{\partial H_x}{\partial y} = -i\omega D_z \, . \tag{V.19}$$

The resolution of these equations, as well as the particular form of matrices Q_ε and Q_μ, will now depend on the problem, in the sense that some types of periodicity and invariance of the device will make some derivatives disappear. For example, in the case of a single grating with a direction of the rulings parallel to the Oz axis, $\frac{\partial}{\partial z} = i\gamma_0$. In addition, as the grating is periodic with respect to the x direction, the derivatives with respect to x will be expressed explicitly. We will then obtain the equations established in Chapter IV for an isotropic material in in-plane or conical mounting. Later in Chapter VIII, we will apply the equations for anisotropic gratings. Two-dimensional photonic crystals could be treated as a stack of identical gratings (Chapter X). For crossed gratings, periodic in x- and z-direction (Chapter IX), the z-derivative is not null, but it can be explicitly expressed, as it is done for the x-derivative. For a three-dimensional periodic photonic crystal, the periodicity along the three dimensions of space eliminates all derivatives, and the problem is reduced to an eigenvalue problem (Chapter X). However, in most cases, periodic in one or two directions only, the modulated region is limited inside a y-interval, let's say between 0 and a, where a is the groove depth, while the x and/or z periodicities or invariance give simple expressions for the x and/or z derivatives. Only the y-derivative remains in the set of Eqs. (V.14)-(V.19). After one- or two-dimensional Fourier analysis, the Fourier components of the field *will only depend on y*, and will be solution of the following equations:

$$\frac{\partial[E_z]}{\partial y}-\frac{\partial[E_y]}{\partial z}=i\omega[B_x] \tag{V.14'}$$

$$\frac{\partial[E_x]}{\partial z}-\frac{\partial[E_z]}{\partial x}=i\omega[B_y] \tag{V.15'}$$

$$\frac{\partial[E_y]}{\partial x}-\frac{\partial[E_x]}{\partial y}=i\omega[B_z] \tag{V.16'}$$

$$\frac{\partial[H_z]}{\partial y}-\frac{\partial[H_y]}{\partial z}=-i\omega[D_x] \tag{V.17'}$$

$$\frac{\partial[H_x]}{\partial z}-\frac{\partial[H_z]}{\partial x}=-i\omega[D_y] \tag{V.18'}$$

$$\frac{\partial[H_y]}{\partial x}-\frac{\partial[H_x]}{\partial y}=-i\omega[D_z], \tag{V.19'}$$

where $[D_j]$ and $[B_j]$, with j = x,y,z, have to be expressed through Eqs. (V.12) and (V.13). In these equations, it is understood that for a device periodic, say, along the x-axis with period d_x, $\partial/\partial x$ must be replaced by the multiplication by $i\alpha$, where α is a diagonal matrix with diagonal elements $(\alpha)_{mn}=\delta_{mn}\alpha_n\equiv\delta_{mn}(\alpha_0+n2\pi/d_x)$. Here α_0 is the x-component of the wavevector of the incident wave and δ_{nm} the Krönecker symbol. The particular case of a x-invariant device can be dealt with by taking $d_x\rightarrow\infty$, so that matrix α becomes a scalar $(\alpha)_{nm}=\alpha_0\delta_{nm}$. If, in addition to the geometrical invariance with respect to x, the incident wavevector is perpendicular to the x-axis, then $\alpha_0=0$, which leads to a total invariance with respect to the x-coordinate, i.e., $\partial/\partial x=0$.

Similarly, the periodicity along z will express the z-derivatives as a multiplication by a diagonal matrix $i\gamma$, where γ has diagonal elements $\gamma_n=(\gamma_0+n2\pi/d_z)$ with γ_0 denoting the z-component of the incident wavevector. Invariance with respect to the z-coordinate will reduce the z-derivative to a multiplication by a scalar $i\gamma_0$, and it will vanish if the incidence is in the Oxy plane.

Finally, the periodicity along y (e.g., three-dimensional photonic crystal) will replace $\partial/\partial y$ by a multiplication by a diagonal matrix $i\beta$ with diagonal elements $i\beta_n=i(\beta_0+n2\pi/d_y)$.

In fact, when programming, the values of α_n, β_n and γ_n are stored as column vectors rather than as almost empty matrices. In case of double periodicity, let us say along x and z directions, the multiplication by the column vector α must be repeated for each component γ_n and vice versa. The same rule holds for a double periodicity along another pair of coordinate axis, and has to be extended to triple periodicity, if necessary.

The resolution of the set (V.14')-(V.19'), which is now indeed a set of first-order classical differential equations with respect to y only, is conducted along the following lines. We only keep as unknown functions the vectors $[E_x]$, $[E_z]$, $[H_x]$ and $[H_z]$, which are respectively solutions of Eqs. (V.16'), (V.14'), (V.19') and (V.17'). To that aim, we eliminate $\left[E_y\right]$ and $\left[H_y\right]$ in the following way.

Eq. (V.12) leads to:

$$\left[D_y\right] = Q_{\varepsilon,yx}[E_x] + Q_{\varepsilon,yy}\left[E_y\right] + Q_{\varepsilon,yz}[E_z]$$

from which we get:

$$\left[E_y\right] = Q_{\varepsilon,yy}^{-1}\left(\left[D_y\right] - Q_{\varepsilon,yx}[E_x] - Q_{\varepsilon,yz}[E_z]\right).$$

In this equation, $\left[D_y\right]$ is expressed in terms of $[H_x]$ and $[H_z]$ only, thanks to Eq. (V.18'). We then get:

$$\left[E_y\right] = Q_{\varepsilon,yy}^{-1}\left\{\frac{i}{\omega}\left(\frac{\partial}{\partial z}[H_x] - \frac{\partial}{\partial x}[H_z]\right) - Q_{\varepsilon,yx}[E_x] - Q_{\varepsilon,yz}[E_z]\right\}. \tag{V.20}$$

A similar work for H_y leads to:

$$\left[H_y\right] = Q_{\mu,yy}^{-1}\left\{-\frac{i}{\omega}\left(\frac{\partial}{\partial z}[E_x] - \frac{\partial}{\partial x}[E_z]\right) - Q_{\mu,yx}[H_x] - Q_{\mu,yz}[H_z]\right\}. \tag{V.21}$$

Of course, the previously stated convention concerning the interpretation of the x and z partial derivatives still applies. We thus obtain a set of first-order differential equations of the form:

$$\frac{d}{dy}\begin{pmatrix}[E_x]\\ [E_z]\\ [H_x]\\ [H_z]\end{pmatrix} = M\begin{pmatrix}[E_x]\\ [E_z]\\ [H_x]\\ [H_z]\end{pmatrix}, \tag{V.22}$$

where M is a known matrix with elements given by:

$$
\begin{aligned}
M_{11} &= -i\alpha Q_{\varepsilon,yy}^{-1} Q_{\varepsilon,yx} - iQ_{\mu,zy} Q_{\mu,yy}^{-1} \gamma \\
M_{12} &= -i\alpha Q_{\varepsilon,yy}^{-1} Q_{\varepsilon,yz} + iQ_{\mu,zy} Q_{\mu,yy}^{-1} \alpha \\
M_{13} &= i\omega Q_{\mu,zy} Q_{\mu,yy}^{-1} Q_{\mu,yx} - i\frac{\alpha}{\omega} Q_{\varepsilon,yy}^{-1} \gamma - i\omega Q_{\mu,zx} \\
M_{14} &= i\omega\ Q_{\mu,zy} Q_{\mu,yy}^{-1} Q_{\mu,yz} + i\frac{\alpha}{\omega} Q_{\varepsilon,yy}^{-1} \alpha - i\omega\ Q_{\mu,zz} \\
M_{21} &= -i\gamma Q_{\varepsilon,yy}^{-1} Q_{\varepsilon,yx} + iQ_{\mu,xy} Q_{\mu,yy}^{-1} \gamma \\
M_{22} &= -i\gamma Q_{\varepsilon,yy}^{-1} Q_{\varepsilon,yz} - iQ_{\mu,xy} Q_{\mu,yy}^{-1} \alpha \\
M_{23} &= -i\omega\ Q_{\mu,xy} Q_{\mu,yy}^{-1} Q_{\mu,yx} - i\frac{\gamma}{\omega} Q_{\varepsilon,yy}^{-1} \gamma + i\omega\ Q_{\mu,xx} \\
M_{24} &= -i\omega\ Q_{\mu,xy} Q_{\mu,yy}^{-1} Q_{\mu,yz} + i\frac{\gamma}{\omega} Q_{\varepsilon,yy}^{-1} \alpha + i\omega\ Q_{\mu,xz} \\
M_{31} &= -i\omega\ Q_{\varepsilon,zy} Q_{\varepsilon,yy}^{-1} Q_{\varepsilon,yx} + i\frac{\alpha}{\omega} Q_{\mu,yy}^{-1} \gamma + i\omega\ Q_{\varepsilon,zx} \qquad \text{(V.22')} \\
M_{32} &= -i\omega\ Q_{\varepsilon,zy} Q_{\varepsilon,yy}^{-1} Q_{\varepsilon,yz} - i\frac{\alpha}{\omega} Q_{\mu,yy}^{-1} \alpha + i\omega\ Q_{\varepsilon,zz} \\
M_{33} &= -i\alpha Q_{\mu,yy}^{-1} Q_{\mu,yx} - iQ_{\varepsilon,zy} Q_{\varepsilon,yy}^{-1} \gamma \\
M_{34} &= -i\alpha Q_{\mu,yy}^{-1} Q_{\mu,yz} + iQ_{\varepsilon,zy} Q_{\varepsilon,yy}^{-1} \alpha \\
M_{41} &= i\omega\ Q_{\varepsilon,xy} Q_{\varepsilon,yy}^{-1} Q_{\varepsilon,yx} + i\frac{\gamma}{\omega} Q_{\mu,yy}^{-1} \gamma - i\omega\ Q_{\varepsilon,xx} \\
M_{42} &= i\omega\ Q_{\varepsilon,xy} Q_{\varepsilon,yy}^{-1} Q_{\varepsilon,yz} - i\frac{\gamma}{\omega} Q_{\mu,yy}^{-1} \alpha - i\omega\ Q_{\varepsilon,xz} \\
M_{43} &= -i\gamma Q_{\mu,yy}^{-1} Q_{\mu,yx} + iQ_{\varepsilon,xy} Q_{\varepsilon,yy}^{-1} \gamma \\
M_{44} &= -i\gamma Q_{\mu,yy}^{-1} Q_{\mu,yz} - iQ_{\varepsilon,xy} Q_{\varepsilon,yy}^{-1} \alpha\,.
\end{aligned}
$$

At the boundaries of the modulated region, the field can be expressed by Rayleigh expansions. The numerical solution of the most general diffraction problem has to be matched to these expansions. The diffraction problem is then reduced to a boundary-value problem similar to the one described in Section IV.3 for conical diffraction by a grating. As usual, the numerical contamination during the integration process is avoided by the use of the S-matrix propagation algorithm. The combination of Eqs. (V.22)-(V.22') and S-matrix propagation algorithm results in a new, fast-converging formulation of the Maxwell equations in a truncated Fourier space, which has been used with great success to obtain numerical results when studying light diffraction by various periodic objects.

V.1.4 Extension of normal and tangent unit vectors outside the diffracting surface

If the diffracting device is contained in a limited y-domain and its surface is defined by equation f(x,y,z) = 0, the unit vector $\vec{N}$ is given by:

$$\vec{N} = \frac{\overrightarrow{\text{grad}}\, f}{\left|\overrightarrow{\text{grad}}\, f\right|}.$$

If the diffraction surface can be expressed as y = g (x,z) we then obtain:

$$\vec{N} = \frac{\left(-\frac{\partial g}{\partial x},\ 1,\ -\frac{\partial g}{\partial z}\right)}{\sqrt{1+\left(\frac{\partial g}{\partial x}\right)^2+\left(\frac{\partial g}{\partial z}\right)^2}}. \tag{V.23}$$

This expression does not depend on y and is used $\forall(x,z)$ to define N_x, N_y and N_z. Similar to what was pointed out in Section IV.2.5, it means that the unit vector of the normal at a point M(x,y,z) inside the modulated area is parallel to the unit vector of the normal on the diffracting surface at point $M_0(x, g(x,z), z)$. $\vec{T}_1$ and $\vec{T}_2$ are then derived from the components of $\vec{N}$. We choose $\vec{T}_1$ in the xOy plane :

$$\vec{T}_1 = \left(\frac{N_y}{\sqrt{N_x{}^2+N_y{}^2}},\ \frac{-N_x}{\sqrt{N_x{}^2+N_y{}^2}},\ 0\right) \tag{V.24}$$

and thus:

$$\vec{T}_2 = \vec{T}_1 \times \vec{N} = \left(\frac{-N_x N_z}{\sqrt{N_x{}^2+N_y{}^2}},\ \frac{-N_y N_z}{\sqrt{N_x{}^2+N_y{}^2}},\ \frac{N_y{}^2+N_x{}^2}{\sqrt{N_x{}^2+N_y{}^2}}\right). \tag{V.25}$$

Since all the components of $\vec{T}_1$, $\vec{T}_2$ and $\vec{N}$ are y-independent, if the diffracting material is homogeneous, the Fourier components of all elements of matrix C_ε (see Eq. (V.11)) will have to be computed only once in the modulated area, and not at each ordinate y. Moreover, only one matrix inversion will be necessary to compute $[\![C_\varepsilon]\!]^{-1}$, which appears in the definition of matrix Q_ε.

V.1.5 The particular case of classical gratings made of isotropic, non-magnetic material

It is interesting to check how the general theory simplifies in the classical grating problem that has been previously addressed in Sections IV.2.2 and IV.3. For a nonmagnetic material, $\bar{\bar{\mu}}$ reduces to a scalar equal to the vacuum permittivity, so that matrix Q_μ reduces to $\mu_0 \mathbb{1}$, where $\mathbb{1}$ is the unit square matrix. As a result, almost half of the right-hand terms in Eqs. (V.22') vanish. Moreover, $[\varepsilon]$ reduces to a scalar: $\varepsilon_{i,j} = \varepsilon(x,y)\delta_{i,j}$.

Since the direction of the rulings coincides with the z-axis, we choose $\vec{T}_2$ in the tangential plane so that $\vec{T}_2 = \vec{z}$, where $\vec{z}$ is the unit vector of the z-axis. The normal vector is given by $\vec{N} = (N_x, N_y, 0)$ and its components are proportional to $\overrightarrow{\text{grad}}\, f(x,y)$, f(x,y) = 0 being the equation of the groove profile. From the equation $\vec{T}_2 \times \vec{T}_1 = \vec{N}$ or $\vec{T}_1 = \vec{N} \times \vec{T}_2$, we obtain: $\vec{T}_1 = (N_y, -N_x, 0)$. The determination of matrix C_ε requires the calculation of $\vec{N} \bullet \bar{\bar{\varepsilon}} \bullet \vec{N}$, $(\vec{N} \bullet \bar{\bar{\varepsilon}}) \times \vec{T}_2$ and $(\vec{N} \bullet \bar{\bar{\varepsilon}}) \times \vec{T}_1$. We obtain that

$$\vec{N} \bullet \bar{\bar{\varepsilon}} \bullet \vec{N} = \varepsilon(x,y)\ \vec{N} \bullet \mathbb{1} \bullet \vec{N} = \varepsilon(x,y)$$

$$(\vec{N} \bullet \bar{\bar{\varepsilon}}) \times \vec{T}_2 = \varepsilon(x,y)(N_y, -N_x, 0)$$

and

$$(\vec{N} \bullet \bar{\bar{\varepsilon}}) \times \vec{T}_1 = \varepsilon(x,y)(0,0,-1).$$

Thus the two matrices required to determine Q_ε read:

$$C_\varepsilon = \begin{pmatrix} N_y & N_x/\varepsilon & 0 \\ -N_x & N_y/\varepsilon & 0 \\ 0 & 0 & 1 \end{pmatrix} \text{ and } \varepsilon C_\varepsilon = \begin{pmatrix} \varepsilon N_y & N_x & 0 \\ -\varepsilon N_x & N_y & 0 \\ 0 & 0 & \varepsilon \end{pmatrix}.$$

Since N_x and N_y are defined outside the grating surface in *a continuous way*, as explained in Section V.1.4, Laurent's rule is used to calculate the corresponding Toeplitz matrices:

$$[\![C_\varepsilon]\!] = \begin{pmatrix} [\![N_y]\!] & [\![\frac{1}{\varepsilon}]\!][\![N_x]\!] & 0 \\ [\![-N_x]\!] & [\![\frac{1}{\varepsilon}]\!][\![N_y]\!] & 0 \\ 0 & 0 & \mathbb{I} \end{pmatrix} \tag{V.26}$$

and

$$[\![\varepsilon\ C_\varepsilon]\!] = \begin{pmatrix} [\![\varepsilon]\!][\![N_y]\!] & [\![N_x]\!] & 0 \\ -[\![\varepsilon]\!][\![N_x]\!] & [\![N_y]\!] & 0 \\ 0 & 0 & [\![\varepsilon]\!] \end{pmatrix}. \tag{V.27}$$

Inverting matrix $[\![C_\varepsilon]\!]$ given in Eq. (V.26) gives:

$$[\![C_\varepsilon]\!]^{-1} = \begin{pmatrix} [\![N_y]\!] & -[\![N_x]\!] & 0 \\ [\![\frac{1}{\varepsilon}]\!]^{-1}[\![N_x]\!] & [\![\frac{1}{\varepsilon}]\!]^{-1}[\![N_y]\!] & 0 \\ 0 & 0 & \mathbb{I} \end{pmatrix},$$

so that:

$$Q_\varepsilon = [\![\varepsilon C_\varepsilon]\!] \bullet [\![C_\varepsilon]\!]^{-1} =$$

$$\begin{pmatrix} [\![\varepsilon]\!][\![N_y^2]\!] + \left[\!\!\left[\frac{1}{\varepsilon}\right]\!\!\right]^{-1}[\![N_x^2]\!] & -\left([\![\varepsilon]\!] - \left[\!\!\left[\frac{1}{\varepsilon}\right]\!\!\right]^{-1}\right)[\![N_x N_y]\!] & 0 \\ -\left([\![\varepsilon]\!] - \left[\!\!\left[\frac{1}{\varepsilon}\right]\!\!\right]^{-1}\right)[\![N_x N_y]\!] & [\![\varepsilon]\!][\![N_x^2]\!] + \left[\!\!\left[\frac{1}{\varepsilon}\right]\!\!\right]^{-1}[\![N_y^2]\!] & 0 \\ 0 & 0 & [\![\varepsilon]\!] \end{pmatrix}.$$

The derivation of this equation has assumed the commutativity of some Toeplitz matrices. This point will be justified in Appendix II.

As can be seen, $Q_{\varepsilon,yz} \equiv 0 \equiv Q_{\varepsilon,zy}$; thus from Eqs. (V.22'), $M_{12} = 0 = M_{22} = M_{33} = M_{34}$. Moreover, $Q_{\varepsilon,xz} \equiv 0 \equiv Q_{\varepsilon,zx}$. If we remember that the device is z-invariant, the diagonal matrix γ in Eqs. (V.22') reduces to a scalar γ_0, equal to the z-component of the incident wavevector. It is then easy to check that elements M_{ij} derived from Eqs. (V.22') reduce to those shown in Eq. (IV.35) for conical diffraction mountings. Further simplification occurs for in-plane mountings, for which $\gamma_0 = 0$. We then obtain

$$M = \tag{V.28}$$

$$i\begin{pmatrix} -\alpha Q_{\varepsilon,yy}^{-1} Q_{\varepsilon,yx} & 0 & 0 & \frac{\alpha}{\omega} Q_{\varepsilon,yy}^{-1}\alpha - \omega\mu_0 \\ 0 & 0 & \omega\mu_0 & 0 \\ 0 & \omega Q_{\varepsilon,zz} - \frac{\alpha^2}{\omega\mu_0} & 0 & 0 \\ \omega\left(Q_{\varepsilon,xy} Q_{\varepsilon,yy}^{-1} Q_{\varepsilon,yx} - Q_{\varepsilon,xx}\right) & 0 & 0 & -Q_{\varepsilon,xy} Q_{\varepsilon,yy}^{-1}\alpha \end{pmatrix}$$

Eq. (V.22) then leads to

$$\frac{d}{dy}[E_z] = M_{23}[H_x] = i\omega\mu_0[H_x]$$

$$\frac{d}{dy}[H_x] = M_{32}[E_z] = \left(i\omega[\![\varepsilon]\!] - i\frac{\alpha^2}{\omega\mu_0} \right)[E_z],$$

which leads to the basic propagation equations for TE polarization, i.e., to Eqs. (I.17) and (I.18) written in Fourier space. We also obtain:

$$\begin{aligned} \frac{d[E_x]}{dy} &= -i\alpha Q_{\varepsilon,yy}^{-1} Q_{\varepsilon,yx}[E_x] + \left(\frac{i\alpha}{\omega} Q_{\varepsilon,yy}^{-1}\alpha - i\omega\mu_0 \right)[H_z] \\ \frac{d[H_z]}{dy} &= -i\omega\left(Q_{\varepsilon,xx} - Q_{\varepsilon,xy} Q_{\varepsilon,yy}^{-1} Q_{\varepsilon,yx} \right)[E_x] - iQ_{\varepsilon,xy} Q_{\varepsilon,yy}^{-1}\alpha[H_z]. \end{aligned} \qquad \text{(V.29)}$$

Remarking that $N_y^2 = c^2$ and $N_x^2 = 1 - c^2$, then $Q_{\varepsilon,ij}$ can be expressed in terms of the notations of Section IV.2.2. We obtain:

$$Q_{\varepsilon,xx} = [\![\varepsilon]\!][\![c^2]\!] + \left[\!\left[\frac{1}{\varepsilon}\right]\!\right]^{-1}[\![1-c^2]\!] = \frac{\Delta[\![c^2]\!]}{\omega^2\mu_0} + \left[\!\left[\frac{1}{\varepsilon}\right]\!\right]^{-1} = A + \left[\!\left[\frac{1}{\varepsilon}\right]\!\right]^{-1}$$

$$Q_{\varepsilon,yy} = [\![\varepsilon]\!][\![1-c^2]\!] + \left[\!\left[\frac{1}{\varepsilon}\right]\!\right]^{-1} c^2 = \frac{-\Delta}{\omega^2\mu_0}[\![c^2]\!] + [\![\varepsilon]\!] = -A + [\![\varepsilon]\!]$$

$$Q_{\varepsilon,yx} = B = Q_{\varepsilon,xy}.$$

Elementary algebraic calculations show that Eqs.(V.29) are identical to Eqs. (IV.21) and (IV.22) that govern the TM case of polarization. The conclusion is that the general equations reduce to the basic equations that lead to fast convergence on the particular case of classical isotropic nonmagnetic gratings.

Reference

V.1. E. Popov and M. Nevière, "Maxwell equations in Fourier space: fast converging formulation for diffraction by arbitrary shaped, periodic, anisotropic media," J. Opt. Soc. Am. A 18, 2886-2894, 2001

Chapter VI

Rigorous Coupled-Wave (RCW) Method

This method was widely popularized by Moharam and Gaylord at the beginning of the 1980s. They called it "rigorous coupled-wave analysis" [I.20]. The qualifier "rigorous" was used to distinguish it from the previously developed "coupled-wave theory" [VI.1] and [VI.2] and to point out the fact that it does not introduce any hypothesis or simplification in Maxwell equations. But its denomination is somehow confusing in the sense that the previously developed differential theory is as rigorous as it is; moreover they are both "coupled-wave" theories. Li has called it "coupled-wave method" [I.44], but recently proposed to call it "Fourier modal method" [VI.3] or "modal method by Fourier expansion" [VI.4]. But the qualifier "modal" may introduce confusion with the classical modal theory [I.16] and [I.17]. Thus, in this book, we will use the denomination of "rigorous coupled-wave theory" (RCW) despite its inconveniences, because it is the most widely used denomination. Indeed, the first presentation of the RCW theory can be found in the early work of Peng et al. [I.18], who refered to the work of Lewis and Hessel [VI.5]; an independent work was also published in a Russian proceeding in 1977 [VI.6]. However, the method was not widely used before the beginning of the 1980s. This is due to the fact that it is restricted to particular situations in which matrix M in Eq. (V.22) is independent of y as it happens for lamellar (or rectangular groove) gratings. This may occur in slanted volume holographic gratings after a convenient choice of the coordinates [VI.6], too. This type of profile was not common before the advent of binary optics and the ion-etching technique. Both Peng, Tamir and Bertoni on one side [I.18], and Moharam and Gaylord on the other side [VI.7] assumed that it was possible to extend its validity to arbitrary profiles via a staircase approximation, in view of having a general approach to analyze any kind of gratings. However, this hypothesis, which may seem quite natural, was never seriously checked. Until recently, the convergence of the staircase approximation was not analyzed in detail, and turns out to fail in TM polarization, as will be shown in Section VI.5.1.2. For surface relief gratings, the RCW method is then limited to lamellar profiles; it is then really performant. It replaces the numerical integration of Eq. (V.22) by the search of eigenvalues and eigenvectors of matrix M. However, it does not avoid the numerical contamination by growing exponential functions, since the eigenvalues are computed with a finite accuracy. In order to avoid the inversion of the global T matrix of a deep grating, the S-matrix algorithm has to be used too. Moreover, since the RCW method uses the same Fourier basis as the differential method, it suffered for a long time from the slow convergence of the Fourier series of the field when analyzing metallic gratings lighted with TM polarized waves. How this point was resolved will be explained in Section VI.2.

VI.1 Original version of the RCW theory

We illustrate the theory on the case of a lamellar grating (Fig. VI.1) illuminated in the cross-section plane, so that the diffraction problem decouples into the two TE and TM fundamental cases of polarization.

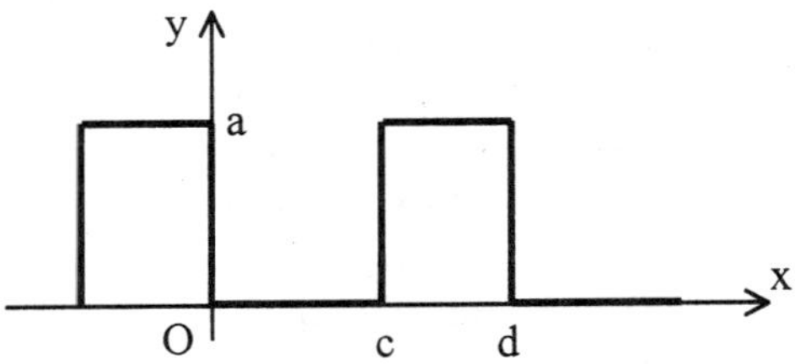

Fig. VI.1. Schematic representation of a lamellar grating.

Using the notations introduced in Section II.1.2, the propagation equation (Eqs. (I.17) and (I.18) written in Fourier space) for TE polarization and Eqs. (II.21) and (II.22) for TM one can be cast into the same form:

$$\frac{d\left[F^{TE,TM}(y)\right]}{dy} = M^{TE,TM}(y)\left[F^{TE,TM}(y)\right]. \qquad \text{(VI.1)}$$

For TE polarization, vector $\left[F^{TE}(y)\right]$ contains the two blocks made of the Fourier components of E_z and H_x. It is worth noticing that the tilde sign no longer appears. $\left[F^{TM}(y)\right]$ contains the two blocks of Fourier components of H_z and E_x in the TM case. Matrices $M^{TE,TM}$ are submatrices of the general matrix M defined in Chapter V and are given by:

$$M^{TE,TM} \equiv \begin{pmatrix} 0 & M_{12}^{TE,TM} \\ M_{21}^{TE,TM} & 0 \end{pmatrix} \qquad \text{(VI.1')}$$

with block elements:

$$\begin{aligned} M^{TE}_{1,2,n,m} &= i\omega\mu_0\delta_{n,m} \\ M^{TM}_{1,2,n,m} &= -i\omega\varepsilon_{n-m} \\ M^{TE}_{2,1,n,m} &= \frac{\alpha_n^2}{i\omega\mu_0}\delta_{n,m} + i\omega\varepsilon_{n-m} \\ M^{TM}_{2,1,n,m} &= \frac{i}{\omega}\alpha_n\left(\frac{1}{\varepsilon}\right)_{n-m}\alpha_m - i\omega\mu_0\delta_{n,m} \end{aligned},$$

For the rectangular-groove geometry, $\varepsilon(x,y)$ is indeed independent of y inside the modulated area. Thus ε_{n-m} and $\left(\frac{1}{\varepsilon}\right)_{n-m}$ are independent of y, too, and matrix $M^{TE,TM}(y)$ is indeed a constant matrix $M^{TE,TM}$. If, for sake of simplicity, we drop the superscrips TE and TM, it is then well known that the solution of Eq. (VI.1) can be explicitly expressed in terms of an exponential function of matrix M:

$$[F(y)] = \exp(My)[F(0)]. \tag{VI.2}$$

Eq. (VI.2) generalizes the usual solution that is known when the dimension of matrix M is unity. The exponential of a matrix M is defined by the entire series:

$$\exp(M) = 1 + M + \frac{(M)^2}{2!} + \ldots\ldots + \frac{(M)^k}{k!} + \ldots\ldots \tag{VI.3}$$

with an infinite radius of convergence. Since the entire series of matrices converges provided the modulus of the eigenvalues are less then the radius of the convergence of the series, the convergence is verified and Eq. (VI.3) defines $\exp(M)$, $\forall M$. If R is the corresponding diagonal matrix made with the different eigenvalues ρ_m, we have $M = QRQ^{-1}$, where Q is a square matrix made with the eigenvectors of M as columns put in the same sequence as the sequence of eigenvalues on the diagonal of R. As a result:

$$\begin{aligned} M^2 &= \left(QRQ^{-1}\right)\left(QRQ^{-1}\right) = QR^2Q^{-1} \\ &\ldots \\ M^k &= QR^kQ^{-1} \end{aligned},$$

so that Eq. (VI.3) leads to $\exp(M) = Q\exp(R)Q^{-1}$. Matrix exp(My) can be easily obtained from Eq. (VI.3); if $R_{n,m} = \rho_m \delta_{n,m}$, then $(\exp Ry)_{n,m} = \exp(\rho_m y)\delta_{n,m}$. As a result, Eq. (VI.2) leads to:

$$[F(y)] = Q\Phi(y)Q^{-1}[F(0)], \tag{VI.4}$$

where $\Phi(y)$ is a diagonal matrix with elements $\exp(\rho_m y)\delta_{n,m}$.

The determination of the eigenvalues and eigenvectors can be done via standard subroutines existing in all computer libraries. Thus Eq. (VI.4) provides F(a) as soon as F(0) is given. The diffraction problem is then resolved via the shooting method described in Section II.1.2.2.

An interesting point of the RCW theory is that the form of matrix M in Eq. (VI.1) ensures that its eigenvalues appear in pairs $\pm\rho_m$. Thus the field inside a rectangular groove is represented as a series of "upgoing" and downgoing" components, corresponding to eigenmodes whose y-dependence is clearly known from Eq. (VI.4). It is readily seen that the $\exp(\rho_m y)$ functions with $\mathrm{Re}(\rho_m) > 0$ will diverge if $y \to \infty$. Thus, if the groove depth a is large enough, some terms of the form $\exp(\rho_m a)$ will take large values and the numerical contamination discussed in Sections II.2.2 and III.5 will appear. It can be avoided using the S-matrix propagation algorithm described in Section III.6. A peer analysis of this algorithm, when applied to RCW theory or modal method, is done in Ref. [I.38].

Although we have presented the theory on the simpler case of a rectangular profile, a parallelogrammic grating, also called "slanted lamellar grating," which turns out to be equivalent to a slanted phase grating, can be analyzed with the RCW theory. Slanted gratings have a specific property that inside the modulated area the permittivity $\varepsilon(x,y)$ only depends on the scalar product $\vec{K}_s \cdot \vec{r}$ where $\vec{r} = (x,y,z)$ and $\vec{K}_s$ is the grating vector along the direction normal to the slanted walls or fringes (Fig. VI.2). It means that $\varepsilon(x,y)$ is constant in a region inside the modulated area lying on a plane perpendicular to $\vec{K}_s$. As a result, if $y \in [0,a]$, $\varepsilon(x,y) = \varepsilon(x',0)$ provided $x = x' - y\tan\delta$, where δ is the angle between the y-axis and the slanted walls. Expanding $\varepsilon(x',0)$ as:

$$\varepsilon(x',0) = \sum_{n=-\infty}^{+\infty} \varepsilon_n \exp(inKx'),$$

where ε_n are constant coefficients, we thus get:

$$\varepsilon(x,y)=\varepsilon(x',0)=\varepsilon(x+y\tan\delta,0)=\sum_{n=-\infty}^{+\infty}\varepsilon_n e^{inK(x+y\tan\delta)}.$$

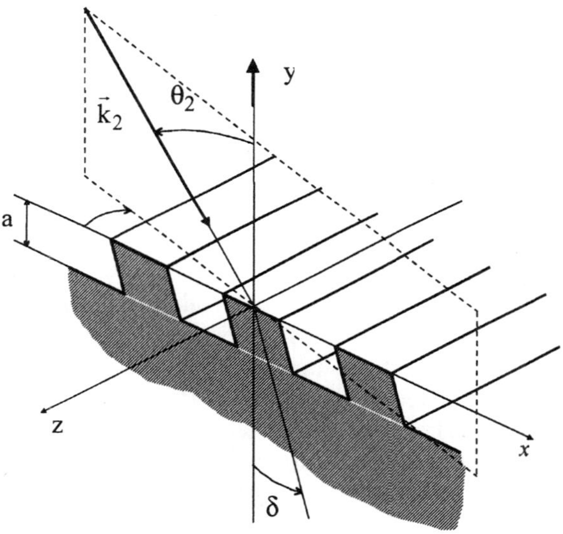

Fig. VI.2. Schematic representation of a slanted lamellar grating.

Defining $K_y = K\tan\delta$, we finally obtain:

$$\varepsilon(x,y)=\sum_{n=-\infty}^{+\infty}\varepsilon_n\, e^{in\left(Kx+K_y y\right)}. \tag{VI.5}$$

This expression suggests to represent all field components in the modulated area, satisfying Eq. (I.29') by:

$$u(x,y)=\sum_{n=-\infty}^{+\infty}\hat{u}_n(y)\,e^{i\left(\alpha_n x+nK_y y\right)}, \tag{VI.6}$$

which means that we express the pseudo-Fourier components $u_n(y)$ of u(x,y) by:

$$u_n(y) = \hat{u}_n(y)e^{inK_y y} \tag{VI.7}$$

Eqs. (VI.5) and (VI.6) put into the propagation Eqs. (I.20) and (I.21) lead to:

$$\frac{d\hat{\hat{H}}_n}{dy} = -inK_y\hat{\hat{H}}_n - i\omega \sum_{m=-\infty}^{+\infty} \varepsilon_{n-m}\hat{\hat{E}}_m \tag{VI.8}$$

$$\frac{d\hat{\hat{E}}_n}{dy} = -inK_y\hat{\hat{E}}_n + \frac{i}{\omega}\alpha_n \sum_{m=-\infty}^{+\infty} \alpha_m \left(\frac{1}{\varepsilon}\right)_{n-m} \hat{\hat{H}}_m - i\omega\mu_0\hat{\hat{H}}_n, \tag{VI.9}$$

where $\hat{\hat{E}}_n$ and $\hat{\hat{H}}_n$ are derived from H_n and E_n using Eq. (VI.7). As can be shown, the $\exp(inK_y y)$ functions have vanished. The differential set to integrate thus has coefficients independent of y and the RCW method applies.

The RCW method has been widely used [VI.7] and [VI.8], and is very efficient when applied in TE polarization case. It is also very efficient on thick phase gratings with continuous variations of refractive index, provided it is combined with the S-matrix propagation algorithm, for both TE and TM polarization. An alternative approach to deal with the exponentially growing terms has been proposed [VI.9], which is equivalent to the S-matrix propagation algorithm.

However, the RCW theory suffers from the slow convergence of the Fourier series of the field, in the same way as the classical differential method did. This is quite natural, since both theories use the same Fourier representation.

VI.2 Heuristic convergence improvement in TM polarization

Two independent teams [I.42] and [I.43] have at the same time (1996) proposed to replace Eq. (II.23) or Eq. (VI.1) of the TM case by Eqs. (IV.5) and (IV.6), i.e., to change $[\![\varepsilon]\!]$ into $\left[\!\!\left[\frac{1}{\varepsilon}\right]\!\!\right]^{-1}$ and $\left[\!\!\left[\frac{1}{\varepsilon}\right]\!\!\right]$ into $[\![\varepsilon]\!]^{-1}$ in the expression of blocks $M_{1,2}$ and $M_{2,1}$. This heuristic change produced the spectacular improvement of the convergence of the Fourier series of the field illustrated in Fig. VI.3. The example concerns a lamellar dielectric grating with refractive index 2.5, the superstrate is vacuum, the modulation ratio a/d is unity and the fill-in ratio c/d is 0.4. The period d is 1 μm, the wavelength λ is 0.6328 μm and incidence is 30°. The "error" is defined in Section IV.2.4. As can be seen, 21 Fourier components (N = 10) lead to an error of 10^{-4}, while 61 coefficients (N = 30) only give an error of 10^{-2} in the classical formulation. Better improvements could even be found on metallic gratings.

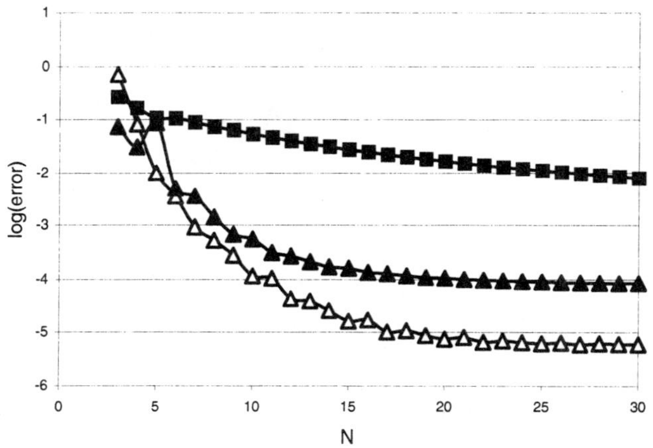

Fig. VI.3. Convergence of the previous and new versions of the RCW theory for the case of a dielectric lamellar grating with high contrast. Squares, old version of the RCW theory for TM polarization; open triangles, new version, TM polarization; solid triangles, TE polarization (after [I.45]).

The reasons for that improvement have been clearly explained by Li [I.44]. Now, they seem quite natural in view of the discussion in Section IV.2.3 that Eqs. (IV.5) and (IV.6) are the form taken for a lamellar profile by the general Eqs. (IV.21) and (IV.22) derived with the FFF method. As a consequence, it can easily be predicted that Eqs. (IV.5) and (IV.6), which work fairly well for rectangular lamellae, fail on a slanted lamellar (or parallelogrammic) grating, as well as on a classical lamellar grating lighted in conical diffraction, as explained in the next section. Confirmation of this prediction for the slanted grating is illustrated in Fig. VI.4.

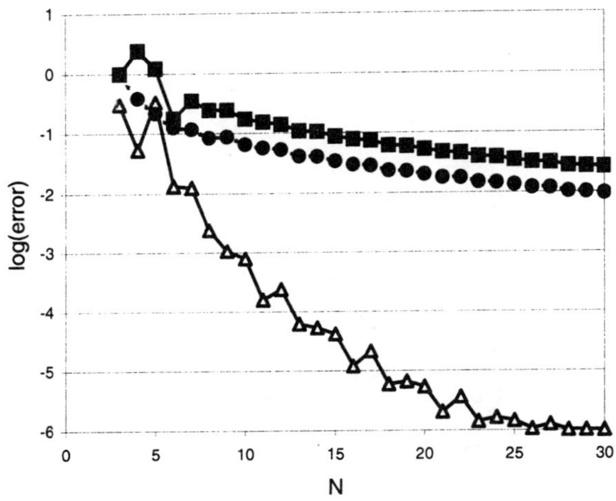

Fig. VI.4. Convergence of the previous (■) and new versions (Δ) of the differential theory, and the Lalanne and Morris [I.42] equations (•) for a slanted (45°) lamellar grating. The grating material is the same as in Fig. VI.3. Solid circles are obtained with Eqs. (IV.5) and (IV.6) (after [I.45]).

VI.3 Lamellar grating in conical diffraction

For a rectangular profile lighted in conical diffraction mounting, the correct equations to be used are Eqs. (IV.34) and (IV.35). For that profile, matrix Q_ε greatly simplifies into:

$$Q_\varepsilon = \begin{pmatrix} \left[\!\left[\frac{1}{\varepsilon} \right]\!\right]^{-1} & 0 & 0 \\ 0 & [\![\varepsilon]\!] & 0 \\ 0 & 0 & [\![\varepsilon]\!] \end{pmatrix}$$

and is y-independent. Matrix M in Eq. (IV.35) reduces to:

$$M = i\begin{pmatrix} 0 & 0 & -\alpha[\![\varepsilon]\!]^{-1}\dfrac{\gamma_0}{\omega} & \dfrac{\alpha}{\omega}[\![\varepsilon]\!]^{-1}\alpha - \omega\mu \\ 0 & 0 & \omega\mu - \dfrac{\gamma_0^2}{\omega}[\![\varepsilon]\!]^{-1} & \gamma_0[\![\varepsilon]\!]^{-1}\dfrac{\alpha}{\omega} \\ \dfrac{\alpha\gamma_0}{\omega\mu} & \omega[\![\varepsilon]\!] - \dfrac{\alpha^2}{\omega\mu} & 0 & 0 \\ \dfrac{\gamma_0^2}{\omega\mu} - \omega\left[\!\left[\dfrac{1}{\varepsilon}\right]\!\right]^{-1} & -\dfrac{\gamma_0\alpha}{\omega\mu} & 0 & 0 \end{pmatrix} \qquad \text{(VI.10)}$$

and Eq. (IV.34) gives:

$$\begin{aligned} \frac{d[E_x]}{dy} &= M_{13}[H_x] + M_{14}[H_z] \\ \frac{d[E_z]}{dy} &= M_{23}[H_x] + M_{24}[H_z] \\ \frac{d[H_x]}{dy} &= M_{31}[E_x] + M_{32}[E_z] \\ \frac{d[H_z]}{dy} &= M_{41}[E_x] + M_{42}[E_z] \end{aligned} \qquad \text{(VI.11)}$$

where elements M_{ij} (i, j = 1-3) are y-independent. The solution of Eq. (VI.11) can be expressed in terms of eigenvectors and eigenvalues, similar to what was done in Eq. (VI.4). An interesting remark is that the set of equations (VI.11) does not decouple into the classical TE and TM polarizations sets, even for the simple geometry of rectangular profile. On the other hand, one can verify that if the incident wavevector is in the cross-section plane of the grating $(\gamma_0 = 0)$, M_{13}, M_{24}, M_{31} and M_{42} vanish and the problem decouples into the TE and TM classical polarization cases.

However, even for conical diffraction $(\gamma_0 \neq 0)$, it is possible to decouple Eqs. (VI.11) into two independent sets, each having half of the size of Eqs. (VI.11). This requires separating the electric and magnetic field vectors into two polarizations [VI.10], transverse-electric or magnetic *with respect to the x-axis*, characterized, respectively, by $E_x = 0$ or $H_x = 0$. Let us denote the two cases as (e) and (h). When $E_x = 0$ (case (e)), the first of Eqs. (VI.11) permits to eliminate $\left[H_z^{(e)}\right]$, whereas in (h) case it is possible to eliminate $\left[E_z^{(h)}\right]$. Thus the set of equations takes the form of two decoupled systems:

$$\frac{d\left[E_z^{(e)}\right]}{dy} = \left(M_{23} - M_{24}M_{14}^{-1}M_{13}\right)\left[H_x^{(e)}\right]$$
$$\frac{d\left[H_x^{(e)}\right]}{dy} = M_{32}\left[E_z^{(e)}\right] \qquad \text{(VI.11')}$$

in (e) case, and

$$\frac{d\left[H_z^{(h)}\right]}{dy} = \left(M_{41} - M_{42}M_{32}^{-1}M_{31}\right)\left[E_x^{(h)}\right]$$
$$\frac{d\left[E_x^{(h)}\right]}{dy} = M_{14}\left[H_z^{(h)}\right] \qquad \text{(VI.11")}$$

in (h) case.

This procedure reduces by half the size of the matrix, whose eigenvalues and eigenvectors are searched to express the field inside the lamellae region. However, the price to pay is the introduction of two matrix inversions and four multiplications, which puts under question the advantages of Eqs. (VI.11') and (VI.11") when compared with Eqs. (VI.11). In addition, the field matching at the bottom and top of lamellae with the Rayleigh expansion above and below the grating region couples again (e) and (h) case.

On the contrary, when using the modal method, which does not use Fourier representation of the field inside the modulated region, Eqs. (VI.11') and (VI.11") are much easier to apply [VI.10], as the matrix $[\![\varepsilon]\!]$ reduces to scalars (different inside the groove and the lamella regions), so that the matrix operations in Eqs. (VI.11') and (VI.11") reduce to scalar operations, which can be handled analytically.

VI.4 Slanted grating used in conical diffraction

A further extension the RCW theory is to adapt it to both slanted gratings and conical diffraction [VI.11a]. We have to follow the lines stated in Section IV.3 and to use Eq. (IV.33) to express matrix Q_ε. When we introduce the change of functions stated by Eq. (VI.7), the y-derivatives introduce new matrix elements:

$$\frac{du_n}{dy} = \left(\frac{d\hat{\hat{u}}_n}{dy} + inK_y \hat{\hat{u}}_n \right) \exp(inK_y y) .$$

Again, $\exp(inK_y y)$ are eliminated in the calculations, so that Eq. (IV.34) is transformed into:

$$\frac{d}{dy} \begin{pmatrix} \left[\hat{\hat{E}}_x\right] \\ \left[\hat{\hat{E}}_z\right] \\ \left[\hat{\hat{H}}_x\right] \\ \left[\hat{\hat{H}}_z\right] \end{pmatrix} = \hat{\hat{M}} \begin{pmatrix} \left[\hat{\hat{E}}_x\right] \\ \left[\hat{\hat{E}}_z\right] \\ \left[\hat{\hat{H}}_x\right] \\ \left[\hat{\hat{H}}_z\right] \end{pmatrix} . \tag{VI.12}$$

The blocks of matrix $\hat{\hat{M}}$ are simply related to those of matrix M given in Eq. (IV.35) by:

$$\hat{\hat{M}}_{i,j} = M_{i,j} + L , \tag{VI.13}$$

where L is a diagonal matrix with elements $(-inK_y)$. The numerical implementation of Eqs. (VI.12) and (VI.13) has been done in Ref. [VI.11b].

In order to illustrate the possibilities of the RCW theory improved by the FFF method, when applied to slanted gratings used in conical diffraction, we choose a lamellar grating etched in gold substrate with complex refractive index 0.22 + i6.71. A linearlyy polarized plane wave is incident at an angle $\theta_2 = 30^{\circ}$ in a plane that makes a 30° angle with the z-axis. The superstrate is air. The electric field vector is in the plane of incidence. The grooves of the lamellar grating are slanted at 45°, the fill-in ratio or duty cycle of the grating is 0.5, the groove depth a is 0.3 μm, the groove spacing d is 1 μm and the wavelength is 0.8 μm. Fig. VI.5 and Fig. VI.6 show the convergence of the efficiencies as a function of the truncation order N. We present two figures for the same grating in order to show in detail the oscillations that occur at N < 30. Increasing N from 2 to 30, Fig. VI.5 shows the convergence of the zeroth order diffraction efficiencies computed by RCW method with the FFF based formulation (filled circles and solid curve), computed with the conventional formulation that does not used FFF method (filled triangles and dash-dotted curve), and with the "inverse" formulation that replaces $[\![\varepsilon]\!]$ by $\left[\!\left[\frac{1}{\varepsilon}\right]\!\right]^{-1}$ (crosses and dotted curve).

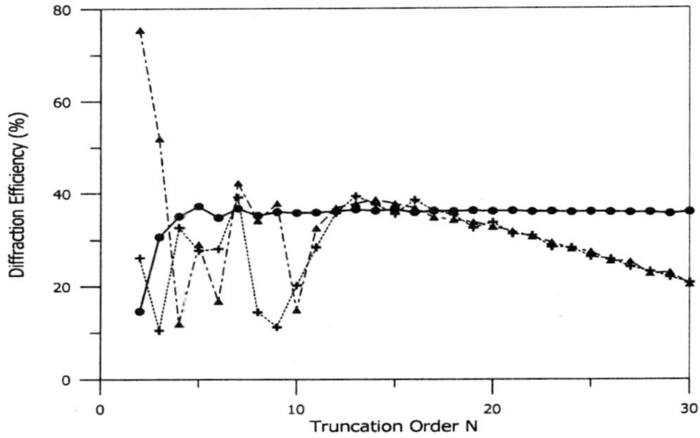

Fig. VI.5. Convergence of the zeroth-order diffraction efficiencies computed by RCW method with the FFF-based formulation (filled circles and solid curve), with the conventional formulation (filled triangles and dash-dotted curve), and with the "inverse" formulation (crosses and dotted curve) (after [VI.11]).

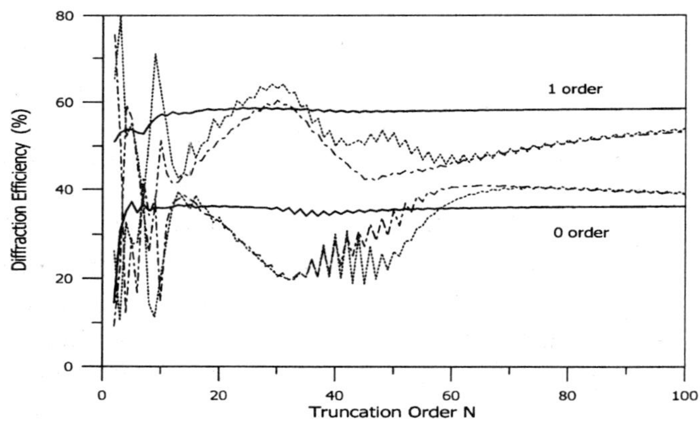

Fig. VI.6. Convergence of the zeroth- and minus-first-order diffraction efficiencies computed by RCW method with the FFF-based formulation (solid curve), with the conventional formulation (dash-dotted curve), and with the "inverse" formulation (dotted curve) (after [VI.11]).

The numerical results for the convergence of the zeroth- and minus-first-order diffraction efficiency are shown in Fig. VI.6 when N increases from 2 to 100. For both diffracted orders, the efficiencies computed with FFF-based formulation converge remarkably quickly as N increases. For the conventional and "inverse" formulation the results of calculations oscillate and converge very slowly; in fact, they do not begin converging until N > 60. Fig. VI.6 shows that the FFF formulation with N = 9 provides more accurate results than the convential and the "inverse" formulation with N = 101.

VI.5 Extension of the RCW theory to arbitrary profiles

After having published the RCW theory for slanted-fringe phase gratings [I.20], Moharam and Gaylord tried to extend the applicability of the method to arbitrary-shaped surface relief gratings, as previously suggested by Peng, Tamir and Bertoni [I.18]. To that end, they approximated the grating profile by a staircase, which turns the diffraction problem into the study of a stack of lamellar gratings to which the RCW theory applies. They assumed [I.20] that "if the individual planar gratings are sufficiently thin, any grating profile can be analyzed to an arbitrary level of accuracy." This hypothesis appeared so natural to the authors that no mathematical justification was ever presented, as well as no numerical study was done to clarify its domain of validity. When wrong results were obtained, they were merely attributed to numerical artifacts, or poor implementation of the theory. Only recently a detailed study was done [VI.12] which proves that the staircase approximation cannot be used to investigate arbitrarily shaped relief gratings in TM polarization. On the other hand, the method gives good results in TE polarization. Both mathematical and physical arguments will be presented to explain the failure and the success.

In order to avoid numerical artifacts, we will present a comparative study of three different methods, based on the differential theory. These are the classical differential method, improved using the fast Fourier factorization (FFF) technique, the RCW method and the modal method [I.16]. The choice is linked with the fact that the three methods are similar: solution of Maxwell differential equations is searched by projecting the solution on some basis of functions of x (see Fig. VI.7), periodic along the grating surface. Then the field dependence in the vertical (y) direction along the groove height is found using different techniques:

1. The classical differential method uses Fourier basis in x with numerical integration of a finite set of ordinary differential equations in y.

2. The RCW method also uses Fourier basis in x, but the grating is lamellar. This method avoids numerical integration by using an eigenvalue/eigenvector technique to find the solution of the set of differential equations.

3. The modal method also applies to lamellar gratings; it projects the field components on a basis of functions of x, called modes, which are rigorous solutions of the Maxwell equations, and boundary conditions along x (modes), assuming infinite height of the grooves along y (see Section I.1.2

and [I.16]). The solution of the diffraction problem is found in this basis by matching a superposition of the modes (with unknown coefficients) with the Rayleigh expansion in the substrate and the superstrate.

It is important to remark that, when applied to a given grating, consisting of a smooth profile represented with a fixed number of slices in the staircase approximation, the differential and the RCW methods give identical results, even when the number of the Fourier components is not sufficient to correctly represent the field. This fact shows that if a problem exists, it is not due to the techniques used to integrate the set of equations. However, the results obtained for the discretized (staircase-like) and non-discretized profile may differ significantly as will be shown in what follows.

VI.5.1 Convergence rate for a sinusoidal aluminum grating

VI.5.1.1 TE convergence

We begin by studying the staircase approximation in TE polarization (electric field vector parallel to the grooves). The grating under investigation is made of aluminum and has a sinusoidal profile with period d = 0.5 μm and groove depth h = 0.2 μm. This is a typical grating widely used in lasers and spectroscopy, because it is characterized by almost perfect blazing (very high diffraction efficiency) in TM polarization in a wide spectral range in the visible.

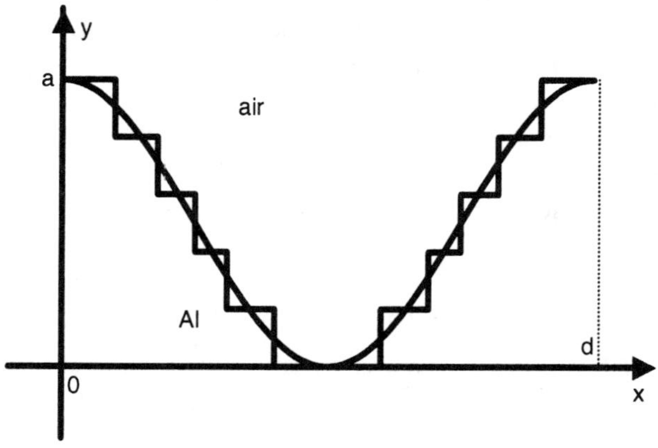

Fig. VI.7 Schematic representation of a sinusoidal profile together with a 5-stair ($\tilde{M} = 5$) approximation of the same profile (after VI.12]).

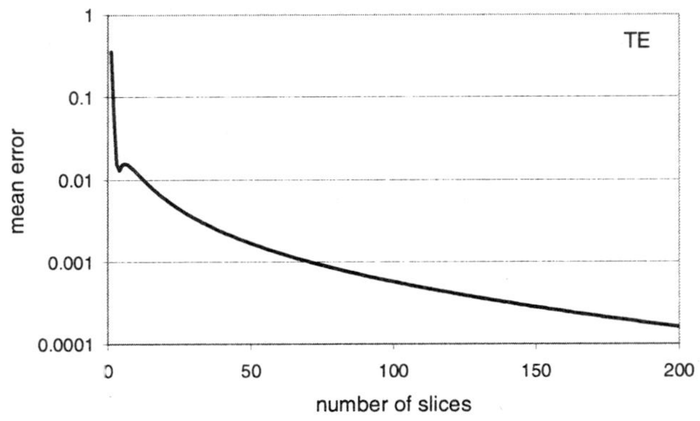

Fig. VI.8 Mean error in the reflected diffraction order efficiencies as a function of the number of slices $\tilde{M}$ in the staircase approximation (Fig. VI.7) of an aluminum grating with a sinusoidal groove profile, period d = 0.5 μm, groove depth a = 0.2 μm, aluminum refractive index ν_{Al} = 1.3 + i7.6, illuminated at 40° incidence with TE polarized light with wavelength $\lambda = 0.6328$ μm (after VI.12]).

The grating is illuminated at an angle θ_2 = 40° with light having wavelength λ = 0.6328 μm; thus two diffracted orders propagate in air. The aluminum complex refractive index is ν_{Al} = 1.3 + i 7.6. In what follows we are mainly interested in the convergence rate of the different methods as a function of the truncation parameter N, which characterizes the number of the Fourier components (or modes) used in the field expansion along x. However, to start with, it is interesting to know the number $\tilde{M}(= M - 2)$ of slices which are necessary to correctly represent the sinusoidal profile (Fig. VI.7). As a rule of the thumb, one can expect that the characteristic dimensions of the stairs must be less than λ/50. And indeed, Fig. VI.8 points out that a sufficient accuracy (>1%) is already obtained with approximately 15 slices.

The figure represents the mean error, defined in Section IV.2.4. In the calculations, the truncation parameter N is fixed equal to 90, a value sufficient to provide convergent results for both differential and RCW method, when applied to the same discretized profile. The error decreases monotonically and becomes close to 10^{-4} for $\tilde{M}$ > 200. The reference values to estimate the error are calculated using two independent methods, namely the integral method [VI.13] and the method of fictitious sources [VI.14].

The convergence with respect to the truncation parameter N for the three different methods is presented in Fig. VI.9. The differential method is applied to the smooth sinusoidal profile, while the modal and the RCW methods are applied to the staircase profile with $\tilde{M} = 200$ slices. The results of the modal method are not presented for large N, because the corresponding computer code deals with the more general case of crossed gratings in conical diffraction, which needs more computational resources for the same values of N.

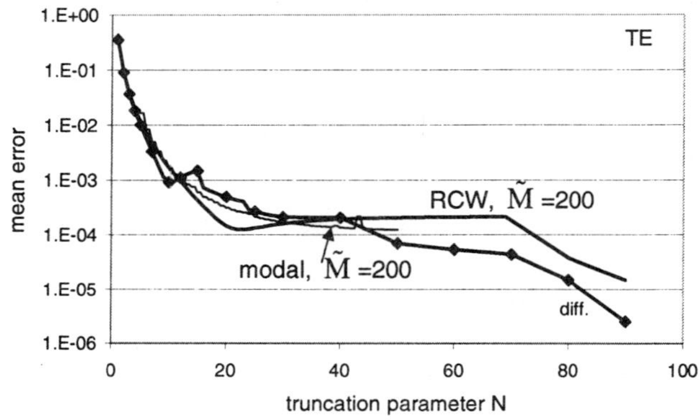

Fig. VI.9 Convergence of the mean error of the grating with parameters given in Fig. VI.8 as a function of the truncation parameter N, the number of the Fourier components or modes being equal to 2N+1; TE polarization. The RCW method (heavy solid curve) and the modal method (thin line) are used with $\tilde{M} = 200$ slices, the differential method (curve "diff.") is applied to the smooth sinusoidal profile (after VI.12]).

For this polarization we can conclude that the three methods lead to the same error for a given truncation, and that a relative error of less than 0.1% is obtained with N = 10, which is a good indication for short computation times.

VI.5.1.2 TM convergence

In TM polarization, the zeroth-order efficiency is very close to zero, so that the relative error is an unreliable characteristic. In what follows it is thus preferable to study the convergence rates of the minus-first-order efficiency. Fig. VI.10 presents the calculated efficiency as a function of N for the different methods and different numbers of slices $\tilde{M}$. Again, the differential method is applied to the real smooth sinusoidal profile. When applied to the staircase profile, it gives the same results as the RCW method. For the smooth profile (curve "diff."), the asymptotic value is approached within 1% as soon as N = 11 (23 Fourier components). Using 13 layers in the S-matrix propagation algorithm and 4 steps of integration per layer, the calculation time is one second on a PC Pentium III (800 MHz) computer.

(a)

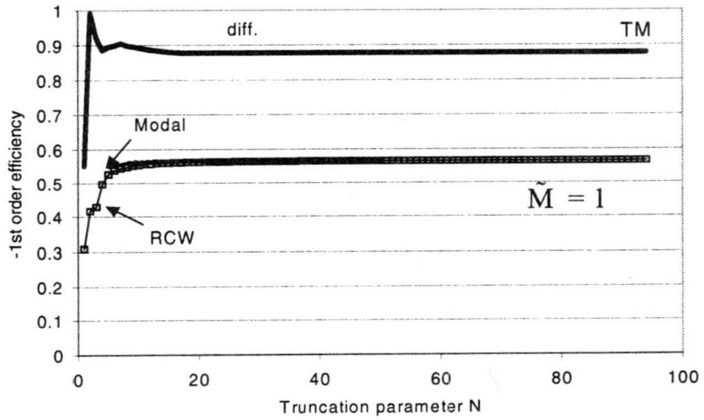

(b)

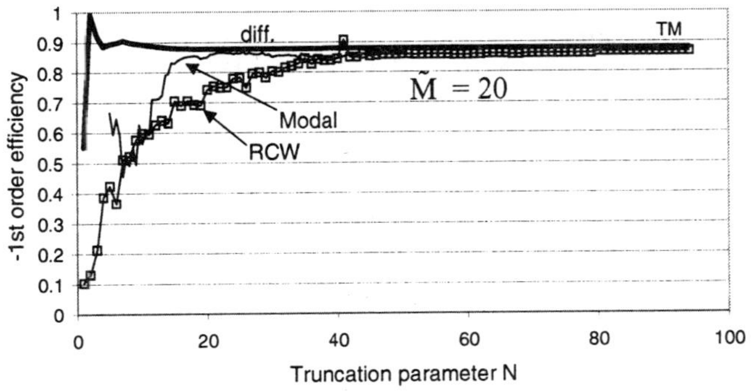

Fig. VI.10 Convergence of the minus-first-order efficiency in TM polarization of the RCW and modal method (indicated on the figure) for a single-step ($\tilde{M} = 1$) lamellar grating (a), and a staircase grating with $\tilde{M} = 20$ (b), as compared to the convergence of the differential method for a smooth sinusoidal profile (curve "diff."). The grating parameters are given in Fig. VI.8 (after VI.12]).

When $\tilde{M} = 1$, the sinusoidal grating is replaced by a lamellar one. Fig. VI.10 (a) shows that the RCW and the modal method give results having the same convergence rate as a function of N. The convergence rate for the discretized profile is the same as for the smooth sinusoidal profile (curve "diff."), although the asymptotic efficiency significantly differs for the two profiles, which is not surprising. This shows that the single-step approximation is rather poor and that the number of slices $\tilde{M}$ must be increased. When $\tilde{M}$ is increased to about 20 (Fig. VI.10 (b)), the asymptotic value of the efficiency approaches to within 1% the efficiency of the smooth profile (almost in the same way as in TE polarization, where $\tilde{M} = 15$ was sufficient). However, the number of Fourier coefficients (or modes, for the modal method), required to approach the asymptotic value increases for both the RCW and the modal methods. In order to obtain the same accuracy (error < 1%), it is necessary to increase N up to 20 for the modal method and up to 50 for the RCW method (i.e., 101 Fourier spatial harmonics), requiring 30 s calculation time, which has to be compared to 1 s necessary for the differential method when modeling the smooth sinusoidal profile with the same accuracy.

The situation becomes worse for $\tilde{M} = 200$. As reported in [VI.12], even N = 100 is not sufficient for the RCW method to converge (to stress again, the same results are obtained when applying the differential method to the staircase profile). In order to explain this surprising deterioration of the convergence rate, it is necessary to study the field map inside the grooves.

VI.5.2 Near-field maps

VI.5.2.1 TM field maps

In order to understand the poor convergence of all the methods for staircase profiles in TM polarization, it is sufficient to compare the near-field maps for a discretized profile with 20 slices, with the field map of the smooth sinusoidal profile [VI.12]. The field maps for a staircase profile were obtained using the RCW method (Fig. VI.11), while the differential method was used for the sinusoidal profile (Fig. VI.12). In TM polarization the magnetic field vector is parallel to the grooves (i.e., to the z-axis) and is a continuous function even across the grating surface. The electric field has x and y components, E_x and E_y, respectively.

A 3-D view of $|E_x|^2$ and $|E_y|^2$ as functions of x and y for $\tilde{M} = 20$ is shown in (Fig. VI.11) with a zoom inside the groove region where the field is stronger. Well-pronounced peaks of $|E_x|^2$ and $|E_y|^2$ are observed close to the edges.

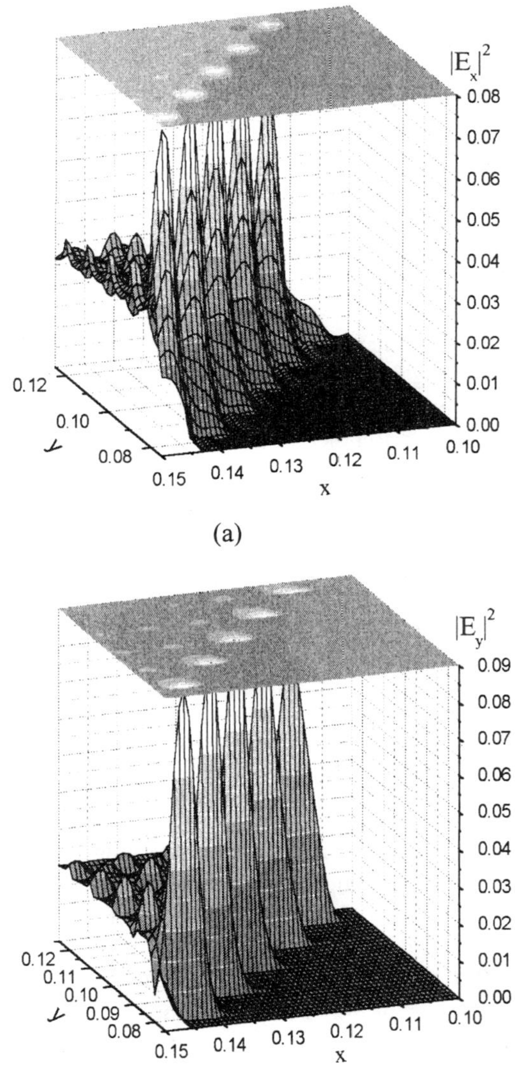

Fig. VI.11 Spatial field distribution in the vicinity of several steps inside a groove of a 20-step staircase profile, used to approximate the sinusoidal grating under study. TM polarization, (a) $|E_x|^2$, (b) $|E_y|^2$ (after VI.12]).

Strong maxima near the step ridges are observed, which could be expected for a metallic grating having sharp ridges, due to the charge accumulation at the ridges. However, the maxima of $|E_x|^2$ and $|E_y|^2$ do not occur at the same location. This is due to the fact that the two components of the electric field vector are discontinuous across the different segments of the profile: $|E_x|^2$ is continuous across the horizontal part (parallel to the x-axis). Its jump across the vertical segments (which are due to the large jump of the refractive index) allows for the existence of the sharps peaks on the vertical segments, while the continuity along the y-axis and the fact that inside the metal the electric field is weak limits $|E_x|^2$ in the vicinity of the horizontal segments. The opposite is true for $|E_y|^2$, which explains why its maxima occur on the horizontal segments. Anyway, the maxima of the two components occur close to the ridge, as already discussed.

It must be stressed that the existence of these sharp peaks is not due to Gibbs phenomena arising from representation of the discontinuous functions E_x and E_y by truncated Fourier series. The convergence of the field maps has been checked and no visible difference was found when going from 161 to 321 Fourier components.

The fact that close to the ridges of the staircase profile one observes regions with field enhancement explains the deterioration of the convergence rate of the methods with increasing the number of slices: the greater the numbers of stairs, the closer and thinner the maxima and the greater their number, thus the greater the number of Fourier components required to correctly represent the field in a form of truncated series. In the modal method, this leads to the increase of the number of modes inside the grooves, although not in such extend as for the RCW method, probably because the modes are not equidistant spectrally, as the Fourier harmonics are.

The field enhancement close to the ridges occurs when using any of the three methods and is a real physical effect, due to the fact that the smooth profile is replaced with a staircase-like profile having ridges. To observe the difference, it is sufficient to compare the previous figures with the field map of the sinusoidal (non-discretized) grating, presented in Fig. VI.12, obtained using the differential theory. A closer view (Fig. VI.12(b)) of the same region as was presented in Fig. VI.11 shows no peaks close to the profile, and the field inside the groove (on the left part of Fig. VI.12(b)) is the same as the background field (outside the peak region) inside the grooves in Fig. VI.11(a) (approximately equal to 0.30). The weak variations of the field in Fig. VI.12(b) come from the Gibbs phenomena, as the calculations were made using 201 Fourier components. The absence of sharp and narrow peaks of the field for the smooth profile explains the fast convergence of the differential method with respect to the number of Fourier components. As previously discussed, 23 Fourier components are sufficient.

(a)

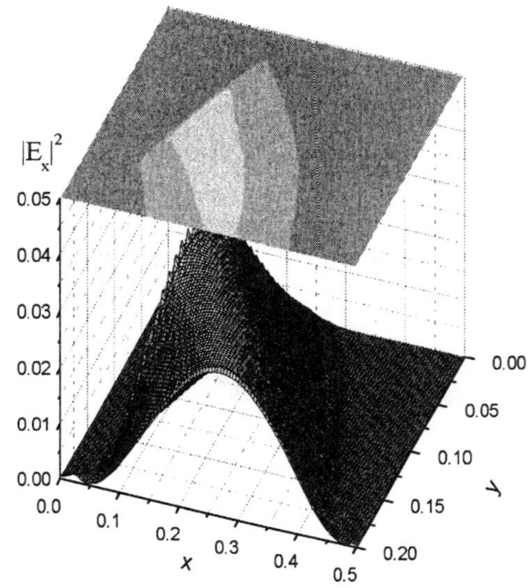

(b)

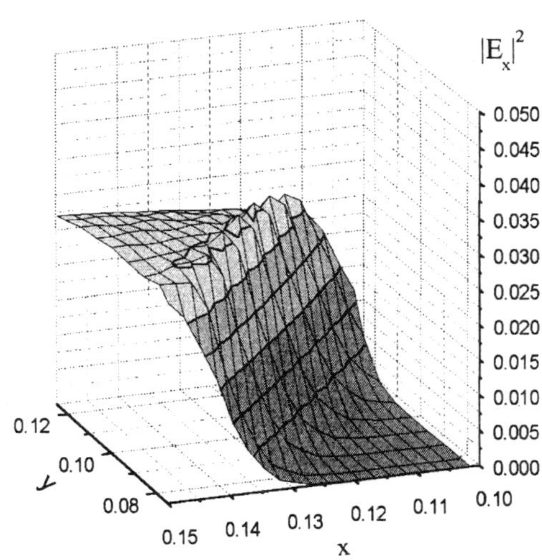

Fig. VI.12 Spatial distribution of $|E_x|^2$ calculated for non-discretized sinusoidal profile in TM polarization (after VI.12]). (a) The entire groove, and (b) the same region as presented in Fig. VI.11.

VI.5.2.2 TE field maps

In order to explain the fast convergence of the three methods in TE polarization, independently of the number of slices used in the staircase approximation, again it is worth studying the field maps and, in particular, it is sufficient to only look at the region of the groove where the field is the strongest.

In TE polarization, both electric and magnetic fields are continuous across the grating surface, because we consider non-magnetic media. Peaks are also found in the maps of the field components, but they rise much lower from the smooth background as compared with the TM case.

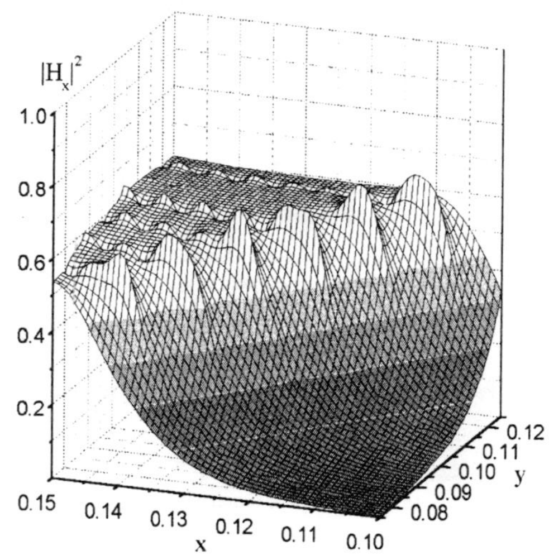

Fig. VI.13 The distribution of $|H_x|^2$, inside the groove for a 20-step staircase approximation of the sinusoidal grating in TE polarization. Same groove region as in Fig.VI.11a (after VI.12]).

This is due to the fact that the two components of the magnetic field (as well as the only component, the transverse one, of the electric field) are continuous functions when crossing the profile and their variations are not so rapid as in the TM case. This fact allows for averaging the peaks when the number of slices is increased so that the step dimensions decrease. The effect could be observed for the 20-stairs approximation, as shown in Fig. VI.13, where the height of the peaks is much lower when compared to the TM case. Thus when increasing $\tilde{M}$ in the TE case, it is not necessary to increase the number of Fourier harmonics in the field representation, which explains the better convergence rate with respect to N, even when $\tilde{M}$ is large, already observed in Fig. VI.9.

VI.5.3 Behavior of the staircase approximation in the limit of an infinite number of stairs

The question that naturally arises when looking at Fig.VI.11 is whether a field "homogenization" occurs when the number of slices is increased, for TM polarization. By "homogenization," we mean an averaging of the peaks which could occur when their number is high enough so that they begin to overlap. The process could result in a smooth surface similar to the one obtained in Fig. VI.12 for TE polarization. Unfortunately, results for $\tilde{M} = 200$ and 500 do not show any homogenization process [VI.12]. Increasing the wavelength λ by a multiplicative factor of 20, so that the ratio between the slice thickness and the wavelength is $1/1300$, does not reduce the height of the peaks observed in Fig. VI.11 [VI.12], too.

The lack of field "homogenization" observed in TM polarization can be explained by looking at the propagation equations. When dealing with a staircase-like profile, both differential and RCW method compute the solution of Eqs. (IV.5) and (IV.6), while for a smooth profile Eqs. (IV.21) and (IV.22) have to be used. These equations remain different whatever the number of slices may be . These two sets are both derived from the general Eqs. (V.22)-(V.22'), but for the first set $N_x = 1$, $N_y = 0$, whereas for the second set $N_x = -\sin\theta$ and $N_y = \cos\theta$. For the staircase profile, E_x is continuous across the horizontal segments (where E_y is discontinuous), and E_y is continuous across the vertical segments (where E_x is discontinuous). None of the two situations is valid for a smooth, continuous profile for which $\vec{N}$ lies neither along the x- nor along y-axis, except at a flat bottom and on vertical walls of the groove, if any. The conclusion is that neither the propagation equations, nor the boundary conditions, nor the field component behavior of the staircase profile tend to the ones of the smooth profile when $\tilde{M} \to \infty$, whatever the truncation order N may be.

Of course, the situation is quite different in TE polarization. The basic propagation equation (II.2') is the same for both staircase-like and smooth profile, as well as the behavior of the unknown field component E_z which is purely tangential in both cases. When $\tilde{M} \to \infty$, the coefficients ε_{n-m} for each elementary lamellar step tend to $\varepsilon_{n-m}(y)$ of the smooth profile. It can thus be expected that the field map of the discretized profile tends to the field map of the original one. The same conclusion is found if we use the set of Eqs. (I.17) and (I.18) written in Fourier space and if we remember that H_x is continuous across any profile since $\mu = \mu_0$ everywhere.

One idea for improving the convergence of the staircase approximation analyzed via the RCW method would be, when the number $\tilde{M}$ of steps is large enough, to replace Eqs. (IV.5) and (IV.6) by Eqs. (IV.21) and (IV.22) written at constant ordinates y_j, for example, the ordinate at the medium height of each step. Numerical studies [VI.12] have shown that indeed for large values of $\tilde{M}$ ($\tilde{M} \sim 200$), the results then given by the RCW theory on the staircase profile closely approach those of the smooth profile obtained with the differential theory. But one has to keep in mind that such a process is then equivalent to using the differential method improved by the FFF technique with the worst integration algorithm, i.e., the rectangular rule with equidistant points of integration. It is well-known that standard integration algorithms (e.g., Runge-Kutta or Adams-Moulton techniques) do much better.

As a side effect, the fact that when $\tilde{M}$ is high enough, Eqs. (IV.21) and (IV.22) give better results than Eqs. (IV.5) and (IV.6) enables us to understand a paradoxical result found by Lalanne [VI.15]. After having announced that for lamellar profiles Eqs. (IV.5) and (IV.6) work better than the equation of the classical differential theory, Eqs. (IV.3) and (IV.4), he noticed that in the low-modulation limit $(a/d \to 0)$ eqs. (IV.3, 4) work better than Eqs. (IV.5) and (IV.6). Indeed, in that limit $N_x \to 0$ and $N_y \to 1$ almost everywhere on the profile. Thus Eqs. (IV.21) and (IV22), which have proved to work better than Eqs. (IV.5) and (IV.6) when $a \to 0$, reduce to Eqs. (IV.3) and (IV.4), which is consistent with the Lalanne's observations.

The conclusion is that the RCW method is well suited for lamellar profiles, or for staircase profiles having a low number of steps, or for gratings consisting of rectangular rods, for which they lead to short computation time. For arbitrary profiles the staircase approximation leads to good results in TE polarization, but to poor results in TM case, linked with the introduction of artificial sharp maximums in the local field near the ridges, which do not vanish when $\tilde{M} \to \infty$. The differential theory should then be preferred.

References

VI.1. D. Marcuse, *Theory of Dielectric Optical Waveguides*, (Academic Press, New York, 1974)

VI.2. H. Kogelnik, "Theory of Dielectric Waveguides," in *Integrated Optics*, T. Tamir, ed. (Springer-Verlag, New York, Heidelberg, Berlin, 1975), pp 13-81

VI.3. L. Li, "New formulation of the Fourier modal method for crossed surface-relief gratings," J. Opt. Soc. Am. A14, 2758-2767, 1997

VI.4. L. Li, "Improvement of the coordinate transformation method for surface-relief gratings with sharp edges," J. Opt. Soc. Am. A13, 2247-2255, 1996

VI.5. L. R. Lewis and A. Hessel, "Propagation characteristics of periodic arrays of dielectric slabs," IEEE Trans. Microwave Theory Tech., MTT-19, 276-286, 1971

VI.6. E. I. Krupitsky and B. C. Chernov, "Rigorous analysis of arbitrarily slanted volume holographic gratings" (in Russian), Proc. IX All-Union School of Holography, Leningrad, 1977, pp 84-95

VI.7. T. K. Gaylord and M. G. Moharam, "Analysis and Applications of Optical Diffraction by Gratings," Proceeding of the IEEE, 73, 894-937, 1985

VI.8. E. Glystis and T. Gaylord, "Three-dimensional (vector) rigorous coupled wave analysis of anisotropic grating diffraction," J. Opt. Soc. Am. A7, 1399-1419, 1990

VI.9. M. Nevière, E. Popov, R. Reinisch, and G. Vitrant, *Electromagnetic Resonances in Nonlinear Optics*, (Gordon and Breach Science Publishers, London, 2000), pp. 90-99

VI.10. L. Li, "A modal analysis of lamellar diffraction gratings in conical mountings," J. Mod. Opt. 40, 553-557, 1993

VI.11a. L. Li, "Oblique-coordinate-system-based Chandezon method for modeling one-dimensionally periodic, multilayer, inhomogeneous, anisotropic gratings, " J. Opt. Soc. Am. A 16, 2521-2531, 1999

VI.11b. B. Chernov, M. Nevière, and E. Popov, "Fast Fourier factorization method applied to modal analysis of slanted lamellar gratings in conical mountings," Opt. Commun. 194, 289-297, 2001

VI.12. E. Popov, M. Nevière, B. Gralak, and G. Tayeb, " Staircase approximation validity for arbitrary-shaped gratings," J. Opt. Soc. Am. A 19, 33-42, 2002

VI.13. D. Maystre, "Integral methods", in *Electromagnetic theory of Gratings*, R. Petit, ed. (Springer-Verlag Berlin, 1980), ch. 3

VI.14. G. Tayeb, "The method of fictitious sources applied to diffraction gratings," special issue on "Generalized Multipole Techniques (GMT)" of

Applied Computational Electromagnetic Society (ACES) Journal, 9, 90-100, 1994

VI.15. P. Lalanne, "Convergence performance of the coupled-wave and the differential method for thin gratings," J. Opt. Soc. Am. A14, 1583-1591, 1997

Chapter VII
Coordinate Transformation Methods

Coordinate transformations have been widely used in the past, in various domains of physics, to resolve boundary-value problems. Conformal mappings [VII.1] are particularly well suited to analyze problems of electrostatic or fluid mechanics, which satisfy the same Laplace equation. Their application to the Helmholtz equation (i.e., to electromagnetics) seems to be more recent to our best knowledge. They were used at the end of the sixties to study waveguide problems [VII.2-VII.6] and were applied to diffraction problems at the beginning of the seventies [I.21].

The idea is to transform a complicated boundary (e.g., a grating profile) into a simpler one (e.g., a plane), with the hope that the basic equation of the problem will remain resolvable in the new space. The Laplace equation is perfect for that aim, since it is unchanged throughout a conformal mapping. The Helmholtz equation is somehow modified, but not too much, and the decision of using a coordinate transformation assumes that the benefit coming from the simplification of the boundary will be bigger than the trouble arising from the change in the equation. Whether a coordinate transformation should be conformal or not depends on the problem. Conformal mapping preserves the angles, which may be useful when a normal derivative must vanish at a boundary; however, their expression may be difficult to find. Thus, simpler nonconformal transformations may be of interest, too.

This chapter will present applications of the two kinds of transformations. First, the conformal mapping technique, initially developed to study perfectly conducting gratings, before being extended to finite conductivity gratings [VII.7-VII.9]; second, a nonconformal coordinate transformation [I.22], which is able to tackle both kinds of gratings.

VII.1 The conformal mapping technique for perfectly conducting gratings lighted in the cross-section plane

VII.1.1 The boundary-value problem in the original plane

The diffraction grating is illustrated in Fig. I.1, but for convenience in what follows, we will choose $d = 2\pi$. The substrate has infinite conductivity. The unknown function u(x,y) represents either E_z in the TE polarization case or H_z in the TM one. In both cases, u verifies the same Helmholtz equation in medium 2. Only one boundary condition applies, i.e., the vanishing of the tangential component of $\vec{E}$ at the metal surface, which results in the vanishing of u or of its normal derivative depending on polarization (the continuity of the tangential component of $\vec{H}$ produces a surface current which is unknown, so that this boundary condition does not help in resolving the problem). As a result, the unknown function u(x,y) is solution of the boundary-value problem:

$$\left(\Delta + k_2^2\right)u(x,y) = 0 \qquad \text{if } y > g(x) \tag{VII.1}$$

$$u(x+2\pi, y) = u(x,y)\exp(i\alpha_0 2\pi) \tag{VII.2}$$

$$u - u^{(i)} = \sum_{n=-\infty}^{+\infty} B_n^{(2)} \exp\left[i(\alpha_n x + \beta_n^{(2)} y)\right] \qquad \text{when } y \to \infty \tag{VII.3}$$

$$u\left(x, g(x)\right) = 0 \qquad \text{in TE polarization} \tag{VII.4}$$

$$\left.\frac{du(x,y)}{dN}\right|_{y=g(x)} = 0 \qquad \text{in TM polarization.} \tag{VII.5}$$

In view of simplifying the writing of Eqs. (VII.4) and (VII.5), we look for a mapping that will link the profile to a straight line, namely the X-axis, and which does not disturb Eqs. (VII.1)-(VII.3) too much.

VII.1.2 Search for a family of conformal mappings

Let us consider a simply connected area $\mathcal{D}$ in the complex w-plane. The area $\mathcal{D}$ contains the origin Ω and is bounded by a sufficiently regular curve (C). In the W-plane, we consider the area $\mathcal{D}$' bounded by the circle (C') having center Ω' and unit radius. According to the general theorems of conformal mappings, it is possible to find a mapping that maps $\mathcal{D}$ onto $\mathcal{D}$' and links origins Ω and Ω' and boundaries (C) and (C'). This implies the existence of a function w(W) analytic for $|W| < 1$, continuous for $|W| \leq 1$, which is equal to zero if and only if $W = 0$ and such that if $W = 1$, w(W) is a point on (C).

We come back to the mapping of the grating profile onto the X-axis by setting:

$$W = \exp(i\mathcal{Z}) \tag{VII.6}$$

$$w = \exp(iz)\,, \tag{VII.7}$$

where $\mathcal{Z} = X + iY$ and $z = x + iy$. The mapping defined by Eq. (VII.6) links the domain $\mathcal{T}$'$(Y \geq 0)$ illustrated in Fig. VII.2 with domain $\mathcal{D}$' presented in Fig. VII.1, as shown in Fig. VII.3.

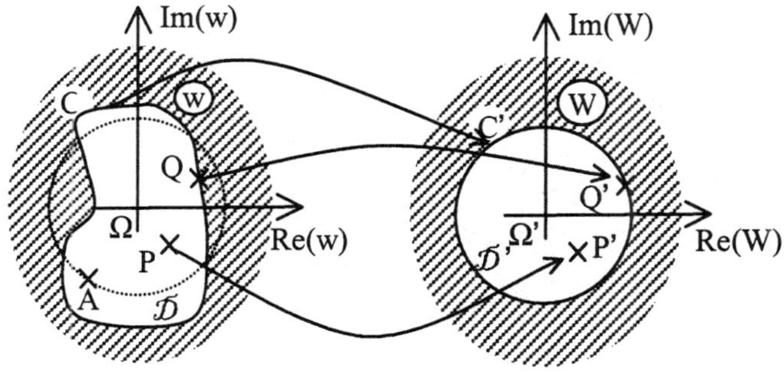

Fig. VII.1.Correspondence between the simply connected area 𝒟 and the interior of the circle 𝒟'.

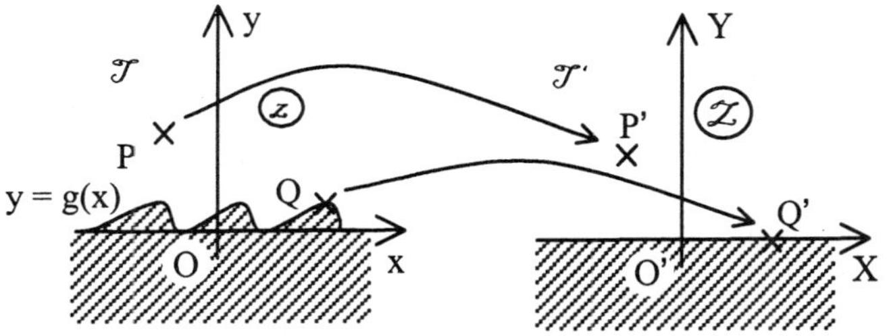

Fig. VII.2. Correspondence between area 𝒯 $(y \geq g(x))$ and half-plane 𝒯'$(Y \geq 0)$.

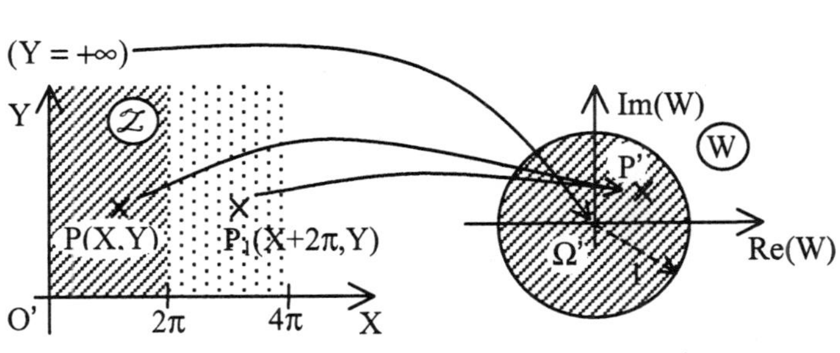

Fig. VII.3. Correspondence between the 𝒵 and W-complex planes.

In a similar way, Eq. (VII.7) links domain 𝒯$(y \geq g(x))$ to domain 𝒟 in Fig. VII.1, as illustrated in Fig. VII.4. The result is that it exists an analytic function 𝒵(z) which links areas 𝒯$(y \geq g(x))$ and 𝒯'$(Y \geq 0)$ in the complex planes z and 𝒵, as shown in Fig. VII.2.

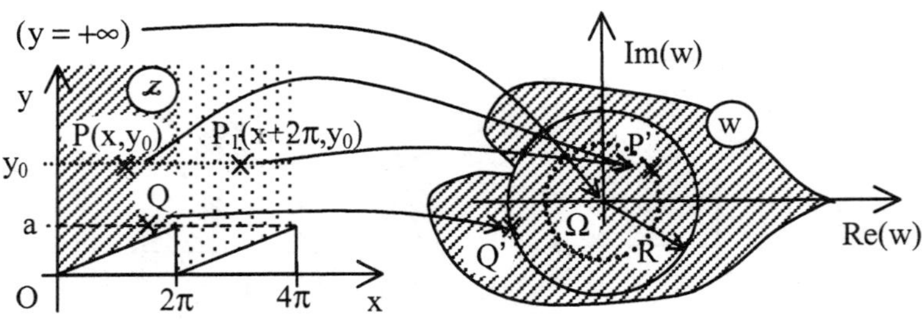

Fig. VII.4. Correspondence between the z and w complex plane. R = exp(-a).

Since the function w(W) is analytic in $\mathcal{D}$ ', we can represent it by Taylor's expansion around the origin Ω', thus:

$$w(W) = \sum_{n=1}^{\infty} c_n W^n. \tag{VII.8}$$

This series converges in the interior of its circle of convergence, whose radius is the distance from the origin Ω' to the closest singular point of w(W). As these singularities are necessarily localized at the exterior of $\mathcal{D}$ ' , the series (VII.8) is then convergent in all this area.*

We thus have:

$$\frac{w(W)}{W} = \sum_{n=0}^{\infty} c_{n+1} W^n$$

which can be written in another form by introducing coefficients b_n that can be expressed in terms of c_n :

$$\ln \frac{w(W)}{W} = i \sum_{n=0}^{\infty} b_n W^n.$$

The desired family of conformal mappings is then expressed as:

---

* This reasoning could not be invoked if we were looking for the mapping in the form:

$$W(w) = \sum_{n=1}^{\infty} c'_n w^n. \tag{VII.8'}$$

In fact, if the function W(w) possesses a singularity in a point A (Fig. VII.1) closer to Ω than some point of $\mathcal{D}$, this series diverges in all the part of the area $\mathcal{D}$ outside the circle with center Ω and radius ΩA, thus we can no longer use this series to represent the required function W(w).

$$z = \mathcal{Z} + i\sum_{n=0}^{\infty} b_n \exp(in\mathcal{Z}) \tag{VII.9}$$

The way of finding coefficients b_n corresponding to a given profile will be detailed in Section VII.1.4. Before that it is necessary to study how the initial boundary-value problem is modified during the coordinate transformation.

VII.1.3 The boundary-value problem in the transformed plane

The bijectivity of the conformal mapping between all points of domain $\mathcal{T}$ and $\mathcal{T}'$ warranties the existence of the function $\mathcal{Z}(z)$, inverse of the function $z(\mathcal{Z})$ defined in Eq. (VII.9). The consequence is the existence of the two functions X(x,y) and Y(x,y) which satisfy Cauchy-Riemann conditions:

$$\frac{\partial X}{\partial x} = \frac{\partial Y}{\partial y} \tag{VII.10}$$

$$\frac{\partial X}{\partial y} = -\frac{\partial Y}{\partial x}. \tag{VII.11}$$

If u(x,y) is transformed into U(X,Y) throughout the conformal mapping, the theorems of derivation of composite functions give:

$$\frac{\partial u(x,y)}{\partial x} = \frac{\partial U(X,Y)}{\partial X}\frac{\partial X}{\partial x} + \frac{\partial U(X,Y)}{\partial Y}\frac{\partial Y}{\partial x} \tag{VII.12}$$

$$\begin{aligned}\frac{\partial^2 u(x,y)}{\partial x^2} = {} & \frac{\partial^2 U}{\partial X^2}\left(\frac{\partial X}{\partial x}\right)^2 + \frac{\partial^2 U}{\partial X \partial Y}\frac{\partial Y}{\partial x}\frac{\partial X}{\partial x} + \frac{\partial U}{\partial X}\frac{\partial^2 X}{\partial x^2} + \\ & \frac{\partial^2 U}{\partial Y \partial X}\frac{\partial X}{\partial x}\frac{\partial Y}{\partial x} + \frac{\partial^2 U}{\partial Y^2}\left(\frac{\partial Y}{\partial x}\right)^2 + \frac{\partial U}{\partial Y}\frac{\partial^2 Y}{\partial x^2}\end{aligned} \tag{VII.13}$$

Writing similar expressions for $\frac{\partial u(x,y)}{\partial y}$ and $\frac{\partial^2 u(x,y)}{\partial y^2}$ we thus obtain:

$$\Delta u(x,y)+k^2u(x,y)=\frac{\partial^2 U}{\partial X^2}\left[\left(\frac{\partial X}{\partial x}\right)^2+\left(\frac{\partial X}{\partial y}\right)^2\right]+$$

$$\frac{\partial^2 U}{\partial Y^2}\left[\left(\frac{\partial Y}{\partial x}\right)^2+\left(\frac{\partial Y}{\partial y}\right)^2\right]+\frac{\partial^2 U}{\partial X\partial Y}\left[2\frac{\partial X}{\partial x}\frac{\partial Y}{\partial x}+2\frac{\partial X}{\partial y}\frac{\partial Y}{\partial y}\right]+ \qquad \text{(VII.14)}$$

$$\frac{\partial U}{\partial X}\left[\frac{\partial^2 X}{\partial x^2}+\frac{\partial^2 Y}{\partial x^2}\right]+\frac{\partial U}{\partial Y}\left[\frac{\partial^2 X}{\partial y^2}+\frac{\partial^2 Y}{\partial y^2}\right]+k^2U(X,Y)$$

in which we have assumed the identity $\frac{\partial^2}{\partial X\partial Y}=\frac{\partial^2}{\partial Y\partial X}$.

Cauchy-Riemann conditions immediately lead to $\Delta X=0=\Delta Y$, so that the terms with $\frac{\partial U}{\partial X}$ and $\frac{\partial U}{\partial Y}$ in Eq. (VII.14) vanish. Moreover, Eqs. (VII.10) and (VII.11) show that the coefficient of $\frac{\partial^2 U}{\partial X\partial Y}$ is null, too, and that the coefficients of $\frac{\partial^2 U}{\partial X^2}$ and $\frac{\partial^2 U}{\partial Y^2}$ are equal. If we remark that:

$$\frac{d\mathcal{Z}}{dz}=\frac{\partial \mathcal{Z}}{\partial x}=\frac{\partial X}{\partial x}+i\frac{\partial Y}{\partial x},$$

we get: $\left|\frac{\partial \mathcal{Z}}{\partial z}\right|^2=\left|\frac{\partial X}{\partial x}\right|^2+\left|\frac{\partial Y}{\partial x}\right|^2$, and Eq. (VII.1) is transformed into:

$$\frac{\partial^2 U(X,Y)}{\partial X^2}+\frac{\partial^2 U(X,Y)}{\partial Y^2}+k_2^2\left|\frac{dz}{d\mathcal{Z}}\right|^2 U(X,Y)=0. \qquad \text{(VII.15)}$$

Defining a new function K(X,Y) by $K^2(X,Y)=k_2^2\left|\frac{dz}{d\mathcal{Z}}\right|^2$, the propagation equation writes:

$$\Delta U(X,Y)+K^2(X,Y)U(X,Y)=0. \qquad \text{(VII.16)}$$

It is worth noticing that, in the new (XO'Y) plane, the unknown function satisfies a propagation equation which, in both TE and TM polarization, is identical to the propagation Eq. (I.19), which holds for finite conductivity gratings in TE polarization. Indeed, as it was found for $k^2(x,y)$, $K^2(X,Y)$ is rigorously periodic along the X-axis (with period 2π), as can be verified by looking at Eq. (VII.9). Moreover, since

$$\frac{dz}{d\mathcal{Z}} = 1 - \sum_{n=1}^{\infty} nb_n \exp[in(X+iY)],$$

it is seen that $K^2(X,Y)$ tends to k_2^2 as $\exp(-2Y)$ when $Y \to +\infty$. As a result, $K^2(X,Y)$ can be replaced by k_2^2 as soon as $Y = 2\pi$, with an error equal to $-b_1 \times 3.5 \times 10^{-6}$. Let us now study how the other basic Eqs. (VII.2)-(VII.5) of the initial boundary-value problem are modified.

From Eq. (VII.9) we derive:

$$x = X - \operatorname{Im}\left\{\sum_{n=0}^{\infty} b_n \exp(inX)\right\}\exp(-nY) \tag{VII.17}$$

$$y = Y + \operatorname{Re}\left\{\sum_{n=0}^{\infty} b_n \exp(inX)\right\}\exp(-nY). \tag{VII.18}$$

If we keep Y unchanged while we change X into $(X+2\pi)$, Eq. (VII.18) shows that y is unchanged while Eq. (VII.17) shows that x is changed into $x+2\pi$. The reciprocal result is immediately established by adding 2π to the two sides of Eq. (VII.17). The conclusion is that in the new plane, the pseudoperiodicity property is preserved:

$$U(X+2\pi, Y) = U(X,Y)\exp(i\alpha_0 2\pi) \tag{VII.19}$$

Concerning the outgoing-wave condition, Eq. (VII.3), we simply remark that Eq. (VII.9) implies that $z \to Z$ when $y \to \infty$. As a result:

$$U - U^{(i)} = \sum_{n=-\infty}^{+\infty} B_n^{(2)} \exp\left[i(\alpha_n X + \beta_n^{(2)} Y)\right] \tag{VII.20}$$

when $Y \to \infty$. Since we have admitted that $K^2(X,Y)$ may be replaced by k_2^2 when $Y > 2\pi$, Eq. (VII.20) will be used as soon as Y reaches the matching distance Y_a, shown to be equal to the grating period, i.e., to 2π.

The transformation of the boundary condition is obvious in TE polarization. Eq. (VII.4) simply gives:

$$U(X,0) = 0 \tag{VII.21}$$

in TE polarization. To obtain the transformed boundary condition in TM polarization, we calculate $\dfrac{\partial U(X,Y)}{\partial Y}$:

$$\frac{\partial U(X,Y)}{\partial Y} = \frac{\partial u(x,y)}{\partial x}\frac{\partial x}{\partial Y} + \frac{\partial u(x,y)}{\partial y}\frac{\partial y}{\partial Y}.$$

Eqs. (VII.10) and (VII.11) give:

$$\frac{\partial U(X,Y)}{\partial Y} = \frac{\partial u(x,y)}{\partial x}\left(-\frac{\partial y}{\partial X}\right) + \frac{\partial u(x,y)}{\partial y}\frac{\partial x}{\partial X}.$$

For $y = g(x)$, which implies $Y = 0$, this expression can be transformed introducing the components N_x and N_y of the unit vector $\vec{N}$ normal to the grating profile:

$$\left.\frac{\partial U(X,Y)}{\partial Y}\right|_{Y=0} = \left.\frac{\partial u(x,y)}{\partial x}\right|_{y=g(x)} N_x \left[\left(\frac{\partial y}{\partial X}\right)^2 + \left(\frac{\partial x}{\partial X}\right)^2\right]^{1/2} +$$

$$\left.\frac{\partial u(x,y)}{\partial y}\right|_{y=g(x)} N_y \left[\left(\frac{\partial y}{\partial X}\right)^2 + \left(\frac{\partial x}{\partial X}\right)^2\right]^{1/2} =$$

$$\left|\frac{dz}{d\mathcal{Z}}\right| \left.\overrightarrow{\text{grad}}u(x,y)\right|_{y=g(x)} \bullet \vec{N} = \left|\frac{dz}{d\mathcal{Z}}\right| \left.\frac{du(x,y)}{dN}\right|_{y=g(x)}.$$

As a result, the boundary condition $\left.\dfrac{du(x,y)}{dN}\right|_{y=g(x)} = 0$, Eq. (VII.5), leads to:

$$\left|\frac{d\mathcal{Z}}{dz}\right| \left.\frac{\partial U(X,Y)}{\partial Y}\right|_{Y=0} = 0.$$

Assuming that $\left|\frac{d\mathcal{Z}}{dz}\right| \neq 0$, a condition that may be violated in some points on the profile (i.e., on a null mesure set), Eq. (VII.5) will thus be replaced in the new plane by:

$$\left.\frac{\partial U(X,Y)}{\partial Y}\right|_{Y=0} = 0. \tag{VII.22}$$

Eqs. (VII.16) and (VII.19)-(VII.22) constitute the boundary-value problem in the new plane. It has the following physical interpretation: A perfectly conducting grating is equivalent to a phase grating with thickness equal to the period, deposited over a perfectly conducting plane mirror. The resolution of the diffraction problem can then be continued with the classical differential method developed for TE polarization. Since the same propagation equation applies to TE and TM polarization, the present problem is not concerned with the FFF method improvement, as already mentioned for phase gratings. On the other hand, the S-matrix propagation algorithm may be used, although it is not so often necessary as for real metallic gratings with finite conductivity, due to the fact that smaller values of the truncation parameter N are required to obtain convergent results.

In view of reducing computation time, it is useful to note that since the TE and TM boundary-value problem only differ by the boundary condition at Y = 0, Eqs. (VII.21) and (VII.22), the numerical integration can be done once only, and can be used for both polarizations, provided it is started from the top ($Y = Y_a$) of the domain and conducted decreasing the Y ordinate. In that case, we first construct (2N+1) particular solutions $\left[\hat{U}(Y)\right]_p$ corresponding to outgoing or evanescent reflected waves by starting the integration with:

$$\hat{U}(Y_a)_{n,p} = \exp\left[i\beta_n^{(2)} Y_a\right]\delta_{n,p}$$

$$\left.\frac{d\hat{U}_{n,p}}{dY}\right|_{Y=Y_a} = i\beta_n^{(2)} \exp\left[i\beta_n^{(2)} Y_a\right]\delta_{n,p}.$$

We then construct one more particular solution $\left[\hat{U}^{(i)}(Y)\right]$ corresponding to the incident field with the starting values:

$$\hat{U}_n^{(i)}(Y_a) = \exp\left[-i\beta_n^{(2)} Y_a\right]\delta_{n,0}$$

$$\left.\frac{d\hat{U}_n^{(i)}}{dy}\right|_{Y=Y_a} = -i\beta_n^{(2)} \exp\left[-i\beta_n^{(2)} Y_a\right] \delta_{n,0}.$$

The linearity of the problem implies that the Rayleigh coefficients are found by solving:

$$\hat{U}_n^{(i)}(0) + \sum_{m=-N}^{+N} B_n^{(2)} \hat{U}_{n,m}(0) = 0 \qquad \text{(VII.23)}$$

for TE polarization, and

$$\left.\frac{d\hat{U}_n^{(i)}}{dY}\right|_{Y=0} + \sum_{m=-N}^{+N} B_n^{(2)} \left.\frac{d\hat{U}_{n,m}}{dY}\right|_{Y=0} = 0 \qquad \text{(VII.24)}$$

for TM polarization. Thus the resolutions of TE and TM polarization case only differ by the matrix to invert at the end on the common integration process.

VII.1.4 Determination of the conformal mapping associated with a given grating profile

Two situations have to be distinguished, the case in which a period of the grating profile consists of a few number of segments of straight lines, and the case of a completely arbitrary profile.

VII.1.4.1 Grating profiles consisting of segments of straight lines

Triangular, lamellar and trapezoidal profiles are the most common examples. In these cases, the classical Schwarz-Chritoffel transformation [VII.1]:

$$\frac{dz}{d\mathcal{Z}} = \prod_j (\mathcal{Z} - a_j)^{\mu_j}, \qquad \text{(VII.25)}$$

where a_j and μ_j are constant numbers, transforms the real axis into a polygon with j summits at finite or infinite distance. We extend this transformation to a broken periodic line by defining a new transformation given by:

$$\frac{dz}{dZ} = \prod_j \left\{1 - \exp\left[i(Z - a_j)\right]\right\}^{\mu_j}. \qquad \text{(VII.26)}$$

As can be easily verified, the right-hand member vanishes for

$$X = a_j + p2\pi, \qquad p \text{ integer and}$$
$$Y = 0.$$

Let us study the argument $\arg\left(\frac{dz}{d\mathcal{Z}}\right)$ of $\frac{dz}{d\mathcal{Z}}$ when Y = 0 (i.e., y = g(x)) and X is varied between a_j and a_{j+1}:

$$\arg\left(\frac{dz}{d\mathcal{Z}}\right) = \sum_j \mu_j \arg\left\{1 - \exp\left[i(X - a_j)\right]\right\}$$
$$= \sum_j \mu_j \arg\left\{1 - \cos(X - a_j) - i\sin(X - a_j)\right\} = \sum_j \mu_j \arg\left\{1 + i\tan\frac{\pi + X - a_j}{2}\right\}$$
$$= \sum_j \mu_j \tan^{-1}\left(\tan\frac{\pi + X - a_j}{2}\right) = \sum_j \mu_j \left(\frac{\pi + X - a_j}{2}\right).$$

Since $\frac{dz}{d\mathcal{Z}} = \frac{dz}{dX} = \frac{dx}{dX} + i\frac{dy}{dX}$, we also have:

$$\arg\left(\frac{dz}{d\mathcal{Z}}\right) = \tan^{-1}\left[\frac{dg(x)/dX}{dx/dX}\right] = \tan^{-1}\left(\frac{dg}{dx}\right).$$

If we want the point M(x,y) to describe a segment of straight line while X is varied between a_j and a_{j+1}, we must have $\frac{dg}{dx}$ = constant, which implies that $\arg\left(\frac{dz}{d\mathcal{Z}}\right)$ = constant. The equation $\sum_j \mu_j \left(\frac{\pi + X - a_j}{2}\right)$ = constant implies:

$$\frac{1}{2}\pi \sum_j \mu_j + \frac{X}{2}\sum_j \mu_j - \frac{1}{2}\sum \mu_j a_j = \text{const.}$$

This equation can be verified $\forall X \in (a_j, a_{j+1})$ if:

$$\sum_j \mu_j = 0. \qquad \text{(VII.27)}$$

As an example, we study the conformal mapping associated to an arbitrary triangular profile with blaze angle θ' and second base angle θ''. Since we have two summits per period, we choose: $a_2 = 0$, a_1 being a parameter to be determined ($0 < a_1 < 2\pi$). We state $\mu_1 = -\mu_2$, and $\mu_j = 0 \quad \forall j \geq 3$. Eq. (VII.26) becomes:

$$\frac{dz}{d\mathcal{Z}} = \left(\frac{1 - \exp\left[i(\mathcal{Z} - a_1) \right]}{1 - \exp(i\mathcal{Z})} \right)^{\mu_1}. \tag{VII.28}$$

Thus, if $Y = 0$,

$$\arg\left(\frac{dz}{d\mathcal{Z}} \right) = \arg\left\{ e^{-i\frac{a_1}{2}} \frac{\sin\dfrac{X - a_1}{2}}{\sin\dfrac{X}{2}} \right\}^{\mu_1} = -\mu_1 \frac{a_1}{2} + \arg\left\{ \frac{\sin\dfrac{X - a_1}{2}}{\sin\dfrac{X}{2}} \right\}^{\mu_1}.$$

When X varies between 0 and a_1, we obtain:

$$\theta' = \mu_1 \left(\pi - \frac{a_1}{2} \right),$$

while when X varies between a_1 and 2π, we get

$$-\theta'' = -\mu_1 \frac{a_1}{2}.$$

Thus μ_1 and a_1 are determined from the groove angles:

$$\mu_1 = (\theta' + \theta'')/\pi$$
$$a_1 = 2\theta''\pi/(\theta' + \theta'').$$

When the parameter μ_1 and a_1 are known, the coefficients b_n in Eq. (VII.9) can be derived by recurrence formulae. To that end, we write:

$$\frac{dz}{d\mathcal{Z}} = \sum_{n=0}^{\infty} b'_n \exp(in\mathcal{Z}), \tag{VII.29}$$

where $b'_0 = 1$ and $b'_n = -nb_n$, $\forall n \geq 1$.

Using the variable $W = \exp(i\mathcal{Z})$, we introduce a function G(W) defined by: $G(W) = \left[\frac{1 - \tau W}{1 - W} \right]^{\mu_1}$ with $\tau = \exp(-ia_1)$. Taking the logarithmic derivative, we obtain:

$$\frac{G'(W)}{G(W)} = \mu_1 \left(\frac{-\tau}{1-\tau W} + \frac{1}{1-W} \right) = \frac{\mu_1 (1-\tau) W}{1-(1+\tau)W+\tau W^2}.$$

Thus G(W) satisfies the differential equation:

$$\left[1-(1+\tau)W+\tau W^2\right]G' = \mu_1 (1-\tau)G \qquad \text{(VII.30)}$$

Comparing Eqs. (VII.28) and (VII.29), we find that G(W) is also represented by its entire series:

$$G(W) = \sum_{n=0}^{\infty} b'_n W^n. \qquad \text{(VII.31)}$$

Substituting this expression into Eq. (VII.30) leads to a recurrence relation:

$$(n+1)b'_{n+1} = \left[\mu_1 (1-\tau) + n(1+\tau)\right] b'_n + \tau(1-n) b'_{n-1}. \qquad \text{(VII.32)}$$

The recurrence process is initialized by taking $b'_0 = 1$ and $b'_1 = \mu_1 (1-\tau)$. When the coefficients b'_n are known, the quantity $K^2(X,Y)$ in Eq. (VII.16) is fully determined and its Fourier components are simply derived.

For lamellar gratings, for example, Eq. (VII.28) has to be replaced by:

$$\frac{dz}{d\mathcal{Z}} = \left(\frac{1-\exp\left[i(\mathcal{Z}-a_2)\right]}{1-\exp(i\mathcal{Z})} \right)^{\mu_1} \left(\frac{1-\exp\left[i(\mathcal{Z}-a_3)\right]}{1-\exp\left[i(\mathcal{Z}-a_1)\right]} \right)^{\mu_1}. \qquad \text{(VII.33)}$$

The study of the argument of $\frac{dz}{d\mathcal{Z}}$ leads to $\mu_1 = 1/2$ and $a_1 - a_2 - a_3 = -\frac{\pi}{2}$. The determination of coefficients a_2 and a_3 requires resolving a set of two coupled integral equations. The details can be found in [VII.11], which presents the application of this conformal mapping to waveguide problems.

VII.1.4.2 Arbitrary-shaped gratings

Such kinds of gratings can be gratings with sinusoidal profiles, grating made by half circles or parts of circles in the form of bumps, etc. In this case, an iterative method has been developed. Coming back to Eq. (VII.8) and Fig. VII.1, we search for the function w(W) that maps the unit radius disc ($\mathcal{D}$') onto the domain $\mathcal{D}$, and which is analytic in $\mathcal{D}'$. As a result, the function

$\zeta(W) = \dfrac{W}{w(W)}$ is also analytic in the same domain. This function is simply related to the conformal mapping that we are looking for. If we introduce function b(W) defined by:

$$b(W) = \sum_{n=0}^{\infty} b_n e^{in\mathcal{Z}}, \qquad \text{(VII.34)}$$

so that $z - \mathcal{Z} = ib(W)$, we find:

$$w = We^{-b(W)}, \quad \text{i.e.,} \quad \zeta(W) = \exp[b(W)]. \qquad \text{(VII.35)}$$

Representing $\zeta(W)$ by its entire series:

$$\zeta(W) = \sum_{n=0}^{\infty} a_n W^n, \qquad \text{(VII.36)}$$

and remarking that $\dfrac{d\zeta(W)}{\zeta(W)} = db(W)$, a simple relation can be established between the coefficients a_n and b_n that we are looking for:

$$a_0 b_n^{n+1} = (n+1)a_{n+1} - \sum_{p=1}^{n} a_p b_{n-p}(n-p+1). \qquad \text{(VII.37)}$$

Thus the iterative process will determine the coefficients a_n, which implies truncating the series in Eq. (VII.36) to an order N' ($N' \sim 61$).

The function $\zeta(W)$ must have two properties. First, it should be analytic in $\mathcal{D}$', second, it should map the boundary (C') of $\mathcal{D}$' onto the boundary (C) of $\mathcal{D}$. Both conditions are easy to satisfy separately, while it is difficult to enforce both properties simultaneously. Indeed, the first condition is automatically satisfied by *any* truncated Taylor series, like the one appearing in Eq. (VII.36), but *in which the summation runs from* 0 *to* N'. Concerning the second property, the function $\zeta(W)$ that has the required mapping property is simply given by:

$$\zeta(W)\big|_{Y=0} = \frac{W\big|_{Y=0}}{w\big|_{y=g(x)}} = \frac{\exp(iX)}{\exp(iz)\big|_{y=g(x)}} = \frac{\exp(iX)}{\exp[ix(X)]\exp\{-g[x(X)]\}}. \qquad \text{(VII.38)}$$

If we remark that X is the argument of the complex variable W, the function $\zeta\left[\exp(iX)\right]$ is obviously periodic with respect to X with period 2π. Thus the coefficients of its exponential Fourier series are simply given by:

$$a_n = \frac{1}{2\pi}\int_0^{2\pi} \frac{\exp(iX)}{\exp\left[ix(X)\right]\exp\left\{-g\left[x(X)\right]\right\}}\exp(-inX)dX, \qquad \text{(VII.39)}$$

where n varies from $-\infty$ *to* $+\infty$. As a result, the function defined by $\sum_{n=-N'}^{+N'} a_n W^n$ will not be analytic in $\mathcal{D}'$. We enforce the two conditions through the following iterative process. We start the process by taking arbitrary values of the $N'+1$ coefficients a_n. The first approximation of $\zeta(W)$ is obviously analytic, but of course, it does not have the mapping property. In order to enforce this property, we first calculate x which is simply equal to arg(w):

$$x = \arg(w) = \arg(W) - \arg(\zeta).$$

Thus, for Y = 0:

$$x = X - \arg\left[\sum_{n=0}^{N'} a_n \exp(inX)\right]. \qquad \text{(VII.40)}$$

This value of x is put into Eq. (VII.38) to enforce the mapping condition. From that value of $\zeta\left[\exp(iX)\right]$, new coefficients a_n are derived from Eq. (VII.39). We take those corresponding to $0 \le n \le N'$ only, and put them into Eq. (VII.36) which gives a second approximation of $\zeta(W)$ that is analytic, but has partially lost its mapping property. This mapping property is again enforced using Eqs. (VII.40), (VII.38) and (VII.39). The iteration is stopped when the smallest coefficient a_n obtained at two consecutive steps only differ by a given accuracy, typically 10^{-5} or 10^{-6} of its value. Coefficients b_n are then derived through Eq. (VII.37). When the profile is symmetrical with respect to the y-axis, which implies that the b_n coefficients are real, another way can be followed. We then obtain, when Y = 0:

$$y(X) = \sum_{n=0}^{N'} b_n \cos(nX) \text{ or}$$

$$g\left[x(X)\right] = \sum_{n=0}^{N'} b_n \cos(nX),$$

in which x(X) is given by Eq. (VII.40). Coefficients b_n can then be obtained by Fourier formulas.

The iterative method has produced convergent results on many usual grating profiles, including on trapezoidal profiles [VII.10] that can be tackled with the Schwarz-Christoffel transformation. It may fail on deep gratings. For sinusoidal profiles, for example, the convergence fails when the modulation a/d is greater than 0.6. The use of Padé approximations [VII.12], however, has increased its range of usefulness so that it was possible to analyze a sinusoidal profile with $a/d = 4$.

VII.1.5 Application to the analysis of the validity of Rayleigh expansion

It is possible to apply the conformal mapping technique, developed above, to study the validity of the Rayleigh expansion, a problem extensively discussed in the literature during the twentieth century. The work of Millar [VII.13a-VII.14c] has established that the solution of Helmholtz and Laplace equations, which satisfy the same boundary conditions, have the same singularities. We thus study the Rayleigh expansion corresponding to the electrostatic problem derived from our diffraction problem by taking $\frac{\omega}{c} = 0$; moreover, we assume normal incidence and a profile symmetric with respect to the y-axis. Since $d = 2\pi$, $\beta_n^{(2)} = i|n|$ and the "static" Rayleigh expansion reads:

$$u = \sum_{n=-\infty}^{+\infty} B_n^{(2)} \exp\left(-|n| y\right) \exp(inx) + 1 .$$

Let us consider the function $\mathcal{Z}(z)$ inverse of $z(\mathcal{Z})$, which defines the conformal mapping, and is equal to $X(x, y) + iY(x, y)$. If $\mathcal{Z}(z)$ were analytic, the function Y should verify the equation $\Delta Y = 0$, as being the imaginary part of an analytic function. In addition, $Y[x, g(x)] = 0$, due to the mapping property.

Function Y would thus be the solution of our electrostatic problem in TE polarization, and could be identified with the static electric field $E(x, y) \equiv u(x, y)$. As a consequence, the validity of Rayleigh expansion for $y \geq g(x)$ is equivalent to:

$$Y(x, y) = \sum_{n=-\infty}^{+\infty} B_n^{(2)} \exp(-|n| y) \exp(inx) + 1 \qquad \text{if } y \geq g(x) . \quad \text{(VII.41)}$$

Since the incidence is normal and the profile symmetric, $B_{-n} = B_n$. Thus Eq. (VII.41) is transformed into:

$$Y(x,y) = 1 + B_0 + 2\sum_{n=1}^{\infty} B_n^{(2)} \exp(-ny)\cos nx \qquad \text{if } y \geq g(x). \quad \text{(VII.42)}$$

This expression would lead to:

$$\mathcal{Z} = z_0 + 2i\sum_{n=1}^{\infty} B_n^{(2)} \exp(inz) \text{ if } y \geq g(x), \text{ with } \mathrm{Im}(z_0) = 1 + B_0 .$$

Thus, if the Rayleigh expansion were valid for $y \geq g(x)$, the function $\mathcal{Z}(z)$ would be analytic $\forall y \geq g(x)$. The study of the validity of Rayleigh expansion is then reduced to the search of the singularities of $\mathcal{Z}(z)$, the inverse function of $z(\mathcal{Z})$, which provides the bijection illustrated in Fig. VII.2. As a result, the singularities of the inverse function would be only found for $y < g(x)$. They would not show up if they were only located at $y < 0$, since in that case, $\mathcal{Z}(z)$ would be analytic $\forall y \geq 0$ and then, it would also be analytic $\forall y \geq g(x)$, so that the Rayleigh expansion would be valid.

VII.1.5.1 Profiles described by a finite number of coefficients b_n

If the number of coefficients b_n in Eq. (VII.9) is finite, the function $z(\mathcal{Z})$ is analytic over the entire open complex plane. In that case, the only singularities of the inverse function $\mathcal{Z}(z)$ are branching points, obtained when $dz/d\mathcal{Z} = 0$.

VII.1.5.1.1 Cycloidal profiles

They are obtained by choosing $b_n = 0 \;\; \forall n \neq 1$. Thus, $z = \mathcal{Z} + ib_1 \exp(i\mathcal{Z})$, so that the parametric equations of the profile, obtained by taking Y = 0, are:

$$x = X - b_1 \sin X$$
$$y = b_1 \cos X .$$

The corresponding curve name "shortened cycloid" is illustrated in Fig. VII.5.

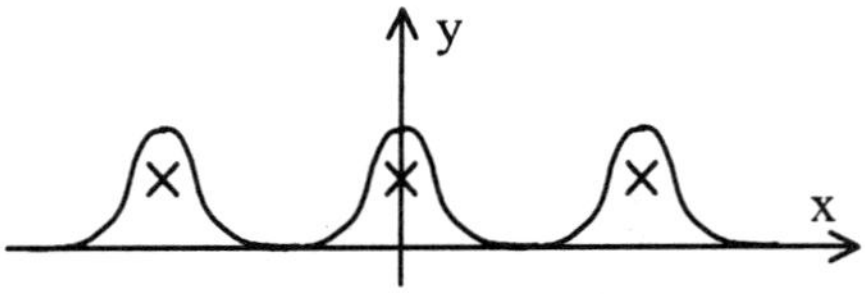

Fig. VII.5. Singularities of $\mathcal{Z}(z)$ for a shortened cycloidal grating.

The derivative $dz/d\mathcal{Z}$ is given by $dz/d\mathcal{Z} = 1 - b_1 \exp(i\mathcal{Z})$. It vanishes if $\mathcal{Z} = i \ln b_1 + 2p\pi$ (p integer), i.e., for values z_0 of z given by $z_0 = i(1 + \ln b_1) + 2p\pi$. Since the superposition of plane waves will converge only if $y > \mathrm{Im}(z_0)$, the Rayleigh hypothesis will be valid only if $\mathrm{Im}(z_0) < -b_1$, which implies $1 + \ln b_1 + b_1 \leq 0$. The solution of the equation is $b_1 = 0.2786$. For higher values of b_1, Rayleigh expansion will not be valid.

If $b_1 = 1$ which gives the pure cycloidal profile shown in Fig. VII.6, the singularities are given by: $z_0 = i + 2p\pi$. They are thus located at the tips, and the superposition of plane waves converges nowhere in the modulated area.

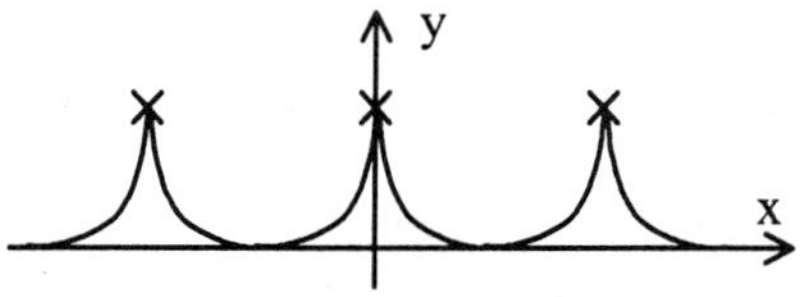

Fig. VII.6. Singularities of $\mathcal{Z}(z)$ for a pure cycloidal grating.

VII.1.5.1.2 Other profiles

If several b_n coefficients are different from zero, e.g., for $n \in [1, J]$ the research of the singularities of Z(z) implies resolving a transcendental equation of the form:

$$1 - \sum_{n=1}^{J} b_n \exp(in\mathcal{Z}) = 0. \qquad \text{(VII.43)}$$

This can be done on a computer and give solutions denoted $\mathcal{Z}_0$. The singularities z_0 are given by $z_0 = \mathcal{Z}_0 + i$.

VII.1.5.2 Triangular profiles

Since this kind of profile requires an infinite number of coefficients b_n, it is impossible to deal with the transcendental equation similar to Eq. (VII.43). We can proceed in a completely different way, by going back to the w and W complex plane. Since the conformal mapping defined by Eq. (VII.9) is equivalent to the one in Eq. (VII.8), the Rayleigh expansion will be valid $\forall y > g(x)$ if the inverse function W(w) is analytic inside the hatched domain illustrated in Fig. VII.4, i.e., if the series in Eq. (VII.8') converges. This entire series converges inside a disc with a the radius R equal to the distance between the center Ω (the image of the point $y \to \infty$) and the nearest singularity of W(w). It diverges outside, which means that the Rayleigh expansion will not be usable at an ordinate smaller than a. More generally, Rayleigh expansion is never valid for profiles having edges oriented toward positive ordinates.

These conclusions hold as far as Helmholtz and Laplace equations have the same singularities. This result from Millar fails for some particular frequencies such as the one implied in the Maréchal and Stroke mounting [VII.15] and some similar situations in which $\left(\frac{\omega}{c}\right)^2$ is an eigenvalue of the Δ-operator of the Helmholtz equation.

VII.2 Perfectly conducting gratings used in conical diffraction mounts

For perfectly conducting gratings, the problem is much simpler to analyze than for finite conductivity gratings (see Section IV.3), in the sense that it reduces to the analysis of two classical in-plane TE and TM polarization cases with a change in the wavelength (or in k_2) [VII.16-VII.18]. We recall that in conical diffraction mount, the incident field is given by:

$$\vec{E}^i = \vec{A} \exp\left[i(\alpha_0 x - \beta_0 y + \gamma_0 z)\right],$$

and that the invariance of the diffracting system along the z coordinate results in the fact that all field components have the same $\exp(i\gamma_0 z)$ dependence. We also recall that as soon as E_z and H_z are known, all other transverse field components can be derived from them by formulae used in waveguide theory [IV.5 and IV.6]. As example, E_x and H_x are given by Eqs. (IV.36) and (IV.37) written in the superstrate, while E_y and H_y are given by:

$$E_y = \frac{1}{k_2^2 - \gamma_0^2}\left(-i\omega\mu\frac{\partial H_z}{\partial x} + i\gamma_0\frac{\partial E_z}{\partial y}\right) \quad \text{(VII.44)}$$

$$H_y = \frac{1}{k_2^2 - \gamma_0^2}\left(i\omega\varepsilon_2\frac{\partial E_z}{\partial x} + i\gamma_0\frac{\partial H_z}{\partial y}\right). \quad \text{(VII.45)}$$

Moreover, E_z and H_z both verify Helmholtz equation if $y \geq g(x)$, which results into:

$$\Delta_T u + \left(k_2^2 - \gamma_0^2\right)u(x,y) = 0 \quad \text{(VII.46)}$$

with $\Delta_T = \frac{\partial^2}{\partial x^2} + \frac{\partial^2}{\partial y^2}$. It means that E_z and H_z verify the same propagation equation as for the in-plane ($\gamma_0 = 0$) TE and TM polarization, except that k_2 has been changed into $\tilde{k}_2$ given by:

$$\tilde{k}_2 = k_2\sqrt{1-\gamma_0^2}. \quad \text{(VII.47)}$$

Concerning the only boundary condition which has to be satisfied for a perfectly conducting material, namely $\vec{N} \times \vec{E} = 0$ if $y = g(x)$, it results into:

$$E_z = 0 \quad \text{if } y = g(x) \quad \text{(VII.48)}$$

$$N_x E_y - N_y E_x = 0 \quad \text{if } y = g(x). \quad \text{(VII.49)}$$

Eqs. (IV.36) and (VII.44) put into Eq. (VII.49) lead to:

$$-\frac{1}{\tilde{k}_2^2}\left[i\omega\mu\left(N_x\frac{\partial H_z}{\partial x} + N_y\frac{\partial H_z}{\partial y}\right) + i\gamma_0\left(N_x\frac{\partial E_z}{\partial y} - N_y\frac{\partial E_z}{\partial x}\right)\right] = 0 \quad \text{if } y = g(x).$$

Since E_z is null not only at $y = g(x)$ but also inside the metal ($y < g(x)$), $\frac{\partial E_y}{\partial y} = 0 = \frac{\partial E_z}{\partial x}$ if $y = g(x)$. The previous equation thus reduces to:

$$\vec{N} \bullet \overrightarrow{\text{grad}} H_z = 0 \quad \text{if } y = g(x). \quad \text{(VII.50)}$$

Eqs. (VII.48) and (VII.50) together with Eq. (VII.46) show that E_z and H_z can be determined by resolving the TE and the TM problem with k_2 replaced by $\tilde{k}_2$. If we designate with $A_n^{(2),TE}$ and $A_n^{(2),TM}$ the Rayleigh coefficients obtained by resolving the two boundary-value problems, the calculation of the Poynting vector leads to the expression of the reflected efficiencies [VII.18]:

$$e_n^{(r)} = \frac{\beta_n^{(2)}}{\beta_0^{(2)}} \left[\sin^2 \delta \left| A_n^{(2),TE} \right|^2 + \cos^2 \delta \left| A_n^{(2),TM} \right|^2 \right], \qquad \text{(VII.51)}$$

where $\beta_n^{(2)}$ are given by Eq. (I.33) in which k_2^2 is replaced by $k_2^2 - \gamma_0^2$, and δ is an angle that defines the polarization state. To that end, we first consider the projection $\vec{k}_2^*$ of the incident wavevector $\vec{k}_2$ onto the cross-section plane xOy, we then construct the vector $\vec{k}_2^+$ which is perpendicular to $\vec{k}_2^*$ in the plane xOy; δ is the angle between $\vec{k}_2^+$ and the polarization vector $\vec{A}$.

A direct consequence of Eq. (VII.51) is an invariance theorem [VII.18] that can be formulated as:

> If the wavelength and/or incidence condition vary in such a way that the projection of the incident wavevector on the cross-section plane of the grating remains constant, the efficiency is unchanged provided the state of polarization remains fixed.

This theorem is of simple use for the two particular $E_\perp$ $(\delta = 0)$ and $H_\perp$ $(\delta = \pi/2)$ cases of polarization, of which the electric and magnetic fields are respectively perpendicular to the grooves; these cases generalize the classical $H_{||}$ (TM) and $E_{||}$ (TE) polarization cases of diffraction in the cross-section plane.

VII.3 Extension of the conformal mapping technique to finite conductivity gratings

The natural idea to extend the conformal mapping technique to finite conductivity gratings is to search for a second conformal mapping that associates regions $Y < 0$ and $y < g(x)$. For some particular profiles (e.g., those with a center of symmetry such that $g(-x) = -g(x)$), the same conformal mapping should work for both superstrate and substrate region. However, even in this easiest situation, two neighboring points in the xOy plane situated on opposite sides of the surface are represented by two points at positions quite distant from each other in the transformed (XOY) plane. Hence the necessity of a matching process that is numerically cumbersome and time consuming.

In order to avoid the inconvenience, R. Depine and J. Simon [VII.9] proposed to use, for $Y \le 0$, a non-conformal mapping defined by:

$$z = \mathcal{Z} + i \sum_{n=0}^{\infty} b_n \exp\left(in\overline{\mathcal{Z}}\right), \qquad \text{(VII.52)}$$

where bar denotes the complex congugate. The mapping is non-conformal since the function $z(\mathcal{Z},\bar{\mathcal{Z}})$ does not satisfy the Cauchy-Riemann conditions and thus is not differentiable.

For $Y = 0$, $\bar{\mathcal{Z}} \equiv \mathcal{Z}$ and the value of z given by Eq. (VII.52) is identical to the one obtained with Eq. (VII.9). It means, first, that two neighboring points with coordinates $(X, O+)$ and $(X, O-)$, give neighboring points on each side of the grating profile. Second, the boundary conditions are transformed in the same way as with Eq. (VII.9). Moreover, it can be verified that the pseudo-periodicity is preserved, as well as the outgoing-wave condition. Since $z \approx \mathcal{Z}$ when $Y \to -\infty$, below some matching negative ordinate $(-Y_a)$, the transformed equation will be replaced by the Helmholtz one. Between $Y = 0$ and $Y = -Y_a$, the propagation equation takes a more complicated form.

VII.3.1 Propagation equation for $Y < 0$

Eq. (VII.52) leads to:

$$x = x(X, Y)$$
$$y = y(X, Y),$$

or to a system of implicit functions:

$$\mathcal{F}_1(x, y, X, Y) = 0$$
$$\mathcal{F}_2(x, y, X, Y) = 0$$

wherefrom the transformed variables can be derived:

$$X = X(x, y) \tag{VII.53}$$
$$Y = Y(x, y), \tag{VII.54}$$

provided the Jacobian **J** of the transformation is not null, i.e., that

$$\mathbf{J} = \frac{\partial(\mathcal{F}_1, \mathcal{F}_2)}{\partial(X, Y)} \neq 0.$$

Let us designate by $U_1(X, Y)$ the solution of the boundary-value problem in the lowest half of the transformed plane $(Y < 0)$, $U_1(x, y) = u_1[x(X, Y), y(X, Y)]$. The calculation already done to establish Eq. (VII.14) shows that the Helmholtz equation for $y < g(x)$ is transformed into:

$$\left(A' \frac{\partial}{\partial X} + B' \frac{\partial}{\partial Y} + C' \frac{\partial^2}{\partial X^2} + D' \frac{\partial^2}{\partial Y^2} + 2E' \frac{\partial^2}{\partial X \partial Y} + K_1' \right) U_1(X, Y) = 0, \quad \text{(VII.55)}$$

where

$$\begin{aligned} &A'(X, Y) = \Delta X \\ &B'(X, Y) = \Delta Y \\ &C'(X, Y) = \left| \overrightarrow{\text{grad}} X \right|^2 \\ &D'(X, Y) = \left| \overrightarrow{\text{grad}} Y \right|^2 \\ &E'(X, Y) = \left(\overrightarrow{\text{grad}} X \right) \cdot \left(\overrightarrow{\text{grad}} Y \right) \\ &K_1' = k_1^2, \end{aligned}$$

and the gradient is calculated with respect to x and y. As observed in Section VII.1.3, if X and Y were satisfying the Cauchy-Riemmann conditions, Eq. (VII.55) would greatly simplify in an equation similar to Eq. (VII.16). However, for $-Y_a < Y < 0$ we have to keep Eq. (VII.55) in its present form.

VII.3.2 Matching the solution at $Y = 0$

In the original plane, the solution $u(x, y)$ of the diffraction problem, denoted by u_2 in the superstrate and u_1 in substrate, must verify the following boundary conditions:

$$u_2 [x, g(x)] = u_1 [x, g(x)] \quad \text{(VII.56)}$$

$$\eta_2 \left. \frac{du_2(x, y)}{dN} \right|_{y=g(x)} = \eta_1 \left. \frac{du_1(x, y)}{dN} \right|_{y=g(x)} \quad \text{(VII.57)}$$

where $\eta_j = 1$ in TE polarization and $\eta_j = (\varepsilon_j)^{-1}$ in TM polarization, with $j = 1, 2$. In the transformed plane, the continuity of the field stated by Eq. (VII.56) leads to the continuity of the transformed field across $Y = 0$.

$$U_2(X, 0) = U_1(X, 0). \quad \text{(VII.58)}$$

Concerning Eq. (VII.57), it has been established in Section VII.1.3 that

$$\left.\frac{du_2(x,y)}{dN}\right|_{y=g(x)} = \left|\frac{dz}{d\mathcal{Z}}\right|^{-1} \left.\frac{\partial U_2(X,Y)}{\partial Y}\right|_{Y=0}. \tag{VII.59}$$

Similar calculation show that:

$$\left.\frac{du_1(x,y)}{dN}\right|_{y=g(x)} = \left|\frac{dz}{d\mathcal{Z}}\right|^{-1} \left[G_x(X)\frac{\partial U_1(X,0)}{\partial X} + G_y(X)\left.\frac{\partial U_1(X,Y)}{\partial Y}\right|_{Y=0} \right] \tag{VII.60}$$

where

$$G_x(X) = \left(\frac{\partial x}{\partial X}\frac{\partial X}{\partial y} - \frac{\partial y}{\partial X}\frac{\partial Y}{\partial x} \right)_{Y=0} \tag{VII.61}$$

$$G_y(X) = \left(\frac{\partial x}{\partial X}\frac{\partial Y}{\partial y} - \frac{\partial y}{\partial X}\frac{\partial Y}{\partial x} \right)_{Y=0}. \tag{VII.62}$$

We stress the fact that the derivative $\frac{dz}{d\mathcal{Z}}$ which occurs in Eqs. (VII.59) and (VII.60) has to be calculated from Eq. (VII.9), while in Eqs. (VII.61) and (VII.62) the derivatives are calculated using Eq. (VII.52). Finally, Eq. (VII.57) is transformed into:

$$\eta_2 \left.\frac{\partial U_2(X,Y)}{\partial Y}\right|_{Y=0} = \eta_1 \left[G_x(X)\frac{\partial U_1(X,0)}{\partial X} + G_y(X)\left.\frac{\partial U_1(X,Y)}{\partial Y}\right|_{Y=0} \right]. \tag{VII.63}$$

VII.3.3 The differential set for $Y < 0$

Normalizing coefficients in Eq. (VII.55) to D', Eq. (VII.55) reads:

$$\left(A\frac{\partial}{\partial X} + B\frac{\partial}{\partial Y} + C\frac{\partial^2}{\partial X^2} + \frac{\partial^2}{\partial Y^2} + 2E\frac{\partial^2}{\partial X \partial Y} + K_1 \right) U_1(X,Y) = 0,$$

where $A = A'/D'$ etc., and we note that, contrary to K_1', K_1 depends on X and Y. Representing $U_1(X,Y)$ by an expansion similar to Eq. (I.29), and the coefficients A, B, C, E and K_1 by their Fourier series thanks to their periodicity in X, the pseudo-Fourier components $U_{1,n}(Y)$ are solutions of the second-order differential equation set

$$\frac{d^2U_{1,n}(Y)}{dY^2} = \sum_{m=-\infty}^{+\infty} W_{nm}(Y)\frac{dU_{1,m}(Y)}{dY} + \sum_{m=-\infty}^{+\infty} V_{nm}(Y)U_{1,m}(Y) \qquad \text{(VII.64)}$$

where the matrix elements are given by:

$$W_{nm}(Y) = -2i\alpha_m E_{n-m}(Y) - B_{n-m}(Y).$$
$$V_{nm}(Y) = \alpha_m^2 C_{n-m}(Y) - i\alpha_m A_{n-m}(Y) - K_{1,n-m}(Y).$$

As usually, the second-order differential equation set can be transformed into a first-order set by defining new functions $R_{1,n}(Y)$ by:

$$R_{1,n}(Y) = \frac{dU_{1,n}(Y)}{dY},$$

which leads to:

$$\frac{dU_{1,n}(Y)}{dY} = R_{1,n}(Y)$$
$$\frac{dR_{1,n}(Y)}{dY} = \sum_{m=-\infty}^{+\infty} V_{n,m}(Y)U_{1,m}(Y) + \sum_{m=-\infty}^{+\infty} W_{n,m}(Y)R_{1,m}(Y), \qquad \text{(VII.65)}$$

so that the matrix of the differential set is equal to $\begin{pmatrix} 0 & \mathbb{1} \\ V & W \end{pmatrix}$.

VII.3.4 The shooting method

At an ordinate $Y = -Y_a$ (with $Y_a \approx d = 2\pi$), an integration process is started by choosing $2N+1$ linearly independent vectors (denoted by subscript p) whose elements verify:

$$\hat{U}_{1,n,p}(-Y_a) = \exp\left[-i\beta_p^{(1)}(-Y_a)\right]\delta_{n,p}$$
$$\hat{R}_{1,n,p}(-Y_a) = -i\beta_p^{(1)}\exp(i\beta_p^{(1)}Y_a)\delta_{n,p}$$
with $n, p \in [-N, +N]$.

An integration algorithm computes $\hat{U}_{1,n,p}(Y)$ and $\hat{R}_{1,n,p}(Y)$ $\forall Y \in \left[-Y_a, 0^-\right]$. Eqs. (VII.58) and (VII.63) give the values of $\hat{U}_{2,n,p}(0^+)$ and $\hat{R}_{2,n,p}(0^+)$, equal to $\left.\frac{d\hat{U}_{2,n,p}}{dY}\right|_{Y=0^+}$. Then the propagation Eq. (VII.16), projected onto the Fourier basis, is integrated between $Y = 0^+$ and $Y = Y_a$. Matching the linear combination $\sum_{p=-N}^{+N} A_p^{(1)} \hat{U}_{2,n,p}(Y_a)$ and its y-derivative with the values derived from Rayleigh expansion (see Eqs. (II.18) and (II.19)) allows us finding the Rayleigh coefficients after a matrix inversion.

VII.3.5 Limitation of the method

On top of the fact that the method presented in Section VII.3 suffers from the same limitation as the conformal mapping technique, i.e., the difficulty of finding the coefficients b_n corresponding to a given arbitrary profile, it may suffer from a second limitation. Indeed, for some deep profiles or particular groove shapes the Jacobian of the transformation (VII.52) may vanish, making the transformation non-bijective. Section VII.4 presents a coordinate transformation, nowadays called C-method, which overcomes the two limitations.

VII.4 The C-method

The C-method proposes to use the non-conformal coordinate transformation:

$$\begin{aligned} X &= x \\ Y &= y - g(x) \end{aligned} \tag{VII.66}$$

which is applied to the entire plane. Its first main advantage is that the transformation is already perfectly determined and that it maps the line $Y = 0$ to the grating profile. The transformed Helmholtz Eq. (VII.55) is then immediately obtained after calculation of the coefficients A', B', etc. We find $A' = \Delta X = 0$, $B' = \Delta Y = -\frac{d^2 g(x)}{dx^2}$, $C' = 1$, $D' = 1 + \left(\frac{dg}{dx}\right)^2$ and $E' = -\left(\frac{dg}{dx}\right)$.
As a result, the propagation equation in both regions $Y > 0$ and $Y < 0$ takes the same form:

$$\left[\frac{\partial^2}{\partial X^2} - 2g'\frac{\partial^2}{\partial X \partial Y} - g''\frac{\partial}{\partial Y} + \left(1+g'^2\right)\frac{\partial^2}{\partial Y^2} + k_j^2\right] U_j(X,Y) = 0, \qquad \text{(VII.67)}$$

where $j = 1, 2$.

This second-order differential equation is rewritten as a pair of first-order equations by defining function $R_j(X,Y)$ by $R_j(X,Y) = -i\frac{\partial U_j}{\partial Y}$, where the factor -i is introduced for convenience in what follows. We thus obtain [VII.19]:

$$\begin{pmatrix} k_j^2 + \frac{\partial^2}{\partial X^2} & 0 \\ 0 & 1 \end{pmatrix} \begin{pmatrix} U_j \\ R_j \end{pmatrix} = \begin{pmatrix} i\left(\frac{\partial}{\partial X} g' + g'\frac{\partial}{\partial X}\right) & 1+g'^2 \\ 1 & 0 \end{pmatrix} \times \frac{1}{i}\frac{\partial}{\partial Y} \begin{pmatrix} U_j \\ R_j \end{pmatrix}. \qquad \text{(VII.68)}$$

As can be seen, the coefficients of the first-order differential set are Y-independent. Similar to what has been done in Chapter VI for the RCW theory (see Eq. (VI.4)), the solution $\begin{pmatrix} U_j \\ R_j \end{pmatrix}$ can thus be searched with an $\exp(i\rho Y)$ dependence, where ρ will be a diagonal matrix whose elements are the eigenvalues of a matrix. As a result, the operator $\frac{\partial}{\partial Y}$ will be replaced by a multiplication by $i\rho$. Since g' is periodic, as well as $g(x)$, it can be represented by its Fourier series. Since $X \equiv x$, the pseudoperiodicity of the field is preserved in the new plane, so that $\frac{\partial}{\partial X} \to i\alpha_m$. In Fourier space, Eq. (VII.68) is transformed into:

$$\begin{pmatrix} \left(\beta^{(j)}\right)^2 & 0 \\ 0 & \mathbb{1} \end{pmatrix} \begin{pmatrix} [U_j] \\ [R_j] \end{pmatrix} = \rho \begin{pmatrix} -(\alpha[\![g']\!] + [\![g']\!]\alpha) & 1 + [\![g']\!][\![g']\!] \\ \mathbb{1} & 0 \end{pmatrix} \begin{pmatrix} [U_j] \\ [R_j] \end{pmatrix} \qquad \text{(VII.69)}$$

where $\left(\beta^{(j)}\right)$ is a diagonal matrix formed by $\beta_m^{(j)}$ defined in Eqs. (I.36) and (I.33), α is the diagonal matrix (defined in Chapter V) with elements α_m, $[\![g']\!]$ is the Toeplitz matrix formed by the Fourier coefficients of $g'(x)$, $[U_j]$

and $\left[R_j\right]$ are column vectors formed by the pseudo-Fourier coefficients of U_j and R_j. As it was shown in [VII.20], the use of $[\![g']\!][\![g']\!]$ instead of the Toeplitz matrix associated to $\left[\!\left[g'^2\right]\!\right]$ ensures fast numerical convergence when the grating profile has sharp edges. The same remark applies to the splitting of the term in $\dfrac{\partial^2}{\partial X \partial Y}$ into 2 parts, which avoids assuming the identity of the crossed second-order derivatives in Eqs. (VII.13) and (VII.14).

After inverting the diagonal matrix on the left-hand side of Eq. (VII.69), we obtain the eigenvalue-eigenvector problem which is the basis of the C-method:

$$\begin{pmatrix} -\left(\beta^{(j)}\right)^{-2}(\alpha[\![g']\!]+[\![g']\!]\alpha) & \left(\beta^{(j)}\right)^{-2}\left(1+[\![g']\!][\![g']\!]\right) \\ \mathbb{1} & 0 \end{pmatrix}\begin{pmatrix}\left[U_j\right] \\ \left[R_j\right]\end{pmatrix}=\frac{1}{\rho}\begin{pmatrix}\left[U_j\right] \\ \left[R_j\right]\end{pmatrix}. \quad \text{(VII.70)}$$

Two eigenvalue problems must be resolved, one in each medium ($j=1,2$). However, the eigenvalue problem is the same for both TE and TM polarization, which represents one of the main interest of the method; only the matching of the solution at the interface $Y=0$ will differ. To that end, U_1 and U_2 are expressed by Eq. (VI.2). The continuity of the tangential component of the field results, for both polarizations, into

$$U_{1,m}(0)=U_{2,m}(0)\,. \qquad \text{(VII.71)}$$

The continuity of the normal derivative (or of $\dfrac{1}{\varepsilon_j}\dfrac{dU_j}{dN}$) requires finding the expression of the tangential component in the cross-section plane of $\vec{E}$ or $\vec{H}$ in the new coordinate system. For example, in TM polarization, the tangential component E_T of $\vec{E}$ in the (x,y) plane, given by [VII.19] $E_T=E_x+g'E_y$ is transformed by expressing E_x and E_y through Maxwell equations. In the (X,Y) plane, we find:

$$E_T=\frac{\sqrt{\mu_0/\varepsilon_0}}{ik_j}\left[g'\frac{\partial U_j}{\partial X}-(1+g'g')\frac{\partial U_j}{\partial Y}\right]. \qquad \text{(VII.72)}$$

Equating the Fourier components of E_T at $Y = 0^+$ and $Y = 0^-$, combined with Eq. (VII.72) lead to a Cramer set of linear equations, the resolution of which gives the diffracted amplitudes.

Since the original papers [I.22], the C-method has been applied with great success to many diffraction grating problems. Popov and Mashev [VII.21] extended it to planar waveguides. Various authors [VII.22-VII.25] extended the method to conical diffraction mountings. Granet et al. [VII.26], Li et al. [VII.27] and Preist et al. [VII.28] allowed the various profiles of a stack of gratings to be different from each other, although keeping the same periodicity. Plumey et al. [VII.29] and Preist et al. [VII.30] allowed overhanging profiles by the use of oblique transformation of coordinates. The extension of the C-method to crossed gratings was done by Granet [VII.31, VII.32] and Harris et al. [VII.33]. The method was then extended to uniaxial [VII.34-VII.36] and biaxial [VII.37] materials. It was extended by the present authors [VII.38] to nonlinear optics, namely to the study of surface enhanced second harmonic generation. Granet et al. [VII.39] allowed the permittivity of the layers to be nonhomogeneous. Numerical improvements were done too, by introducing the S-matrix propagation algorithm for multilayer gratings [VII.24, VII.40], allowing the theory to analyze gratings with sharp edges (at which the derivative g' is not defined) [VII.20], or allowing profiles defined by parametric equations [VII.41]. An extension to cylinders with arbitrary cross-section, which combines both a non-orthogonal coordinate transformation and a numerical integration, was published recently [VII.42]. This short list of references gives an idea of the versatility of the method.

In the original formulation [I.22], the authors used tensorial calculus in non-orthogonal coordinate systems, as was previously done in waveguide theory [VII.43-VII.45]. Since this domain of mathematics is not always taught in the education cursus of physicists and engineers, the theory appeared difficult to understand to many scientists. We chose to present it in a simpler way which avoids the difficulty, as was recently published in [VII.19], although tensorial calculus allows an elegant presentation, which is briefly reproduced here.

In any curvilinear coordinate system the rotational operator is defined through the Levi-Civita permutation indicator ε^{ijk}:

$$(\text{curl}\,\vec{F})^i = \varepsilon^{ijk}\frac{\partial F_k}{\partial X^j} \tag{VII.73}$$

Here X^j denote the transformed coordinates, i.e., $X^1 = x$, $Y^1 = y$, and $Z^1 = z$, and the permutation indicator components are null for any two equal indices, are equal to +1 for "positive" permutation (e.g., 1,2,3) and to −1 for a "negative" permutation (e.g., 1,3,2) of their indices. Eq. (VII.73) must be viewed in terms of covariant and contravariant vector components, which are not equal in a non-orthogonal coordinate system. The transfer from covariant to contravariant components (and vice versa) is made using the metric tensor $\mathfrak{G}$, defined as

$$\mathfrak{G}^{ij} = \frac{\partial X^i}{\partial x}\frac{\partial X^j}{\partial x} + \frac{\partial X^i}{\partial y}\frac{\partial X^j}{\partial y} + \frac{\partial X^i}{\partial z}\frac{\partial X^j}{\partial z}. \tag{VII.74}$$

The electric permittivity ($\bar{\bar{\varepsilon}}$) and magnetic permittivity ($\bar{\bar{\mu}}$) are second-order tensors. In a specific coordinate transformation they are subjected to the laws of transformation of tensors. In isotropic media the transformation law is the same as for the metric tensor $\mathfrak{G}$:

$$\begin{aligned} \mu^{ij} &= \mu_0 \mathfrak{G}^{ij} \\ \varepsilon^{ij} &= \varepsilon\ \mathfrak{G}^{ij} \end{aligned}. \tag{VII.75}$$

Maxwell equations written in a vector form (with covariant vector components) are covariant with respect to coordinate transformations, provided the curl operation is correctly expressed, i.e., by using Eq. (VII.73). It follows that for the specific coordinate system defined by Eq. (VII.66), $\mathfrak{G}$ takes the form:

$$\mathfrak{G}^{ij} = \begin{pmatrix} 1 & -g' & 0 \\ -g' & 1+g'g' & 0 \\ 0 & 0 & 1 \end{pmatrix}. \tag{VII.76}$$

The advantage of this transformation is that it not only maps the grating profile into a plane ($X^2 \equiv Y = \text{const.}$), but also that the covariant transformed field components are the components tangential to this plane, i.e., they are continuous functions of Y. And indeed, one can easily find that the electric field components in the transformed coordinate systems are:

$$E_3 = E_z, \quad \text{and} \quad E_1 = E_T, \tag{VII.77}$$

with E_T expressed in Eq. (VII.72). Similar expressions are valid for H.

In TE polarization, Mawell equations can be written using Eqs. (VII.73) and (VII.75) as:

$$\left|\begin{array}{l} \dfrac{\partial E_3}{\partial X^1} = -i\omega\mu_0\left(\mathcal{G}^{21}H_1 + \mathcal{G}^{22}H_2 \right) \\ \dfrac{\partial E_3}{\partial X^2} = \ i\omega\mu_0\left(\mathcal{G}^{11}H_1 + \mathcal{G}^{12}H_2 \right) \\ \dfrac{\partial H_2}{\partial X^1} - \dfrac{\partial H_1}{\partial X^2} = -i\omega\varepsilon\ \mathcal{G}^{33}E_3 \end{array}\right.$$

The derivative with respect to x (or to $X^1 = x$) results in a multiplication by $i\alpha$. Eliminating the term containing H_2 by using the first equation, one arrives to the same eigenvalue problem, stated in Eq. (VII.70), because the metric tensor $\mathcal{G}$ given by Eq. (VII.76) is independent of Y $(\equiv X^2)$.

TM polarization is treated in the same way. As far as one has only to work with continuous field components, the convergence in TE and TM polarization is the same and does not depend on the refractive index, i.e., highly conducting media do not impose more computational difficulties than dielectrics.

References

VII.1. P. M. Morse and H. P. Feshbach, *Methods of Theoretical Physics*, (McGraw-Hill Book Company, New York, 1953), p. 443

VII.2. H.Y. Yee, “On determination of cutoff frequencies of waveguides with arbitrary cross-section,” Proc. IEEE (Correspondence), 54, 64, 1965

VII.3. M. Chi and P. A. Laura, “Approximate method of determining the cutoff frequencies of waveguides with arbitrary cross-section,” IEEE Trans. on Microwave Theory and Techniques (Correspondence), MTT – 12, 248-249, 1964

VII.4. G. Birkhoff, D. M. Young, and E. H. Zarantonello, “Numerical methods in conformal mapping,” Proceedings of Symposia in Applied Mathematics, vol. IV, New York, McGraw-Hill, pp. 117-140, 1953

VII.5. H. Y. Yee and N. F. Auden, “Uniform waveguides with arbitrary cross-section considered by the point-matching method,” IEEE Trans. On Microwave Theory and Techniques, MTT-13, 847-851, 1965

VII.6. P. A. Laura and A. J. Faulstich, "An application of conformal mapping to the determination of the natural frequency of membranes of regular polygonal shape," Proc. 9th Midwestern Mechanics Conf., University of Wisconsin, Madison, August, 155-163, 1965

VII.7. R. A. Depine and J. M. Simon, "Diffraction grating efficiencies. Conformal mapping method for a good real conductor," Optica Acta, 29, 1459-1473, 1982

VII.8. R. A. Depine and J. M. Simon, "Surface impedance boundary condition for metallic diffraction gratings in the optical and infrared range," Optica Acta, 30, 313-322, 1983

VII.9. R. A. Depine and J. M. Simon, "Diffraction grating efficiencies. An exact differential algorithm valid for high conductivity," Optica Acta 30, 1273-1286, 1983

VII.10. M. Nevière and M. Cadilhac, "Sur une nouvelle formulation du problème de la diffraction d'une onde plane par un réseau infiniment conducteur – cas général," Opt. Commun., 3, 379-383, 1971

VII.11. J. P. Fenélon, "Contribution à l'étude des caractéristiques de guides ouverts ou fermés, chargés périodiquement par des iris ou remplis par un milieu hétérogène," thesis No 24 – 1985, University of Limoges, France (5 September 1985)

VII.12. J. Gilewicz, *Approximants de Padé*, Lecture Notes in Mathematics, vol. 667, (Springer, Berlin, Heidelberg, New York, 1978)

VII.13a. R. F. Millar, "On the Rayleigh assumption in scattering by a periodic surface," Proc. Cambr. Phil. Soc. 65, 773-791, 1969

VII.13b. R. F. Millar, "On the Rayleigh assumption in scattering by a periodic surface. II," Proc. Cambr. Phil. Soc. 69, 217-225, 1971

VII.14a. R. F. Millar, "The location of singularities of two-dimensional harmonic functions. I: Theory," SIAM J. Math. Anal. 1, 333-344, 1970

VII.14b. R. F. Millar, "The location of singularities of two-dimensional harmonic functions. II: Applications," SIAM J. Math. Anal. 1, 345-353, 1970

VII.14c.R. F. Millar, "Singularities of two-dimensional exterior solutions of the Helmholtz equation," Proc. Camb. Phil. Soc. 69, 175-188, 1971

VII.15. A. Maréchal and G. W. Stroke, "Sur l'origine des effets de polarization et de diffraction dans les réseaux optiques," Compt. Rend. Acad. Sci. (Paris), 249, 2042-2044, 1959

VII.16. M. Nevière and M. Cadilhac, "Sur la diffraction d'une onde électromagnétique plane par un réseau infiniment conducteur, lorsque le vecteur d'onde incident n'est pas orthogonal aux sillons," Opt. Commun. 4, 13-18, 1971

VII.17. D. Maystre and R. Petit, "Sur la diffraction d'une onde plane par un réseau infiniment conducteur," Opt. Commun. 4, 97-101, 1971

VII.18. R. Petit and D. Maystre, "Applications des lois de l'électromagnétisme à l'étude des réseaux," Revue de Physique appliquée, 7, 427-441, 1972

VII.19. L. Li, J. Chandezon, G. Granet, and J. P. Plumey, "Rigorous efficient grating-analysis method made easy for optical engineers," Appl. Opt. 38, 304-313, 1999

VII.20. L. Li and J. Chandezon, "Improvement of the coordinate transformation method for surface-relief gratings with sharp edges," J. Opt. Soc. Am. A 13, 2247-2255, 1996

VII.21. E. Popov and L. Mashev, "Rigorous electromagnetic treatment of planar corrugated waveguides," J. Opt. Commun. 7, 127-131, 1986

VII.22. E. Popov and L. Mashev, "Conical diffraction mounting generalization of a rigorous differential method," J. Opt. (Paris) 17, 175-180, 1986

VII.23. S. J.Elston, G. P. Bryan-Brown, and J. R. Sambles, "Polarization conversion from diffraction grating," Phys. Rev. B 44, 6393-6400, 1991

VII.24. L. Li, "Multilayer-coated diffraction gratings: differential method of Chandezon et al. revisited," J. Opt. Soc. Am. A 11, 2816-2828, 1994, Errata 13, 543, 1996

VII.25. J. P. Plumey, G. Granet, and J. Chandezon, "Differential covariant formalism for solving Maxwell's equations in curvilinear coordinates: oblique scattering from lossy periodic surfaces," IEEE Trans. Antennas Propag. 43, 835-842, 1995

VII.26. G. Granet, J. P. Plumey, and J. Chandezon, "Scattering by a periodically corrugated dielectric layer with non-identical faces," Pure Appl. Opt. 4, 1-5, 1995

VII.27. L. Li, G. Granet, J. P. Plumey, and J. Chandezon, "Some topics in extending the C method to multilayer-coated gratings of different profiles," Pure Appl. Opt. 5, 141-156, 1996

VII.28. T. W. Preist, N. P. K. Cotter, and J. R. Sambles, "Periodic multilayer gratings of arbitrary shape," J. Opt. Soc. Am. A 12, 1740-1748, 1995

VII.29. J. P. Plumey, B. Guizal, and J. Chandezon, "Coordinate transformation method as applied to asymmetric gratings with vertical facets," J. Opt. Soc. Am. A 14, 610-617, 1997

VII.30. T. W. Preist, J. B. Harris, N. P. Wanstal, and J. R. Sambles, "Optical response of blazed and overhanging gratings using oblique Chandezon transformations," J. Mod. Opt. 44, 1073-1080, 1997

VII.31. G. Granet, "Analysis of diffraction by crossed gratings using a non-orthogonal coordinate system," Pure Appl. Opt. 4, 777-793, 1995

VII.32. G. Granet, "Analysis of diffraction by surface-relief crossed gratings with use of the Chandezon method: application to multilayer crossed gratings," J. Opt. Soc. Am. A 15, 1121-1131, 1998

VII.33. J. B. Harris, T. W. Preist, J. R. Sambles, R. N. Thorpe, and R. A. Watts, "Optical response of bigratings," J. Opt. Soc. Am. A 13, 2041-2049, 1996

VII.34. J. B. Harris, T. W. Preist, and J. R. Sambles, "Differential formalism for multilayer diffraction gratings made with uniaxial materials," J. Opt. Soc. Am. A 12, 1965-1973, 1995

VII.35. J. B. Harris, T. W. Preist, E. L. Wood, and J. R. Sambles, "Conical diffraction from multicoated gratings containing uniaxial materials," J. Opt. Soc. Am. A 13, 803-810, 1996

VII.36. M. E. Inchaussandague and R. A. Depine, "Polarization conversion from diffraction gratings made of uniaxial crystals," Phys. Rev. E 54, 2899-2911, 1996

VII.37. M. E. Inchaussandague and R. A. Depine, "Rigorous vector theory for diffraction from gratings made of biaxial crystals," J. Mod. Opt. 44, 1-27, 1997

VII.38. E. Popov and M. Nevière, "Surface-enhanced second-harmonics generation in nonlinear corrugated dielectrics: new theoretical approach," J. Opt. Soc. Am. B 11, 1555-1564, 1994

VII.39. G. Granet, J. Chandezon, and O. Coudert, "Extension of the C method to nonhomogeneous media: application to nonhomogeneous layer with parallel modulated faces and to inclined lamellar gratings," J. Opt. Soc. Am. A 14, 1576-1582, 1997

VII.40. N. P. K. Cotter, T. W. Preist, and J. R. Sambles, "Scattering-matrix approach to multilayer diffraction," J. Opt. Soc. Am. A 12, 1097-1103, 1995

VII.41. G. Granet and J. Chandezon, "The method of curvilinear coordinates applied to the problem of scattering from surface-relief gratings defined by parametric equations: application to scattering from a cycloidal grating," Pure Appl. Opt. 6, 727-740, 1997

VII.42. E. Popov and M. Nevière, "Diffraction by a metallic cylinder coated by a thin absorbing layer: coordinate transformation formalism and applications," J. of Electromagnetic Waves and Applications, 12, 1265-1296, 1998

VII.43. J. Chandezon, G. Granet, and G. Raoult, "Equations de propagation des ondes électromagnétiques dans un système de coordonnées présentant la symétrie de révolution axiale," C. R. Acad. Sci. Paris, B 276, 507-509, 1973

VII.44. J. Chandezon, G. Granet, and G. Raoult, "Propagation des ondes dans les guides cylindriques à génératrice sinusoidal Expression des champs," C. R. Acad. Sci. Paris, B 277, 355-358, 1973

VII.45. J. Chandezon, G. Granet, and G. Raoult, "Principe de résolution approchée des équations de Maxwell dans un système de coordonnées "perturbées"," C. R. Acad. Sci. Paris, B 297, 239-242, 1974

Chapter VIII

Gratings Made of Anisotropic Materials

In classical textbooks [I.47], anisotropic materials are generally assumed to be lossless. When the electric permittivity tensor $\bar{\bar{\varepsilon}}$ is real, the energy balance implies that it should be symmetrical ($\varepsilon_{km} = \varepsilon_{mk}$, with $k, m = x, y, z$). As a result, it can be represented in each coordinate basis by a real symmetric matrix. Such a matrix can be diagonalized in a specific coordinate system to present real eigenvalues. It means that there exists a set of orthogonal coordinate axis, called *principal axes* in which $\bar{\bar{\varepsilon}}$ is reduced to a diagonal matrix:

$$\bar{\bar{\varepsilon}} = \begin{pmatrix} \varepsilon_1 & 0 & 0 \\ 0 & \varepsilon_2 & 0 \\ 0 & 0 & \varepsilon_3 \end{pmatrix}.$$

The diagonal elements are called *principal permittivities* and are real numbers, due to absence of absorption. When all principal permittivities are different, the medium is said to be *biaxial*; when $\varepsilon_1 = \varepsilon_2$, it is called *uniaxial*, a fact which generally results from particular symmetries in the crystalline stucture. With losses, $\bar{\bar{\varepsilon}} = \bar{\bar{\varepsilon}}_r + i\bar{\bar{\varepsilon}}_i$, and, except under special circumstances, it is impossible to find an orthogonal set of axis that diagonalizes both the real part $\bar{\bar{\varepsilon}}_r$ and the imaginary part $\bar{\bar{\varepsilon}}_i$, and the classical theory of uniaxial and biaxial material cannot be developed. However, perturbative methods allow to extend it to low absorption materials in view of analyzing the phenomenon of dichroism and pleiochroism [I.47].

Uniaxial and biaxial materials are of great interest in the field of Optics. However, modern technology involves the use of anisotropic materials with strong losses, as it occurs in optical information storage. For such materials, whatever the coordinate set may be, $\bar{\bar{\varepsilon}}$ will be represented by a square matrix with complex elements:

$$\bar{\bar{\varepsilon}} = \begin{pmatrix} \varepsilon_{xx} & \varepsilon_{xy} & \varepsilon_{xz} \\ \varepsilon_{yx} & \varepsilon_{yy} & \varepsilon_{yz} \\ \varepsilon_{zx} & \varepsilon_{zy} & \varepsilon_{zz} \end{pmatrix}, \qquad \text{(VIII.1)}$$

and a similar situation could be found for the magnetic permittivity for materials that are magnetically anisotropic ($\vec{B}$ non parallel to $\vec{H}$). Fortunately, this last situation does not occur frequently. Indeed, as shown in [VIII.1, VIII.2], even magneto-optical effects, that result in a rotation of the polarization vector of both transmitted and reflected field known as Faraday and Kerr effect, respectively [VIII 3-VIII.5] can be studied, at least in the optical domain, by taking $\mu = \mu_0$ and $\bar{\bar{\varepsilon}}$ with non-null and imaginary off-diagonal elements. The case of ferrites used in low frequency domain is the only exception that we know which would require a tensorial expression of the magnetic permittivity.

The problem of diffraction by a grating made of an anisotropic lossy medium was tackled as soon as mid 1980s in view of optical information storage [III.6]. An extension of the classical differential method was used, so that the numerical treatment was only successful with low modulated gratings. Nowadays, the analysis of such devices merely appears as a particular case of Chapter V.

VIII.1 Propagation equations

We assume that $\bar{\bar{\mu}} = \mu_0 \delta_{km}$, and that $\bar{\bar{\varepsilon}}$ is given by Eq. (VIII.1). The grating profile is illustrated in Fig. I.1, so that in the Cartesian Oxyz coordinate system $\vec{N} = (N_x, N_y, 0)$, with $N_x = -g'(x)\big/\sqrt{1+g'^2(x)}$, $N_y = 1\big/\sqrt{1+g'^2(x)}$. As in Section V.1.5, we choose $\vec{T}_2 = \vec{z}$, wherefrom $\vec{T}_1 = (N_y, -N_x, 0)$. The determination of matrix C_ε through Eq. (V.11) first requires the calculation of ξ_ε, equal to $\vec{N}\cdot\bar{\bar{\varepsilon}}\cdot\vec{N}$, which reduces to $\left(\vec{N}\cdot\bar{\bar{\varepsilon}}\right)_x N_x + \left(\vec{N}\cdot\bar{\bar{\varepsilon}}\right)_y N_y$. Then we find that

$$\left(\vec{N}\cdot\bar{\bar{\varepsilon}}\right)\times\vec{T}_2 = \left(\left(\vec{N}\cdot\bar{\bar{\varepsilon}}\right)_y,\ -\left(\vec{N}\cdot\bar{\bar{\varepsilon}}\right)_x,\ 0\right)$$

and

$$\left(\vec{N}\cdot\bar{\bar{\varepsilon}}\right)\times\vec{T}_1 = \left(\left(\vec{N}\cdot\bar{\bar{\varepsilon}}\right)_z N_x,\ \left(\vec{N}\cdot\bar{\bar{\varepsilon}}\right)_z N_y,\ -\left(\vec{N}\cdot\bar{\bar{\varepsilon}}\right)_x N_x - \left(\vec{N}\cdot\bar{\bar{\varepsilon}}\right)_y N_y\right),$$

so that matrix C_ε reads:

$$C_\varepsilon = \begin{pmatrix} (\vec{N}\bullet\bar{\bar{\varepsilon}})_y / \xi_\varepsilon & N_x/\xi_\varepsilon & -(\vec{N}\bullet\bar{\bar{\varepsilon}})_z N_x / \xi_\varepsilon \\ -(\vec{N}\bullet\bar{\bar{\varepsilon}})_x / \xi_\varepsilon & N_y/\xi_\varepsilon & (\vec{N}\bullet\bar{\bar{\varepsilon}})_z N_y / \xi_\varepsilon \\ 0 & 0 & 1 \end{pmatrix}. \qquad \text{(VIII.2)}$$

The other matrix $\bar{\bar{\varepsilon}}C_\varepsilon$ required for the FFF method is calculated from Eq. (VIII.2). Then the Toeplitz matrices $[\![C_\varepsilon]\!]$ and $[\![\bar{\bar{\varepsilon}}\bullet C_\varepsilon]\!]$ are calculated with the Laurent's rule, since N_x and N_y are chosen to be continuous functions of x. Inversing $[\![C_\varepsilon]\!]$ on a computer enables us to determine matrix $Q_\varepsilon = [\![\bar{\bar{\varepsilon}}\bullet C_\varepsilon]\!][\![C_\varepsilon]\!]^{-1}$.

In this problem $Q_\mu \equiv \mu_0 \mathbb{1}$ so that the general propagation equation (V.22) then applies with

$$M = \begin{pmatrix} M_{11} & M_{12} & 0 & M_{14} \\ 0 & 0 & M_{23} & 0 \\ M_{31} & M_{32} & 0 & M_{34} \\ M_{41} & M_{42} & 0 & M_{44} \end{pmatrix}$$

given in the form:

$$\begin{aligned}
M_{11} &= -i\alpha Q_{\varepsilon,yy}^{-1} Q_{\varepsilon,yx} \\
M_{12} &= -i\alpha Q_{\varepsilon,yy}^{-1} Q_{\varepsilon,yz} \\
M_{14} &= \frac{i\alpha}{\omega} Q_{\varepsilon,yy}^{-1} \alpha - i\omega\mu_0 \\
M_{23} &= i\omega\mu_0 \\
M_{31} &= -i\omega\left(Q_{\varepsilon,zy} Q_{\varepsilon,yy}^{-1} Q_{\varepsilon,yx} - Q_{\varepsilon,zx}\right) \\
M_{32} &= -i\omega\left(Q_{\varepsilon,zy} Q_{\varepsilon,yy}^{-1} Q_{\varepsilon,yz} - Q_{\varepsilon,zz}\right) \\
M_{34} &= i Q_{\varepsilon,zy} Q_{\varepsilon,yy}^{-1} \alpha \\
M_{41} &= i\omega\left(Q_{\varepsilon,xy} Q_{\varepsilon,yy}^{-1} Q_{\varepsilon,yx} - Q_{\varepsilon,xx}\right) \\
M_{42} &= i\omega\left(Q_{\varepsilon,xy} Q_{\varepsilon,yy}^{-1} Q_{\varepsilon,yz} - Q_{\varepsilon,xz}\right) \\
M_{44} &= -i Q_{\varepsilon,xy} Q_{\varepsilon,yy}^{-1} \alpha
\end{aligned} \qquad \text{(VIII.3)}$$

Of course, in case of magnetic anisotropy, Q_μ would have to be calculated and more terms should be retained in Eq. (V.22') as done in [VIII.6].

VIII.2 Resolving the boundary-value problem

Outside the groove region, the permittivity is constant. Thus the solution obtained inside the groove region can be matched to Rayleigh expansions at $y=0$ and $y=a$. However, the matching process at $y=0$ needs care; in a bulk anisotropic material, it is not obvious to make a distinction between incoming and outgoing waves, since the direction of propagation of energy, i.e., the direction of Poynting vector, does not coincide with the direction of the wavevector, in general. The problem was addressed more then ten years ago, and the interested reader can refer to Tayeb's thesis [III.6] or to the more recent work of Li [VIII.7]. The easiest way when working with a computer is, for sure, to compute the Poynting vector of each Fourier harmonic in the substrate and to observe the sign of its y-component.

Anyway, the grating problem is reduced to the numerical integration in the interval $(0,a]$ of the differential equation set (V.22), with matrix M given by Eq. (VIII.3), taking into account boundary conditions at $y=0$ and $y=a$. As usual, the boundary-value problem is transformed into an initial-value problem through a "shooting method." Similar to what was encountered in Section IV.3 devoted to conical diffraction, we first perform a first series of $(2N+1)$ numerical integrations by choosing at $y=0$:

$$\hat{E}_{z,n,p} = \delta_{n,p}$$
$$\hat{H}_{z,n,p} = 0\,, \text{ where } p \in [1, 2N+1]\,.$$

From these starting values, $\hat{E}_{x,n,p}$ and $\hat{H}_{x,n,p}$ have to be calculated from Maxwell equations, written in the bulk anisotropic medium, at $y=0$. A standard algorithm then integrates Eqs. (V.22) and (VIII.3) up to $y=a$. A second series of $2N+1$ integrations starts at $y=0$ with

$$\hat{E}_{z,n,p} = 0$$
$$\hat{H}_{z,n,p} = \delta_{n,p}\,, \text{ where } p \in [2N+2, 4N+2]$$

and using the corresponding values of $\hat{E}_{x,n,p}$ and $\hat{H}_{x,n,p}$. The Fourier components $E_{x,n}(a)$, $E_{z,n}(a)$, $H_{x,n}(a)$ and $H_{z,n}(a)$ are then matched with the various Rayleigh expansions at y = a in the isotropic superstrate. The Rayleigh coefficients for E_z and H_z are found after an inversion of a $(4N+2)\times(4N+2)$ matrix. The coefficients of E_x and H_x are derived from Maxwell equations written in the isotropic superstrate, i.e., from Eqs. (IV.36) and (IV.37) in which $\gamma_0 = 0$. Numerical contamination is again avoided using the S-matrix propagation algorithm.

VIII.3 Convergence with respect to the number of Fourier components

It is straightforward to verify that if

$$\bar{\bar{\varepsilon}} = \varepsilon\delta_{km}, \tag{VIII.4}$$

Eqs. (VIII.3) reduce to Eqs. (V.28) found for a classical isotropic grating. Thus the computer code developed to analyze gratings made of anisotropic material is supposed to give the same results as the isotropic grating code when $\bar{\bar{\varepsilon}}$ verifies Eq. (VIII.4). More interesting is to show the convergence of the present method compared with the classical one [III.6] as a function of the truncation parameter N. To that end, we consider a sinusoidal grating with $d = 0.6$ μm and $a = 0.5$ μm, lighted by a TM polarized plane wave with wavelength $\lambda_2 = 0.6328$ μm under incidence $\theta_2 = 30^o$. The grating material is cobalt, for which $\bar{\bar{\mu}}_1 = \mu_0 \mathbb{1}$, and $\varepsilon_{1,xx} = \varepsilon_{1,yy} = \varepsilon_{1,zz} = \varepsilon_0(-8.19 + i16.38)$, $\varepsilon_{1,xz} = -\varepsilon_{1,zx} = \varepsilon_0(-0.495 - i0.106)$ $\varepsilon_{1,xy} = \varepsilon_{1,yx} = \varepsilon_{1,yz} = \varepsilon_{1,zy} = 0$. We assume that $\varepsilon_2 = \varepsilon_0$ and $\mu_2 = \mu_0$. Fig. VIII.1 shows the TE and TM diffraction efficiencies as functions of the truncation order N of the Fourier series of the field components. One can see that as soon as $N = 7$ or 8 (15 to 17 Fourier components) the present method gives the efficiencies of this deep anisotropic grating to better than 1%, an accuracy never reached by the conventional differential method. On the other hand, a comparison made for the same anisotropic grating except for keeping the groove depth small ($a = 0.1$ μm), have shown a good agreement between the two methods for this low-modulated profile [VIII.6]. The proof is done that the differential theory is now able to analyze deep anisotropic gratings, among which are magnetooptic gratings. Chiral materials could be included in the theory.

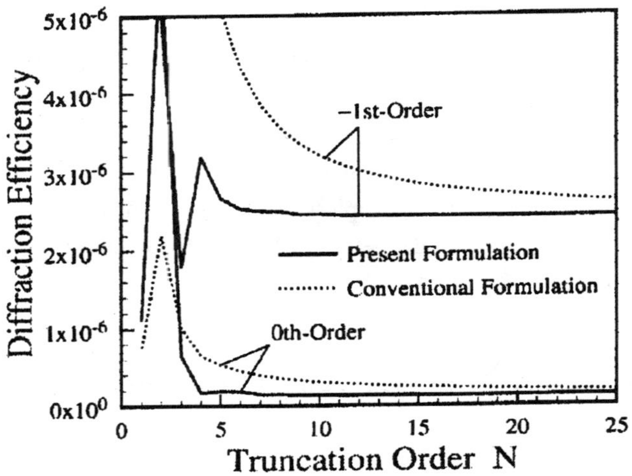

(a) TE diffraction wave

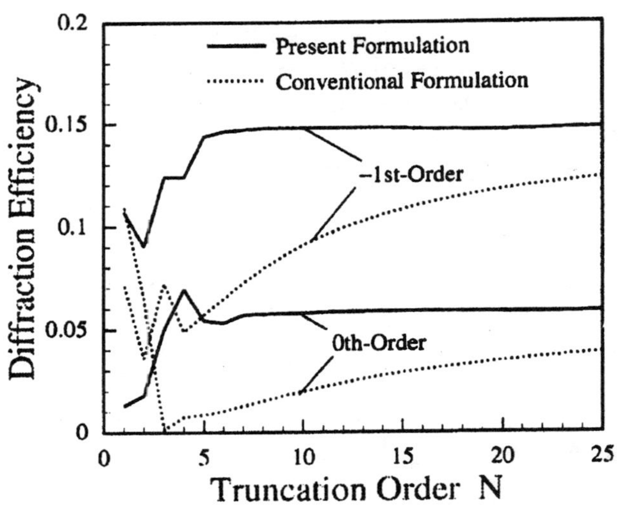

(b) TM diffraction wave

Fig. VIII.1. Comparison of convergence of diffraction efficiencies computed by the conventional and present formulation of the differential theory for a sinusoidal grating. The grating parameters are given in the text (after [VIII.6]).

References

VIII.1. P. S. Pershan, "Magneto-optical effects," J. Appl. Phys. 38, 1482-1490, 1967

VIII.2. L. D. Landau and E. M. Lifshitz, *Electrodynamics of Continuous Media*, (Pergamon Press, Inc., New York, 1960), pp 251-253

VIII.3. P. S. Pershan, "Non-linear optical properties of solids: energy considerations," Phys. Rev. 130, 919-929, 1963

VIII.4. R. P. Hunt, "Magneto-optic scattering from thin solid films," J. Apl. Phys. 38, 1652-1671, 1967

VIII.5. Z. Q. Qiu and S. D. Bader, "Surface magneto-optic Kerr effect (SMOKE)," J. Magn. Magn. Matter. 200, 664-678, 1999

VIII.6. K. Watanabe, R. Petit, and M. Nevière, "Differential theory of gratings made of anisotropic materials," J. Opt. Soc. A 19, 325-334, 2002

VIII.7. L. Li, "Reformulation of the Fourier modal method for surface-relief gratings made with anisotropic materials, " J. Mod. Opt. 45, 1313-1334, 1998

Chapter IX

Crossed Gratings

Crossed gratings are periodic in two different directions, e.g., along the x- and z-axis, although the two directions are not necessarily perpendicular. Their electromagnetic study began at the end of the 1970s [IX.1-IX.4], in view of optimizing the performances of the surface of solar energy absorbers. The methods developed at that time mainly suffered from the limitations in the storage capacity of available computer memories. The authors of [IX.1-IX.4] worked with 40 kilobytes random-access memory, while nowadays most Ph.D. students have at their disposal 256 megabytes or more on a personal computer. If we take into account the increase in the speed of computations by a factor greater than 1000, it is clear that nowadays the electromagnetic study of two-dimensional (2-D) gratings or crossed gratings can be undertaken with much better chances of success. Presently, these devices are expected to have potentialities in information optical storage, or as bandstop filters, since 2-D gratings are the next step toward the general (3-D) photonic crystals. Also, their study may explain some striking effects recently discovered (see Section IX.1.7), or can help in modelling artificial graded-index and anisotropic materials. They also may be used as anti-radar elements (antireflection surfaces) or as display components.

The early studies of crossed gratings used the classical differential method extended to double periodicity [IX.1, IX.2] or developed a new coordinate transformation method [IX.3, IX.4] that reduced the diffraction problem to an eigenvalue-eigenvector technique. The simpler case of perfectly conducting grids was analyzed in [IX.5]. Other theories have been developed in the meantime [VII.32, VII.33 and IX.6-IX.14], most of which have been an extension of the RCW method. With the development of the FFF method (Chapter V), crossed gratings with arbitrary shape appear as a particular case which can be analyzed with Eqs. (V.22) and (V.22'), overcoming the early difficulties of the classical differential method.

IX.1 Theory of crossed gratings made with isotropic materials

A typical crossed grating is represented in Fig. IX.1. Fig. IX.2 defines the orientation of the incident wavevector $\vec{k}_2$ and polarization vector $\vec{A}$ through the use of three angles, Φ, Ψ and δ. Φ is the angle between the incident wavevector $\vec{k}_2$ and the Oy axis normal to the mean plane of the crossed grating, and Ψ is the angle between the projection of $\vec{k}_2$ onto the Oxz plane and the Ox axis, as illustrated in Fig. IX.2. The components of $\vec{k}_2$, denoted by α_0, $-\beta_0$ and γ_0, are given by:

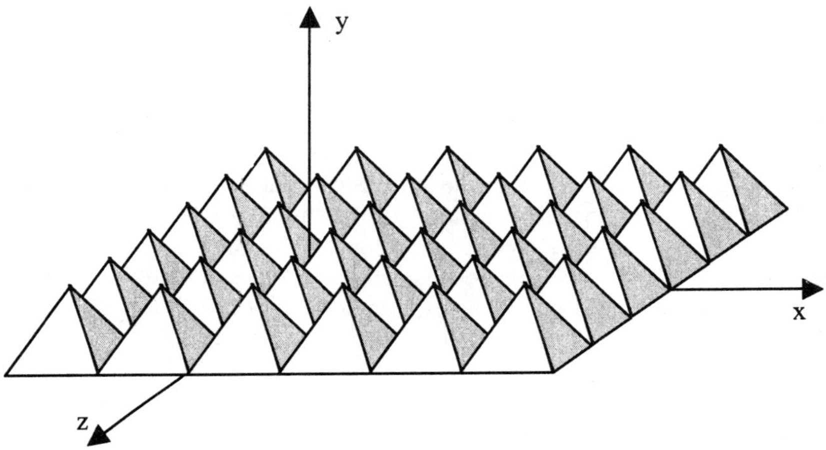

Fig. IX.1. Schematic representation of a pyramidal crossed grating.

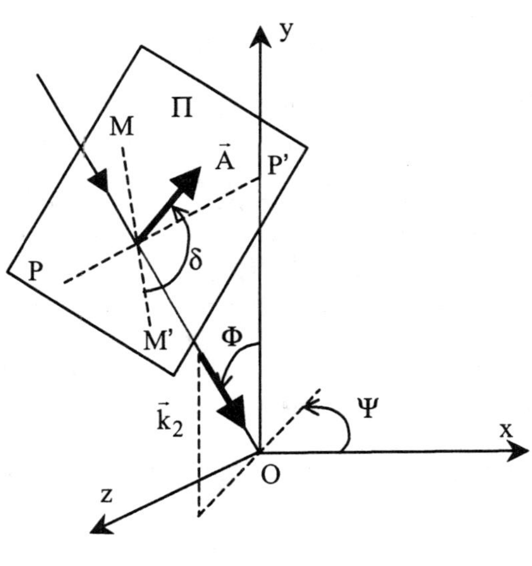

Fig. IX.2. Incidence geometry.

$$\begin{aligned} \alpha_0 &= k_0 \sin\Phi \cos\Psi \\ \beta_0 &= k_0 \cos\Psi \\ \gamma_0 &= k_0 \sin\Phi \sin\Psi . \end{aligned} \tag{IX.1}$$

The vector amplitude of the incident electric field is $\vec{A}$, chosen such that $|\vec{A}| = 1\,\mathrm{V/m}$. Its orientation is determined by an angle δ, located in the plane Π perpendicular to $\vec{k}_2$. PP' is the intersection of plane Π with the plane of incidence defined by $\vec{k}_2$ and $\overrightarrow{Oy}$. MM' is a line perpendicular to PP' in the plane Π. By definition:

$$\delta = \sphericalangle(\overrightarrow{MM'}, \vec{A}).$$

Thus if $\delta = 0^\circ$, $\vec{A}$ is normal to the plane of incidence, and if $\delta = 90^\circ$, $\vec{A}$ lies in the plane of incidence. In the special case of normal incidence, we choose $\Psi = \pi\,\mathrm{rad}$, and δ is then the angle between $\vec{A}$ and $\overrightarrow{Oz}$. With these definitions and recalling that the incident electric field has modulus equal to one, the Cartesian components of $\vec{A}$ read:

$$\begin{aligned} A_x &= \sin\Psi \cos\delta + \sin\delta \cos\Phi \cos\Psi \\ A_y &= \sin\delta \sin\Phi \\ A_z &= \cos\Phi \sin\Psi \sin\delta - \cos\delta \cos\Psi . \end{aligned} \tag{IX.2}$$

The grating surface has d_x period with respect to the x direction and d_z period with respect to z.

IX.1.1 Rayleigh expansion of the field

Due to the x and z periodicity, above the grooves, $\vec{E}$ can be expressed by the sum of the incident field and a double series:

$$\vec{E}(x,y,z) = \vec{E}^{(i)} + \sum_{n=-\infty}^{+\infty} \sum_{m=-\infty}^{+\infty} \vec{B}^{(2)}_{n,m} \exp\left[i\left(\alpha_n x + \beta^{(2)}_{n,m} y + \gamma_m z\right)\right] \tag{IX.3}$$

where

$$\alpha_n = \alpha_0 + nK_x \qquad \text{with } K_x = 2\pi/d_x \tag{IX.4}$$

$$\gamma_m = \gamma_0 + mK_z \qquad \text{with } K_z = 2\pi/d_z \tag{IX.5}$$

$$\beta_{n,m}^{(2)} = \sqrt{k_2^2 - \alpha_n^2 - \gamma_m^2} \quad \text{or} \quad i\sqrt{-\left(k_2^2 - \alpha_n^2 - \gamma_m^2\right)}. \tag{IX.6}$$

Eqs. (IX.4) and (IX.5), which give the direction of the reflected orders, generalize the Fraunhofer equation to crossed gratings.

Eq. (IX.3) combined with Eq. (IX.2) provides the Rayleigh expansions of both E_x and E_z components which appear in Eq. (V.22).

Below the groove region ($y < 0$), the field reads:

$$\vec{E}(x,y,z) = \sum_{n=-\infty}^{+\infty} \sum_{m=-\infty}^{+\infty} \vec{B}_{n,m}^{(1)} \exp\left[i\left(\alpha_n x - \beta_{n,m}^{(1)} y + \gamma_m z\right)\right] \tag{IX.7}$$

where

$$\beta_{n,m}^{(1)} = \sqrt{k_1^2 - \alpha_n^2 - \gamma_m^2} \tag{IX.8}$$

with

$$\operatorname{Im}\left(\beta_{n,m}^{(1)}\right) \geq 0. \tag{IX.9}$$

A similar equation applies for $\vec{H}(x,y,z)$. Inside the groove area $(0 < y < a)$, the permittivity being rigorously periodic is expressed by the double Fourier series:

$$\varepsilon(x,y,z) = \sum_{n=-\infty}^{+\infty} \sum_{m=-\infty}^{+\infty} \varepsilon_{n,m}(y) \exp\left[i\left(nK_x x + mK_z z\right)\right]. \tag{IX.10}$$

If d_x and d_z are of the same order of magnitude, Eqs. (IX.3), (IX.7) and (IX.10) will be truncated at the same order N in the x and z direction. Thus the vectors $[E_x]$, $[E_z]$, etc. will contain $(2N+1)^2$ Fourier components $E_{x,n,m}$, $E_{z,n,m}$, etc. For ε, $(4N+1)^2$ components will then be necessary to calculate the required Toeplitz matrices.

IX.1.2 Propagation equations in the modulated region for an arbitrary-shaped crossed grating

If the grating surface is defined by $y = g(x,z)$ or, more generally, by $f(x,y,z) = 0$, the normal vector components are defined by Eq. (V.23), while vectors $\vec{T}_1$ and $\vec{T}_2$ are given by Eqs. (V.24) and (V.25). Contrary to what occurred for classical gratings for which $N_z = 0$, even in the anisotropic case (see Eq. VIII.2), no null element appears in the construction of matrix C_ε, given by Eq. (V.11) in which $\bar{\bar{\varepsilon}} = \varepsilon(x, y)\mathbb{1}$. The same remark holds for $\bar{\bar{\varepsilon}}C_\varepsilon$, so that, in general, all blocks of $[\![C_\varepsilon]\!]$ and $[\![\bar{\bar{\varepsilon}}C_\varepsilon]\!]$ are different from zero. It is then not so simple to find an explicit expression of $[\![C_\varepsilon]\!]^{-1}$ and then to find an expression of Q_ε by following the general theory of Chapter V. Fortunately, in the special case of isotropic material, matrix Q_ε can be found in a more straightforward way, generalizing what we did in Section IV.3.

Indeed, from Eqs. (IV.32) and (IV.32') we immediately obtain:

$$Q_\varepsilon = [\![\varepsilon]\!] - \left([\![\varepsilon]\!] - \left[\!\!\left[\frac{1}{\varepsilon} \right]\!\!\right]^{-1} \right) [\![\vec{N}\vec{N}]\!].$$

This equation has to be interpreted in a block form as:

$$(Q_\varepsilon)_{mn,ij} = [\![\varepsilon]\!]_{mn}\,\delta_{ij} - \left([\![\varepsilon]\!] - \left[\!\!\left[\frac{1}{\varepsilon} \right]\!\!\right]^{-1} \right)_{mn} [\![\vec{N}_i \vec{N}_j]\!] \qquad \text{with} \begin{cases} i, j = x, y, z \\ m, n \in [-N, N]. \end{cases}$$

In a matrix form, Q_ε is written as:

$$Q_\varepsilon = \begin{pmatrix} [\![\varepsilon]\!] - \Delta[\![N_x^2]\!] & -\Delta[\![N_x N_y]\!] & -\Delta[\![N_x N_z]\!] \\ -\Delta[\![N_x N_y]\!] & [\![\varepsilon]\!] - \Delta[\![N_y^2]\!] & -\Delta[\![N_y N_z]\!] \\ -\Delta[\![N_x N_z]\!] & -\Delta[\![N_y N_z]\!] & [\![\varepsilon]\!] - \Delta[\![N_z^2]\!] \end{pmatrix} \tag{IX.11}$$

$$\text{with } \Delta = [\![\varepsilon]\!] - \left[\!\!\left[\frac{1}{\varepsilon} \right]\!\!\right]^{-1}.$$

This form differs from the classical grating case, Eq. (IV.33), because of the three components of $\vec{N}$, which are, in general, non-null. Since $\mu = \mu_0\mathbb{1}$ everywhere, the propagation equations are given by Eqs. (V.22) and (V.22'), with elements M_{ij} reduced to:

$$
\begin{aligned}
M_{11} &= -i\alpha Q_{\varepsilon,yy}^{-1} Q_{\varepsilon,yx} \\
M_{12} &= -i\alpha Q_{\varepsilon,yy}^{-1} Q_{\varepsilon,yz} \\
M_{13} &= -i\frac{\alpha}{\omega} Q_{\varepsilon,yy}^{-1} \gamma \\
M_{14} &= +i\frac{\alpha}{\omega} Q_{\varepsilon,yy}^{-1} \alpha - i\omega\mu_0 \\
M_{21} &= -i\gamma Q_{\varepsilon,yy}^{-1} Q_{\varepsilon,yx} \\
M_{22} &= -i\gamma Q_{\varepsilon,yy}^{-1} Q_{\varepsilon,yz} \\
M_{23} &= -i\frac{\gamma}{\omega} Q_{\varepsilon,yy}^{-1} \gamma + i\omega\mu_0 \\
M_{24} &= i\frac{\gamma}{\omega} Q_{\varepsilon,yy}^{-1} \alpha \\
M_{31} &= -i\omega Q_{\varepsilon,zy} Q_{\varepsilon,yy}^{-1} Q_{\varepsilon,yx} + i\frac{\alpha\gamma}{\omega\mu_0} + i\omega Q_{\varepsilon,zx} \,. \qquad \text{(IX.12)} \\
M_{32} &= -i\omega Q_{\varepsilon,zy} Q_{\varepsilon,yy}^{-1} Q_{\varepsilon,yz} - i\frac{\alpha\,\alpha}{\omega\mu_0} + i\omega Q_{\varepsilon,zz} \\
M_{33} &= -iQ_{\varepsilon,zy} Q_{\varepsilon,yy}^{-1} \gamma \\
M_{34} &= iQ_{\varepsilon,zy} Q_{\varepsilon,yy}^{-1} \alpha \\
M_{41} &= i\omega Q_{\varepsilon,xy} Q_{\varepsilon,yy}^{-1} Q_{\varepsilon,yx} + i\frac{\gamma\gamma}{\omega\mu_0} - i\omega Q_{\varepsilon,xx} \\
M_{42} &= i\omega Q_{\varepsilon,xy} Q_{\varepsilon,yy}^{-1} Q_{\varepsilon,yz} - i\frac{\gamma\alpha}{\omega\mu_0} - i\omega Q_{\varepsilon,xz} \\
M_{43} &= iQ_{\varepsilon,xy} Q_{\varepsilon,yy}^{-1} \gamma \\
M_{44} &= -iQ_{\varepsilon,xy} Q_{\varepsilon,yy}^{-1} \alpha
\end{aligned}
$$

IX.1.3 Resolution of the boundary-value problem

The boundary-value problem is resolved as usual, with a shooting method similar to the one described in Section IV.3 for conical diffraction. Of course, the S-matrix propagation algorithm has to be included, too. Both procedures, i.e., resolving the boundary-value problem and defining the S matrix for a crossed grating, first require defining a bijection between the (m,n) entry of a $(2N+1)\times(2N+1)$ matrix and the corresponding r^{th} element of a vector with $(2N+1)^2$ components, with r running from 1 to $(2N+1)^2$. If m and n both vary between $-N$ and $+N$, we obtain:

$$r = 1 + (m+N) + (2N+1)(n+N). \tag{IX.13}$$

On the other hand, if r is given, n and m are derived from:

$$\begin{aligned} n &= -N + \mathrm{Int}\left[(r-1)/(N+1)\right] \\ m &= r - 1 - (2N+1)(n+N) - N \end{aligned} \tag{IX.14}$$

where function Int(x) designates the integer part of x.

It is worth giving some details about the transformation of the basic equations through the bijection defined by Eqs. (IX.13) and (IX.14). The double Fourier series components $E_{x,n,m}$ of E_x with $n, m = -N:+N$ are placed inside a block vector $\mathbf{\mathcal{E}}_x$ with components $\mathbf{\mathcal{E}}_{x,r}$, where $r = 1:(2N+1)^2$; the same is made for the other field components. In Eq. (IX.11), the Toeplitz block matrices $[\![\varepsilon]\!]$ now have elements $\varepsilon_{n'-n'',m'-m''}(y)$. They have to be placed inside a square matrix $\mathcal{E}$ with elements $\mathcal{E}_{r',r''} = \varepsilon_{n'-n'',m'-m''}(y)$, where Eq. (IX.13) is used to derive r' from n' and m', and r'' from n'' and m''. The same transformation applies to the other Toeplitz block matrices that appear in Eq. (IX.11). When all these transformations are done, the basic differential set of equations (V.22), with matrix M given by Eq. (IX.12), presents its usual form. It is resolved by starting at $y = 0$ a first series of $(2N+1)^2$ integrations. To this aim, it is necessary to chose $(2N+1)^2$ shooting vectors $\hat{\mathbf{\mathcal{E}}}_z$, defined as:

$$\begin{aligned} \hat{\mathbf{\mathcal{E}}}_{z,r,s} &= \delta_{r,s} \\ \hat{\mathbf{\mathcal{H}}}_{z,r,s} &= 0 \end{aligned} \tag{IX.15}$$

where $s = 1:(2N+1)^2$ is the index numbering the shooting vectors.

From these values of the shooting vectors, the Rayleigh expansions at $y = 0$, Eq. (IX.7), give $\frac{d}{dy}\hat{\mathbf{\mathcal{E}}}_{z,r,s}$ and $\frac{d}{dy}\hat{\mathbf{\mathcal{H}}}_{z,r,s}$. Then $\hat{\mathbf{\mathcal{E}}}_{x,r,s}$ and $\hat{\mathbf{\mathcal{H}}}_{x,r,s}$ are derived from Eqs. (IV.38) extended to a crossed grating (see also Appendix III), which implies changing γ_0 into γ_m and $\beta_n^{(1)}$ into $\beta_{n,m}^{(1)}$. We finally get:

$$\hat{\mathbf{E}}_{x,r,s} = \frac{1}{k_1^2 - \gamma_{m(s)}^2}\left[\omega\mu_0\beta^{(1)}_{n(r),m(s)}\hat{\mathbf{H}}_{z,r,s} - \gamma_{m(s)}\alpha_{n(r)}\hat{\mathbf{E}}_{z,r,s}\right]$$

$$\hat{\mathbf{H}}_{x,r,s} = \frac{1}{k_1^2 - \gamma_{m(s)}^2}\left[-\omega\varepsilon_1\beta^{(1)}_{n(r),m(s)}\hat{\mathbf{E}}_{z,r,s} - \gamma_{m(s)}\alpha_{n(r)}\hat{\mathbf{H}}_{z,r,s}\right] \qquad \text{(IX.16)}$$

where the indices n and m are obtained using Eq. (IX.14). Taking into account Eqs. (IX.15), we obtain:

$$\hat{\mathbf{E}}_{x,r,s} = \frac{-\gamma_{m(r)}\alpha_{m(r)}}{k_1^2 - \gamma_{m(r)}^2}\delta_{rs}$$

$$\hat{\mathbf{H}}_{x,r,s} = \frac{-\omega\varepsilon_1\beta^{(1)}_{m(r),m(r)}}{k_1^2 - \gamma_{m(r)}^2}\delta_{rs} \qquad \text{(IX.17)}$$

The differential set is then integrated up to $y = a$. A second series of $(2N+1)^2$ integrations ($s \in \left[(2N+1)^2 + 1, 2(2N+1)^2\right]$ starts with

$$\hat{\mathbf{E}}_{z,r,s} = 0$$

$$\hat{\mathbf{H}}_{z,r,s} = \delta_{r,s} \qquad \text{(IX.18)}$$

from which we deduce:

$$\hat{\mathbf{E}}_{x,r,s} = \frac{\omega\mu_0\beta^{(1)}_{m(r),m(r)}}{k_1^2 - \gamma_{m(r)}^2}\delta_{rs}$$

$$\hat{\mathbf{H}}_{x,r,s} = \frac{-\gamma_{m(r)}\alpha_{m(r)}}{k_1^2 - \gamma_{m(r)}^2}\delta_{rs} \qquad \text{(IX.19)}$$

The Fourier components $\mathbf{E}_{x,r}(a)$, $\mathbf{E}_{z,r}(a)$, $\mathbf{H}_{x,r}(a)$, $\mathbf{H}_{z,r}(a)$ are computed as linear superpositions of the particular computed values as stated by Eq. (IX.7) and the similar equation for $\vec{H}$. They are then matched with the various Rayleigh expansions at $y = a$. After inversing a $2(2N+1)^2 \times 2(2N+1)^2$ matrix, the Rayleigh coefficients of $\mathbf{E}_{z,r}$ and $\mathbf{H}_{z,r}$ are found at $y = 0$. A matrix product (or a new integration) then gives the Rayleigh coefficients of $\mathbf{E}_{z,r}$ and $\mathbf{H}_{z,r}$ at y = a. These values are used in Eq. (IX.16) to derive the Rayleigh coefficients $\mathbf{E}_{x,r}$ and $\mathbf{H}_{x,r}$ at both $y = 0$ and $y = a$ (see Appendix IV).

IX.1.4 Crossed-grating efficiencies

The previous paragraph provides $B_{x,r}^{(1)}$, $B_{z,r}^{(1)}$, $B_{x,r}^{(2)}$ and $B_{z,r}^{(2)}$. By using Eq. (IX.14), we go back to the double-index (n,m) notations, which gives $B_{x,n,m}^{(2)}$ and $B_{z,n,m}^{(2)}$, x and z components of vectors $\vec{B}_{n,m}^{(2)}$ appearing in Eq. (IX.3). The Poynting vector theorem gives the expression of the absolute reflected efficiencies:

$$e_{n,m}^{(r)} = \frac{1}{\beta_{0,0}^{(2)}\beta_{n,m}^{(2)}}\left\{\left(k_2^2-\gamma_m^2\right)\left|B_{x,n,m}^{(2)}\right|^2+\left(k_2^2-\alpha_n^2\right)\left|B_{z,n,m}^{(2)}\right|^2 + \alpha_n\gamma_m\left[\bar{B}_{x,n,m}^{(2)}B_{z,n,m}^{(2)}+B_{x,n,m}^{(2)}\bar{B}_{z,n,m}^{(2)}\right]\right\}. \quad \text{(IX.20)}$$

For a dielectric grating, the transmitted efficiencies are given by a similar formula, deduced from Eq. (IX.20) by changing $\vec{B}^{(2)} \rightarrow B^{(1)}$ and $\beta_{n,m}^{(2)} \rightarrow \beta_{n,m}^{(1)}$; $\beta_{0,0}^{(2)}$ is unchanged.

IX.1.5 An equivalence rule linking crossed and classical gratings

Some similarities in the absorption behavior of classical gratings lighted by a TM polarized wave and crossed gratings lighted by an unpolarized wave (see p. 271 of [IX.5]) suggest that a simple rule exists to link the two kinds of gratings. The discussion will be restricted to the case of a plane wave falling on a crossed grating under normal incidence, linearly polarized in an arbitrary direction. The grating profile is assumed being the sum of two functions, $g_1(x)$ and $g_2(z)$, with maximum amplitudes (heights) a_1 and a_2, respectively:

$$y = g(x,z) = g_1(x) + g_2(z) \quad \text{(IX.21)}$$

We suppose that $g_1(x)$ and $g_2(z)$ are symmetric functions, so that replacing a_1 by $(-a_1)$ or a_2 by $(-a_2)$ does not change the diffracting structure. It is then possible to expand the $(0,0)^{\text{th}}$ reflected efficiency $e_{g,0,0}^{(r)}$ of the crossed grating into a Taylor series having even powers of a_1 and a_2 only:

$$e_{g,0,0}^{(r)} = R + e_{g_1,0,0}^{\{2\}}a_1^2 + e_{g_2,0,0}^{\{2\}}a_2^2 + e_{g_1,0,0}^{\{4\}}a_1^4 + e_{g_2,0,0}^{\{4\}}a_2^2 + e_{g_{1,2},0,0}^{\{4\}}a_1^2a_2^2 + O(a^6), \quad \text{(IX.22)}$$

where R is the Fresnel reflectance for the unmodulated surface and $a = a_1 + a_2$. If we state $a_1 = 0$, the crossed grating reduces to the second classical grating, the zero order efficiency $e^{(r)}_{g_2,0}$ of which is:

$$e^{(r)}_{g_2,0} = R + e^{\{2\}}_{g_2,0,0} a_2^2 + e^{\{4\}}_{g_2,0,0} a_2^4 + O(a_2^6)\,. \qquad \text{(IX.23)}$$

In a similar way, if $a_2 = 0$, the efficiency of the first grating is:

$$e^{(r)}_{g_1,0} = R + e^{\{2\}}_{g_1,0,0} a_1^2 + e^{\{4\}}_{g_1,0,0} a_1^4 + O(a_1^6)\,. \qquad \text{(IX.24)}$$

Simple algebraic calculations lead to:

$$e^{(r)}_{g,0,0} = \frac{e^{(r)}_{g_1,0} e^{(r)}_{g_2,0}}{R} + \left(e^{\{4\}}_{g_{1,2},0,0} - \frac{e^{\{2\}}_{g_1,0,0} e^{\{2\}}_{g_2,0,0}}{R} \right) a_1^2 a_2^2 + O(a^6)\,,$$

which, for low enough values of groove depths, will be approximated by:

$$e^{(r)}_{g,0,0} = \frac{1}{R} e^{(r)}_{g_1,0} e^{(r)}_{g_2,0}\,. \qquad \text{(IX.25)}$$

In a similar way, we may derive:

$$e^{(r)}_{g,n,0} = \frac{1}{R} e^{(r)}_{g_1,n} e^{(r)}_{g_2,0}$$

$$e^{(r)}_{g,0,m} = \frac{1}{R} e^{(r)}_{g_1,0} e^{(r)}_{g_2,m}\,.$$

It is worth noting that in these equivalence rules, all efficiencies have to be computed without modifying the orientation of the polarization vector. These approximate rules express the efficiency of a class of crossed gratings given by Eq. (IX.21) in terms of the product of the efficiencies of two classical gratings. The rules have been checked numerically and turn out to be useful in many cases for which the groove depth is not too high.

IX.1.6 Theory of hole arrays and grids

Similarly to the classical lamellar gratings, there is a class of crossed gratings, for which the geometry inside the grating region is independent of the y-coordinate, which greatly simplifies the propagation equations and permits using the eigenvalue-eigenvector technique to integrate them. Such are the crossed gratings consisting of cylindrical hole arrays or grids. The cylindrical holes have generatrices parallel to the y-axis, so that the Fourier coefficients of ε are y-independent and, in addition, $N_y = 0$. Matrix Q_ε given by Eq. (IX.11) simplifies into:

$$Q_\varepsilon = \begin{pmatrix} [\![\varepsilon]\!] - \Delta[\![N_x^2]\!] & 0 & -\Delta[\![N_x N_z]\!] \\ 0 & [\![\varepsilon]\!] & 0 \\ -\Delta[\![N_x N_z]\!] & 0 & [\![\varepsilon]\!] - \Delta[\![N_z^2]\!] \end{pmatrix} \tag{IX.26}$$

As a result, matrix M reduces to:

$$M = \begin{pmatrix} 0 & 0 & M_{13} & M_{14} \\ 0 & 0 & M_{23} & M_{24} \\ M_{31} & M_{32} & 0 & 0 \\ M_{41} & M_{42} & 0 & 0 \end{pmatrix}$$

with

$$\begin{aligned}
M_{13} &= -i\frac{\alpha}{\omega}[\![\varepsilon]\!]^{-1}\gamma \\
M_{14} &= i\frac{\alpha}{\omega}[\![\varepsilon]\!]^{-1}\alpha - i\omega\mu_0 \\
M_{23} &= i\left(\omega\mu_0 - \frac{\gamma}{\omega}[\![\varepsilon]\!]^{-1}\gamma\right) \\
M_{24} &= i\frac{\gamma}{\omega}[\![\varepsilon]\!]^{-1}\alpha \\
M_{31} &= i\frac{\alpha\gamma}{\omega\mu_0} - i\omega\Delta[\![N_x N_z]\!]. \\
M_{32} &= i\omega\left([\![\varepsilon]\!] - \Delta[\![N_z^2]\!]\right) - i\frac{\alpha\ \alpha}{\omega\mu_0} \\
M_{41} &= i\frac{\gamma\ \gamma}{\omega\mu_0} - i\omega\left([\![\varepsilon]\!] - \Delta[\![N_x^2]\!]\right) \\
M_{42} &= -i\left(\frac{\gamma\ \alpha}{\omega\mu_0} - \omega\Delta[\![N_x N_z]\!]\right)
\end{aligned} \tag{IX.27}$$

The differential set of Eqs. (V.22) takes the form:

$$\frac{d[E_x]}{dy} = -\frac{i\alpha[\![\varepsilon]\!]^{-1}}{\omega}\gamma[H_x] + i\left(\frac{\alpha}{\omega}[\![\varepsilon]\!]^{-1}\alpha - \omega\mu_0\right)[H_z]$$

$$\frac{d[E_y]}{dy} = i\left(\omega\mu_0 - \frac{\gamma}{\omega}[\![\varepsilon]\!]^{-1}\gamma\right)[H_x] + i\frac{\gamma}{\omega}[\![\varepsilon]\!]^{-1}\alpha[H_z]$$

$$\frac{d[H_x]}{dy} = i\left(\frac{\alpha\ \gamma}{\omega\mu_0} - \omega\Delta[\![N_x N_z]\!]\right)[\![E_x]\!] + \left\{i\omega\left([\![\varepsilon]\!] - \Delta[\![N_z^2]\!]\right) - \frac{i\alpha\ \alpha}{\omega\mu_0}\right\}[\![E_z]\!] \tag{IX.28}$$

$$\frac{d[H_z]}{dy} = i\left\{\frac{\gamma\ \gamma}{\omega\mu_0} - \omega\left([\![\varepsilon]\!] - \Delta[\![N_x^2]\!]\right)\right\}[\![E_x]\!] - i\left(\frac{\gamma\ \alpha}{\omega\mu_0} - \omega\Delta[\![N_x N_z]\!]\right)[\![E_z]\!]$$

Eqs. (IX.28) can be considered as a two-dimensional extension of Eqs. (V.5) and (V.6) valid for ordinary lamellar gratings. They represent a form of truncated Maxwell equations, different from the previously published Li's equations (see Eqs. 33, 34 of [IX.13]). As it was for the classical gratings, the question how to better choose the continuation of $\vec{N}$ still remains open.

In the case of a classical lamellar grating, $N_z \equiv 0$, $N_x^2 = 1$ everywhere and $\gamma = \gamma_0 \mathbb{1}$. If used in an in-plane mount, $\gamma_0 = 0$ and Eqs. (IX.28) then reduce to:

$$\frac{d[E_x]}{dy} = i\left(\frac{\alpha}{\omega}[\![\varepsilon]\!]^{-1}\alpha - \omega\mu_0\right)[H_z] \tag{IX.29}$$

$$\frac{d[E_z]}{dy} = i\omega\mu_0[H_x] \tag{IX.30}$$

$$\frac{d[H_x]}{dy} = \left(i\omega[\![\varepsilon]\!] - i\frac{\alpha\ \alpha}{\omega\mu_0}\right)[E_z] \tag{IX.31}$$

$$\frac{d[H_z]}{dy} = -i\omega\left[\!\!\left[\frac{1}{\varepsilon}\right]\!\!\right]^{-1}[E_x]. \tag{IX.32}$$

Eqs. (IX.30) and (IX.31) are the classical propagation equations, i.e., Eqs. (II.17) and (II.18) written in Fourier space for TE polarization, while Eqs. (IX.29) and (IX.32) represent Eqs. (V.5) and (V.6).

As in the one-dimensional case, the coefficients M_{ij} given in Eq. (IX.27) do not depend on y. The numerical integration of the first-order differential set (IX.28) can then be avoided by using the eigenvalue technique as described in Chapter VI, Eq. (VI.2). To that end, all vectors must be written in monosubscript notation as explained in Section IX.1.3; similar transformation has to be done to all matrices appearing in Eq. (IX.28). The shooting method allows finding all Rayleigh coefficients and thus, deriving the crossed grating efficiencies.

In some particular devices, among which are photonic crystals, hole arrays with periodicity in two non-orthogonal directions are found. Their analysis can be found in [IX.13].

IX.1.7 Extraordinary light transmission of subwavelength hole arrays

A recent paper by Ebbesen et al. [IX.15] has shown experimental evidence of extraordinary light transmission by arrays of holes with diameters ten times smaller than the wavelength. The zero-order transmission spectra presented sharp peaks at particular wavelengths (Fig. IX.3), with intensity several times higher than the hole-surface-to-total-surface ratio, and orders of magnitude greater than predicted by standard aperture theory. Many theoreticians tried to propose an explanation of the phenomenon, which involved the excitation of surface plasmons or of localized resonances. However, due to the lack of an efficient hole-array grating code at that time, all authors of [IX.16-IX.21] worked on a one-dimensional model, in which holes where replaced by infinitely long slits. As shown in [IX.22], such a simplification introduces an efficient channel for light transmission in lamellar gratings, namely the TEM mode, that *does not exist* for hole arrays. Only a two-dimensional analysis, such as the one presented in the previous section, is able to account for this surprising phenomenon.

Inside a square cylindrical hole of the array, any field component can be written [IX.13] as:

$$E_\sigma(x,y,z) = \sum_{n,m,p=-N}^{+N} \left(u_p^+ e^{i\rho_p y} + u_p^- e^{-i\rho_p y} \right) e^{i\alpha_n x + i\gamma_m z} \qquad \text{(IX.33)}$$

where α_n and γ_m are defined in Eqs. (IX.4) and (IX.5), $\sigma = x$ or z, u_p^+ are the upgoing (propagating or decaying in +y direction) modal field amplitudes, u_p^- are the downgoing (propagating or decaying in -y direction) amplitudes, ρ_p are the propagation constants, ($i\rho_p$ being the eigenvalues of matrix M). If the diameter of the holes is smaller than the wavelength, *all* modes inside the holes are evanescent, so that the field intensity below the array is governed [IX.23] by the mode which has the lowest attenuation constant, i.e., by the mode with minimum $\mathrm{Im}(\rho_p)$, with $\mathrm{Im}(\rho_p) > 0$. In what follows, we will simply denote by ρ the corresponding propagation constant.

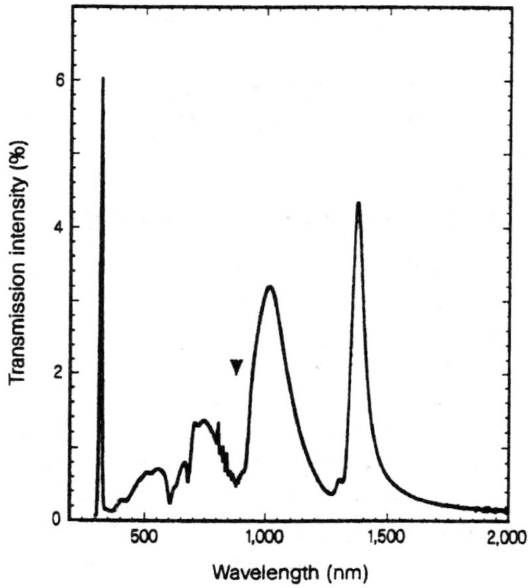

Fig. IX.3. Transmission through a silver film of 0.2 μm thickness deposited on a glass substrate and pierced with holes of 0.2 μm diameter positioned in a rectangular array with period 0.9 μm. Normal incidence (after [IX.15]).

By using a crossed-grating computer code based on the theory presented in Section IX.1.6, the complex values of ρ_p are determined for a square-hole array pierced through a silver plate with $0.2\,\mu\mathrm{m}$ thickness. The x and z period are $0.9\,\mu\mathrm{m}$, and the square length l is $0.25\,\mu\mathrm{m}$. The array is lighted under normal incidence by a linearly polarized wave with $\lambda = 1.37\,\mu\mathrm{m}$, at which the refractive index of silver is $0.1 + i8.94$. The electric field is parallel to one side of the squares. Fig. IX.4 plots as triangles the values of ρ_p. As can be seen, ρ_p lie in two distinct regions of the ρ-complex plane situated close to the imaginary axis or not far from the real one. For a comparison, circles show the propagation constants of the same modes except that the metal is assumed to be perfectly conducting. In that case, the mode propagation can be derived from the standard waveguide theory [IX.24], since no coupling through the metal exists between adjacent holes. Since the fundamental cut-off wavelength is twice the square width, i.e., $0.5\ \mu m$, all modes are purely evanescent. For the fundamental mode, we find:

$$\rho^2 = \frac{4\pi^2}{\lambda^2} - \frac{\pi^2}{l^2} = -129.0973 \mu m^{-2} \quad \Rightarrow \quad \mathrm{Im}(\rho) = 11.362\ \mu m^{-1}.$$

As a result, the wave of the form $u^- \exp(-i\rho y)$ is attenuated from the top to the bottom of the hole by a factor $\exp[-\mathrm{Im}(\rho)a] = 0.00534$.

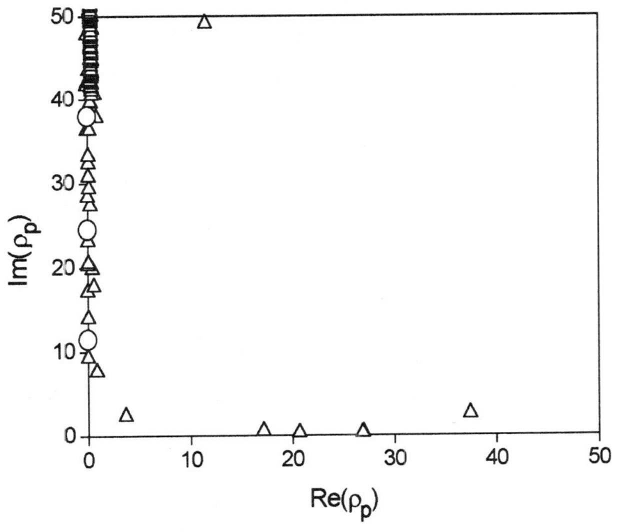

Fig. IX.4. Complex propagation constants of a square-hole array made through a silver and a perfectly conducting plates (after [IX.22]).

Fig. IX.4 shows that for finite conductivity several values of ρ_p with much lower imaginary part can be found, so that their attenuation within the layer thickness $(a = 0.2\ \mu m)$ could be 100 times smaller or more, possibly leading to a 10^4 greater transmittivity. But the final transmittivity enhancement also depends on the coupling strength between the incident field and the various evanescent modes; the same remark also applies to the coupling between these modes and the transmitted field.

In order to determine which mode(s) plays the important role in the transmission, several numerical experiments are possible. By chosing $\lambda = 1.388\,\mu m$, which is close to the most right-hand side maximum in Fig. IX.3, we first computed the mode propagation constants and the transmission, and then started to artificially double the imaginary part of the propagation constant of each mode, one after the other, and again calculate the transmission. The "experiment" revealed that the transmission remains almost the same, except when the propagation constant of one definite mode is changed, for which the transmission falls 40 times, from 7.5% to 0.192%. This was the mode with $\rho_p = 0.02539 + i9.3813$, the mode which is the closest to the imaginary ρ - axis and has the lowest imaginary part.

A second numerical experiment was made to support this conclusion. The transmittivity T computed as a function of the thickness of the silver film is presented in Fig. IX.5 in a semi-logarithmic scale. As can be seen, ln(T) is a linear function of a in the interval 0.2-0.6 μm, which is a proof that the exceptional transmittivity occurs through a single channel (single mode in Eq. (IX.33)). The slope of the line in Fig. IX.5 (approximately 10) well agrees with the imaginary part of $\rho_p = 0.02539 + i9.3813$, determined above.

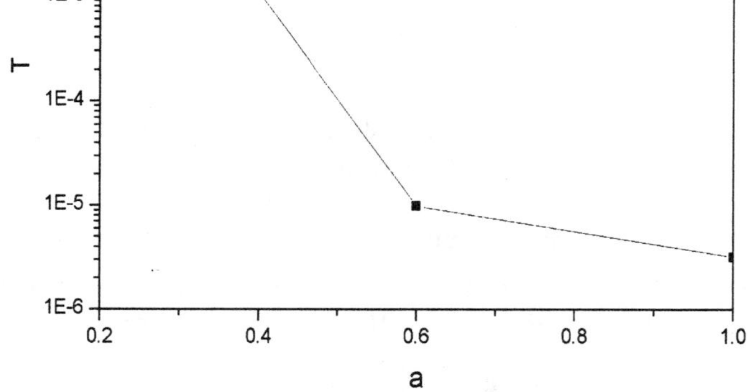

Fig. IX.5. Exponential variation of the transmittivity as a function of the thickness plate (after [IX.25]).

These results show that the finite conductivity of the metal introduces a new channel for light transmission which cannot be predicted by the one-dimensional model. Having proved that only one mode of the entire set of eigenvalues is responsible for the extraordinary transmission through the silver film, the question which remains is why the enhanced transmission is only observed around particular wavelengths, while the channel exists for all wavelengths. To that end, it is helpful to study the poles of the scattering (S) matrix [IX.25] in the complex plane. This requires extending to 2-D-diffracting devices the method previously developed for classical gratings and nowadays known under the name of "polology" (see Chapter XIII), which explained with great success several unexpected phenomena in linear [IX.26] and nonlinear optics [IX.27]. The complex poles of the S-matrix are related with resonance phenomena (guided waves, surface plasmons or cavity resonances, etc.), which can be excited with the help of grating periodicity. This excitation appears at particular wavelength or angle of incidence values, which satisfy what is usually called phase-matching condition (see Section XIII.1.5).

There are several different values of the complex pole of the S-matrix that can be found, the pole situated at the longest wavelength is characterized by its complex wavelength value $\lambda^p = 1.3878 + i\,0.0020227$ and corresponds to the rightmost anomaly in Fig. IX.3. When the hole dimensions are forced to tend to zero, a process easily accessible in a numerical experiment, this pole tends to a value $\lambda^p = 1.369406 + i\,0.0004436$, which is equal to the complex pole corresponding to a plasmon surface wave propagating along a plane (without holes) silver-glass interface. The resonant character of the extraordinary transmission is thus due to the fact that the channel (with field evanescent in the holes) resonantly excites surface plasmons at the lower silver-glass interface thanks to the crossed-grating periodicity. A phenomenological study using both complex poles of the scattering operator and complex zeros of the transmitted amplitude was able to account for this unexpected effect in both qualitative and quantitative ways [IX.25].

It is possible to further enhance the transmission by using as a superstrate a liquid with the same refractive index as the glass substrate in order to excite surface plasmons at both upper and lower surfaces, as reported in [IX.28, IX.29].

IX.2 Crossed gratings made of anisotropic materials

This kind of diffracting device cumulates complexities of both types studied in Chapter VIII and in Section IX.1. Assuming that $\mu = \mu_0$ everywhere, the anisotropy only comes from the permittivity, described in Eq. (VIII.1). The grating surface being defined by the equation $f(x,y,z) = 0$ or by $y = g(x,z)$, all normal vector components are different from zero, so that no simplification occurs in the expression of matrix C_ε given by Eq. (V.11). No null elements of the matrix appear, as was the case of classical gratings made of anisotropic materials (see Eq. VIII.2), but also the method used for crossed gratings made of isotropic material (Section IX.1.2) is not applicable to derive matrix Q_ε, since $\bar{\bar{\varepsilon}}$ does not reduce to a scalar. Thus the general method developed in Chapter V has to be used.

First the components of $\vec{N}$ are calculated through Eq. (V.23), the elements of $\vec{T}_1$ and $\vec{T}_2$ are derived through Eqs. (V.24) and (V.25). Matrix C_ε is then calculated via Eq. (V.11) and the product $\bar{\bar{\varepsilon}} \cdot C_\varepsilon$ is derived. After developing all elements of C_ε and $\bar{\bar{\varepsilon}} \cdot C_\varepsilon$ into double Fourier series similar to Eq. (IX.10), the associated Toeplitz matrices $[\![\bar{\bar{\varepsilon}} \cdot C_\varepsilon]\!]$ and $[\![C_\varepsilon]\!]$ are constructed, and the inverse matrix $[\![C_\varepsilon]\!]^{-1}$ is calculated numerically. The product $[\![\bar{\bar{\varepsilon}} \cdot C_\varepsilon]\!][\![C_\varepsilon]\!]^{-1}$ gives matrix Q_ε and no explicit expression can be provided in this case, contrary to what occurred in Section IX.1.2, see Eq. (IX.11).

The only simplification which occurs in the general theory of Chapter V comes from the fact that $Q_\mu = \mu_0 \delta_{kl}$. Thus the basic propagation equations are given by Eqs. (V.22) and (IX.12), except that in Eqs. (IX.12) the blocks $Q_{\varepsilon,i,j}$ with $i, j = (x, y, z)$ are determined on a computer instead of being provided by Eq. (IX.11).

The incidence conditions are the same as in Section IX.1 and the Rayleigh expansions of the field are given by Eqs. (IX.3) and (IX.7). The boundary-value problem is resolved as explained in Section IX.1.3. It implies using the monosubscript notations given in Eqs. (IX.13) and (IX.14) and applying a shooting method as stated in Eqs. (IX.15)-(IX.19). It is necessary to take care of the outgoing-wave condition by computing the y-component of the Poynting vector for each harmonic, as explained in Section VIII.2. After a matrix inversion, the Rayleigh coefficients are found and the grating efficiencies are given by Eq. (IX.20).

References

IX.1. D. Maystre and M. Nevière, "Electromagnetic theory of crossed gratings," J. Optics (Paris) 9, 301-306, 1978.

IX.2. P. Vincent, "A finite-difference method for dielectric and conducting crossed gratings," Opt. Commun. 26, 293-296, 1978

IX.3. G. H. Derrick, R.C. McPhedran, D. Maystre, and M. Nevière, "Crossed gratings: a theory and its applications," Appl. Phys. 18, 39-52, 1979

IX.4. R.C. McPhedran, G. H. Derrick, M. Nevière, and D. Maystre, "Metallic crossed gratings," J. Optics, Paris 13, 209-218, 1982

IX.5. R.C. McPhedran, G. H. Derrick, and L. C. Botten, "Theory of crossed gratings," in *Electromagnetic Theory of Gratings*, R. Petit ed. (Springer-Verlag, Berlin, 1980), pp. 227-276

IX.6. J. B. Harris, T. W. Preist, J. R. Sambles, R. N. Thorpe, and R. A. Watts, "Optical response of bigratings," J. Opt. Soc. Am. A 13, 2041-2049, 1996

IX.7. D. C. Dobson and J. A. Cox, "An integral equation method for biperiodic diffraction structures,"in *International Conference on the Application and Theory of Periodic Structures*, J. M. Lerner and W. R. McKinney, eds., Proc. SPIE 1545, 106-113, 1991

IX.8. R. Bräuer and O. Bryngdahl, "Electromagnetic diffraction analysis of two-dimensional gratings," Opt. Commun. 100, 1-5, 1993

IX.9. E. Noponen and J. Turunen, "Eigenmode method for electromagnetic synthesis of diffractive elements with three-dimensional profiles," J. Opt. Soc. Am. A 11, 2494-2502, 1994

IX.10. J.-J. Greffet, C. Baylard, and P. Versaevel, "Diffraction of electromagnetic waves by crossed gratings: a series solution," Opt. Lett. 17, 1740-1742, 1992

IX.11. O. P. Bruno and F. Reitich, "Numerical solution of diffraction problems: a method of variation of boundaries. III. Doubly periodic gratings," J. Opt. Soc. Am. A 10, 2551-2562, 1993

IX.12. O. P. Bruno and F. Reitich, "Calculation of electromagnetic scattering via boundary variations and analytic continuation," Appl. Computat. Electromagn. Soc. J. 11, 17-31, 1996

IX.13. L. Li, "New formulation of the Fourier modal method for crossed surface-relief gratings," J. Opt. Soc. Am. A 14, 2758-2769, 1997

IX.14. A.-L. Fehrembach, S. Enoch, and A. Sentenac, "Highly directive light sources using two-dimensional photonic crystal slab," Appl. Phys. Lett. 79, 4280-4282, 2001

IX.15. T. W. Ebbesen, H. J. Lezec, H. F. Ghaemi, T. Thio, and P. A. Wolff, "Extraordinary optical transmission through subwavelength hole arrays," Nature, 391, 667-669, 1998

IX.16. U. Schröter and D. Heitmann, "Surface-plasmon-enhanced transmission through metallic gratings," Phys. Rev. B 58, 15419-15421, 1998

IX.17. M. M. J. Treacy, "Dynamical diffraction in metallic optical gratings," Appl. Phys. Lett. 75, 606-608, 1999

IX.18. J. A. Porto, F. T. Garcia-Vidal, and J. B. Pendry, "Transmission resonances on metallic gratings with very narrow slits," Phys. Rev. Lett. 83, 2845-2848, 1999

IX.19. H. F. Ghaemi, T. Thio, D. E. Grupp, T. W. Ebbesen, and H. J. Lezec, "Surface plasmons enhanced optical transmission through sub-wavelength holes," Phys. Rev. B 58, 6779-6782, 1998

IX.20. T. Lopez-Rios, D. Mendoza, J. J. Garcia-Vidal, J. Sanchez-Dehesa, and B. Pannetier, "Surface shape resonances in lamellar metallic gratings," Phys. Rev. Lett. 81, 665-668, 1998

IX.21. Ph. Lalanne, J. P. Hugonin, S. Astilean, M. Palamaru, and K. D. Möller, "One-mode model and Airy-like formulae for one-dimensional metallic gratings," J. Opt. A: Pure Appl. Opt. 2, 48-51, 2000

IX.22. E. Popov, M. Nevière, S. Enoch, and R. Reinisch, "Theory of light transmission through subwavelength periodic hole arrays," Phys. Rev. B, 62, 16100-16108, 2000

IX.23. A. Sentenac and D. Maystre, "Symmetry of the field transmitted by metallic grids," J. Modern Optics 45, 785-797, 1998

IX.24. R. E. Callen, *Field Theory of Guided Waves*, Second Ed. (IEEE Press, New York, 1991)

IX.25. S. Enoch, E. Popov, M. Nevière, and R. Reinisch, "Enhanced light transmission by hole arrays," J. Optics A, Pure and Applied Optics, to be published

IX.26. M. Nevière, "The homogeneous problem," in *Electromagnetic Theory of Gratings*, R. Petit ed. (Springer-Verlag, Berlin, 1980), ch. 5, 123-157

IX.27. M. Nevière, E. Popov, R. Reinisch, and G. Vitrant, *Electromagnetic Resonances in Nonlinear Optics*, (Gordon and Breach Science Publishers, London, 2000), ch. VI, pp. 120-164

IX.28. A. Krishman, T. Thio, T. J. Kim, H. J. Lezec, T. W. Ebbesen, P. A. Wolf, J. Pendry, L. Martin-Moreno, and J. J. Garcia-Vidal, "Evanescently coupled surface resonance in surface plasmon enhanced transmission," Opt. Commun. 200, 1-7, 2001

IX.29. L. Martin-Moreno, F. J. Garcia-Vidal, H. J. Lezec, K. M. Pellerin, T. Thio, J. B. Pendry, and T. W. Ebbesen, "Theory of extraordinary optical transmission through subwavelength hole arrays," Phys. Rev. Lett. 86, 1114-1117, 2001

Chapter X
Photonic Crystals

Photonic crystals have been the subject of intensive studies over the last 15 years, since the pioneering work of Yablonovich [I.5]. They present a wide band gap, which is almost independent of the incidence angle and, thus, enable light confinement in one, two or three directions. Their periodicity also leads to light diffraction producing grating behavior. It is then tempting to try to cumulate both their diffracting capabilities and their 100% reflectivity for wavelengths inside the band gaps to make gratings with perfect blazing in unpolarized light. With the recent technological developments that allow producing two- or three-dimensional periodic structures with a period shorter than optical wavelengths, the need of a full electromagnetic analysis of such devices becomes evident.

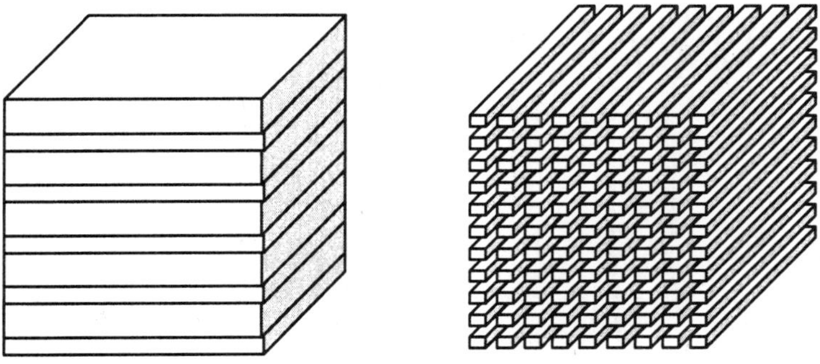

(a) (b)

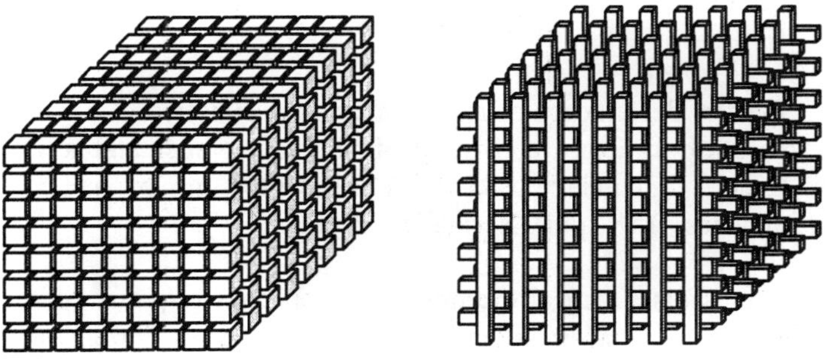

(c) (d)

Fig. X.1. Schematic representation of (a) one-, (b) two- and (c, d) three-dimensional photonic crystals.

A schematic view of one-, two- and three-dimensional photonic crystals is presented in Fig. X.1. Generally, the two- and three-dimensional crystals can consist of objects with arbitrary shapes, not necessarily rectangular as in the figure.

X.1 One-dimensional photonic crystals

A structure characterized by a one-dimensional periodicity (Fig. X.1a) has been well known for quite a long time. And indeed, such are the multilayered dielectric mirrors, used in many domains of physics and optical engineering. If the optical thickness of each layer is equal to a quarter of a given wavelength λ_B, and if two consecutive layers have, respectively, high and low optical indices, the light reflected at the interfaces interferes constructively to form a band of high reflection. Given high enough number of layer pairs, the reflection can reach, at least neglecting the absorption and scattering losses, almost 100%, and a forbidden band in transmission is formed.

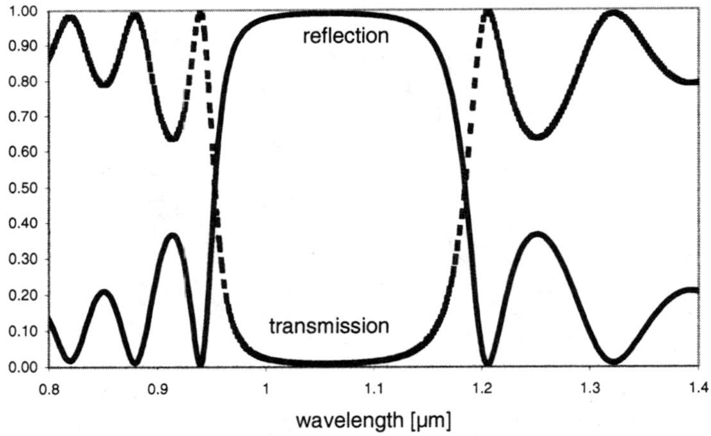

Fig. X.2. Spectral dependence of reflection and transmission of a 21-layer dielectric structure consisting of alternative high and low index layers (high index 1.885, thickness 0.140 μm, low index 1.441, thickness 0.183 μm). Normal incidence.

Fig. X.2 presents a typical example of the spectral response of such a system in reflection and transmission, with $\lambda_B = 1.056\,\mu\text{m}$. Formation of a transmission band-gap is observed between 0.95 and 1.15 μm. The reflection inside the gap increases rapidly with the number of pairs in the structure.

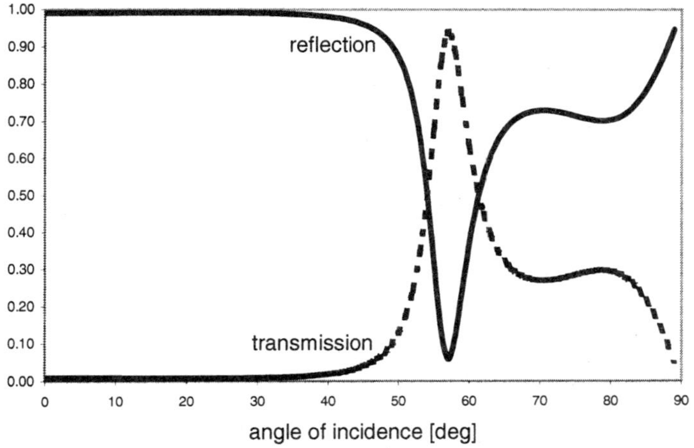

Fig. X.3. Angular dependence of reflection and transmission of the system described in Fig. X.2 at wavelength $\lambda = \lambda_B = 1.056\,\mu m$.

The main problem of 1-D systems is that their response depends on the incident direction, as demonstrated in Fig. X.3.

If one desires to maintain the high reflection (or, in other words, the forbidden transmission) for all angles of incidence, it is necessary to change the symmetry of the system, so that the incident light "sees" the structure almost invariant of the angle. In order to obtain such an independence, it is necessary to use 2-D photonic crystals. An invariance with respect to both incidence and azimutal angles requires 3-D systems. However, before describing such more complicated structures, let us present a useful combination between a one-dimensionally periodic system and a transmission diffraction grating, which could be characterized with very high diffraction efficiency and almost zero absorption losses.

X.2 A grating with a 1-D photonic crystal for chirped pulse amplification for ultrahigh-intensity lasers

X.2.1 The chirped pulse amplification concept

One of the most exciting developments in modern physics is surely the physics of ultrahigh energy opened by the recent development of lasers [X.1]. Their intensities have been increased by more than four orders of magnitude in the last ten years, to reach irradiances about $10^{21}\ W/cm^2$. Such enormous intensities produce field strengths of about $10^{12}\ V/cm$, i.e., hundreds times the natural intra-atomic field. When the first powerful lasers were developed in the sixties, with tera Watts/cm^2 irradiances (10^{12} W/cm^2), they produced fields about one hundredth of the Coulomb field E_{at} to which the ground state electron is subjected in the hydrogen atom ($E_{at} = 5 \times 10^9$ V/cm). This was enough to modify the electronic orbitals in matter and to produce new phenomena – the study of which is called "nonlinear optics." The situation nowadays is completely different from this "bounded" nonlinear optics. Under field strengths of about 10^{12} V/cm, electrons are pulled up from atoms and accelerated to relativistic speeds under accelerations of 10^{21} g (g is Earth's gravity). At these intensities the light pressure is of the order of giga- to tera-bars while the magnetic field reaches a gigagauss. It is fascinating to see that such physical environment of extreme electric and magnetic fields, pressure, acceleration and temperature, which could be only expected in stellar cores or close to black holes, can be produced on a laboratory tabletop. Of course, many new potentialities [X.1] arise from laser technology, among which are high-harmonic-generation, accelerating unstable particles (π mesons), fast ignition in inertial confinement fusion, creating particle-antiparticle pairs from vacuum fluctuations, creating astrophysical conditions inside laboratories, studying new quantum electromagnetism, etc.

This spectacular breakthrough in ultrahigh intensity laser development has become possible thanks to the concept of *chirped pulse amplification* (CPA): a laser produces a nanojoule pulse in the femtosecond (10^{-15} s) regime, the pulse is stretched by a factor of 10^3 to 10^5 into the nanosecond regime, reducing its intensity accordingly. It is then amplified by 10^{11}, i.e., to tens of joules without damage on the amplifier or beam distortion linked with autocollimation. Finally it is recompressed to its initial width, i.e., in the femto second regime. A 100 J energy pulse with a 100 fs duration produces a peta watt (10^{15} W) power beam. When focalized, such a beam may have an irradiance approaching 10^{21} W/cm^2, which is enough to initiate fusion in a deuterium–tritium compressed plasma. Such process, called *fast ignitor concept* [X.1], offers the possibility of high target gain at reduced total drive energy and quite probably brings to a near future massive energy production via nuclear fusion.

Gratings play a key role in chirped pulse amplification. The first attempts [X.2] used the positive group velocity dispersion (in which the red wavelength goes faster than the blue) of a single-mode fiber to temporarily stretch the ultrashort pulse. After amplification, it was recompressed by a pair of parallel diffraction gratings which exhibit a negative group velocity dispersion. However, the two dispersive characteristics did not match exactly and the beam experienced deformations that limited the stretching-compression ratio to about 100. In 1987, the idea arose of making a dispersive system consisting of a telescope of magnification 1 placed between two symmetric, identical gratings [X.1, X.3]. Such a device has exactly the same dispersive function as the compressor using two parallel gratings, but with opposite sign. It means that, in theory, any arbitrary short pulse could be stretched to any pulse duration and then recompressed to its original shape. A stretching factor of $10^3 - 10^4$ was immediately demonstrated and the first tabletop terawatt laser system was produced from a Nd:glass laser. Presently, a stretching up to 10^9 has been demonstrated of 10 fs pulses.

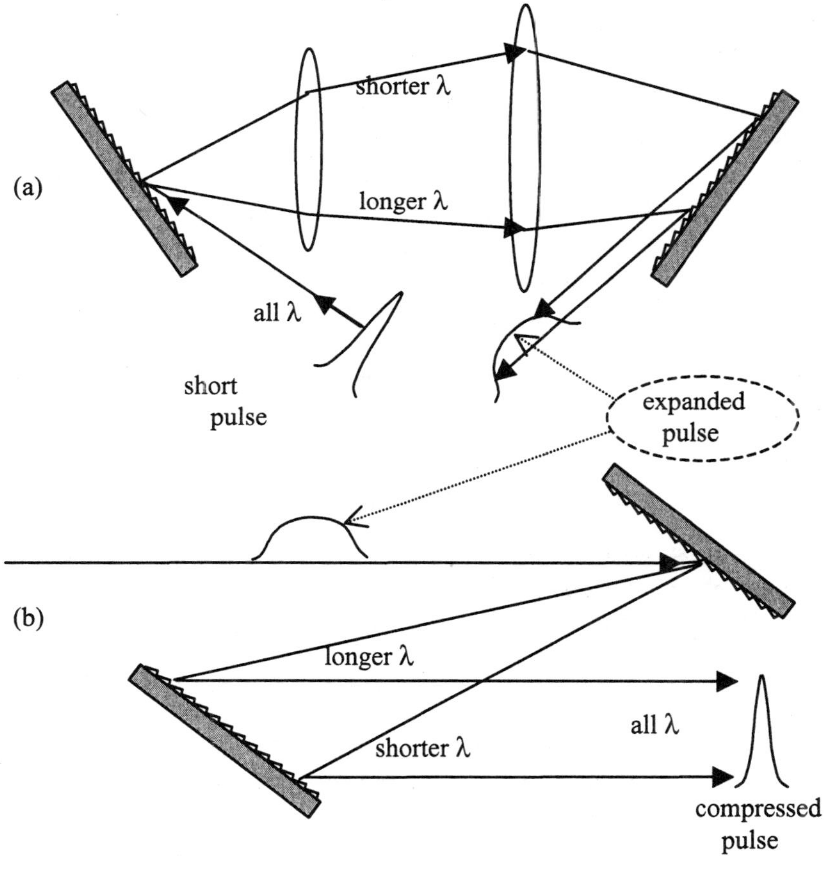

Fig. X.4. Chirped pulse amplification scheme: (a) pulse stretcher and (b) pulse compressor.

Fig. X.4 gives a schematic presentation of both the stretcher (expander) and the compressor. During stretching, the shorter wavelengths of the pulse run a longer path. The shorter the wavelength, the longer is the delay, leading to a temporal expansion of the pulse. The compressor works in an inverse regime, the longer wavelengths are delayed more than the shorter and the pulse is compressed.

In the present state of the art, four identical gratings are thus necessary to achieve the stretching-compression of the laser pulse. Since petawatt systems operate with 46-cm diameter beam, gratings as large as 75 cm must be produced, a true challenge to grating manufacturers. Such gratings must have an absolute -1st-order efficiency $e_{-1}^{(r)}$ as high as possible, since the final power will be proportional to $\left(e_{-1}^{(r)}\right)^4$. They should present zero absorption, and high damage threshold. At these powers, even one percent absorption would result in an immediate destruction of the system.

X.2.2 Design of perfectly blazed reflection gratings

Since absorption must be excluded, metallic gratings are not useful. Although ultrahigh intensity systems use Nd:glass lasers which operate in near infrared ($\lambda \simeq 1.056\ \mu m$) at which good reflectors may present 99% reflectivity, gold gratings would also be damaged by the pulses. Perfect dielectrics must be used [X.4]. The only way to make a perfect reflector with dielectrics is to construct a 1-D photonic crystal as demonstrated in Section X.1, by superposing 10 to 20 bilayers made of a low-index material (SiO_2 with refractive index $\nu_l = 1.441$) and a high-index one (HfO_2 with index $\nu_h = 1.885$). The upper layer in which a grating is ion-etched is a high-index layer. Contrary to metallic gratings for which the electromagnetic energy is concentrated near the surface, in dielectric gratings the electromagnetic field diffracts in the material bulk. As a result, they have a higher damage threshold, as established by measurements [X.5].

A grating for chirped pulse amplification is illustrated in Fig. X.5. The bilayers are all identical, with thicknesses (t_h and t_l) except the pair at the top. Its thicknesses t_1 and t_2 will be optimized separately from the remaining of the stack. The angle of incidence θ_2 in air and the groove period d are imposed by the compressor geometry. It is designed in such a way that only the 0th and the -1st orders will propagate. The thicknesses t_h and t_l are then determined through the following lines.

We first determine the angle of diffraction $\theta_{2,-1}$ of the reflected diffracted order of interest for the central wavelength λ_B of the pulse. From $\theta_{2,-1}$ we deduce the angular deviation A.D. between incident order and -1st one. We then calculate the angle θ' made by the bisector of θ_2 and $\theta_{2,-1}$. The thicknesses t_h and t_l are optimized so that the layers have an optical thickness equal to $\lambda_B/4$ under incidence θ'. If θ'_h and θ'_l are the corresponding refracted angles in high- and low-index layers, which are derived from Snell's law ($\nu_h \sin\theta'_h = \sin\theta' = \nu_l \sin\theta'_l$), the thicknesses t_h and t_l are given by:

$$\begin{aligned} t_h &= \lambda_B/(4\nu_h \cos\theta'_h) \\ t_l &= \lambda_B/(4\nu_l \cos\theta'_l) \end{aligned} . \tag{X.1}$$

Calculations then show that under incidence θ', such a stack approaches 100% reflectivity to about 10^{-6} when the number of bilayers is 20, and 100% is approached to about 10^{-4} with only 10 bilayers.

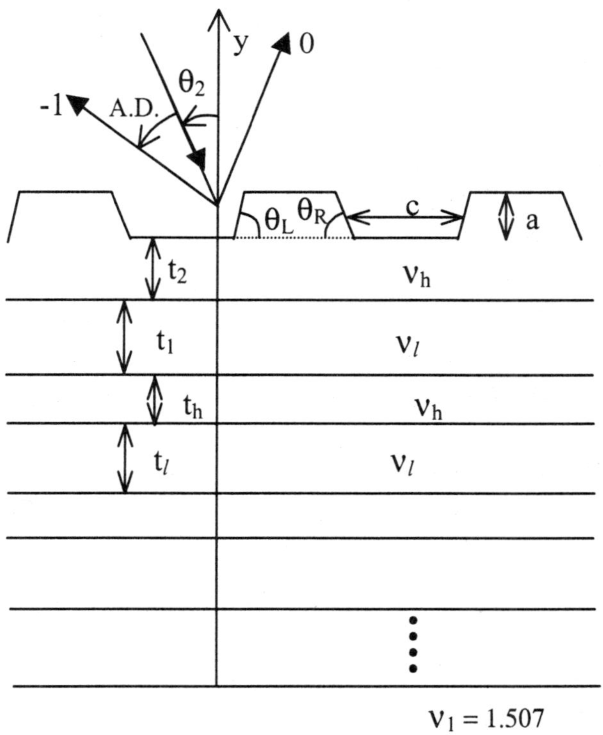

Fig. X.5. Schematic representation of a dielectric grating on a one-dimensional multilayer structure.

The fact that all energy is reflected does not guarantee that it will propagate in the desired dispersive (-1st) order. Grating profile optimization has to be done to reach the goal. Systematic study has shown that sinusoidal profiles are not good candidates. Lamellar profiles have been shown [X.4] to culminate at theoretical 96% efficiency. Triangular profiles introduce sharp edges which could reduce the damage threshold. Trapezoidal profiles have no so sharp edges and propose more degrees of freedom than lamellar ones, namely the slope angles θ_L and θ_R defined in Fig. X.5, in addition to the groove depth a and the groove width c (defined at a half depth). They also represent a more realistic situation, as far as the ion-etching process never gives exactly vertical walls of the lamellae. Typically, angles θ_L and θ_R between 55° and 85° are obtained. On top of this, thicknesses t_l and t_2 are also allowed to be varied.

In order to optimize the grating profile, a computer code based on the classical differential theory, including the S-matrix propagation algorithm (Chapters II and III) is most suitable to optimize the grating profile. Since lasers produce a linearly polarized light, TE polarization can be chosen, so that the fast Fourier factorization is not necessary, which is not the case in TM polarization. The optimization procedure searches for a maximum of a function (namely of $e_{-1}^{(r)}$) of an arbitrary number of parameters following a step-by-step algorithm that does not assume the derivability of the function. In practice, an initial choice of $\theta_L = \theta_R$ and $t_l = t_l$ leaves four independent parameters (a, c, θ_L, t_2). The code finds the nearest maximum of $e_{-1}^{(r)}(a, c, \theta_L, t_2)$ in the vicinity of any starting point. Many possible starting points have to be tried in the physical range of the various parameters. With a number of layers equal to 40 or 41, several perfect blazing conditions were found [X.6], producing -1st order efficiencies reaching unity within an unmeasurable discrepancy (10^{-6} to 10^{-4}).

When the deviation between the incident and the diffracted beams is allowed to be varied, providing a degree of freedom to vary the groove spacing d, additional perfect blazing conditions could occur.

X.2.3 Petawatt laser with a perfectly blazed multilayer dielectric diffraction grating

A study was conducted to produce gratings for the petawatt laser project of the French Commissariat à l'Energie Atomique (CEA) [X.6]. The geometry of the compressor imposed a 10° deviation (A.D.) between the incident beam and the -1st order. The blaze wavelength ($\lambda_B = 1.056\ \mu m$) and the angular deviation determined the groove frequency to be 1740 grooves/mm ($d = 0.57471\ \mu m$) and the incidence angle to be either 62.2548^o (for $A.D. = -10^o$) or 72.2548^o (for $A.D. = +10^o$). The angular deviation is positive when the incident beam is more grazing than the -1st order. Both values give the same position of the bisector of the two directions. Thus Eq. (X.1) led to $t_h = 0.154678\ \mu m$ and $t_l = 0.238432\ \mu m$. In addition, it was chosen that $t_1 = t_l$. Assuming the base angles θ_L and θ_R equal to 55°, and a modulation a/d equal to 0.35, optimizing t_2 and c gave the optimum values $t_{2,opt} = 0.25$ µm and $c_{opt} = 0.308984$. With a stack of 31 layers (15 bilayers plus one extra layer of HfO_2 on top), the optimized efficiency reached 0.999565. Fig. X.6 shows the evolution of the -1st reflected efficiency around the central wavelength. The bandwidth is wider for $D = +10^o$ (larger angle of incidence). The efficiency remains higher than 99% over a range of 8.6 nm, and higher than 98.5% over 11 nm. The conclusion is that, for this compressor, optimizing t_2 and c only already provides an excellent solution.

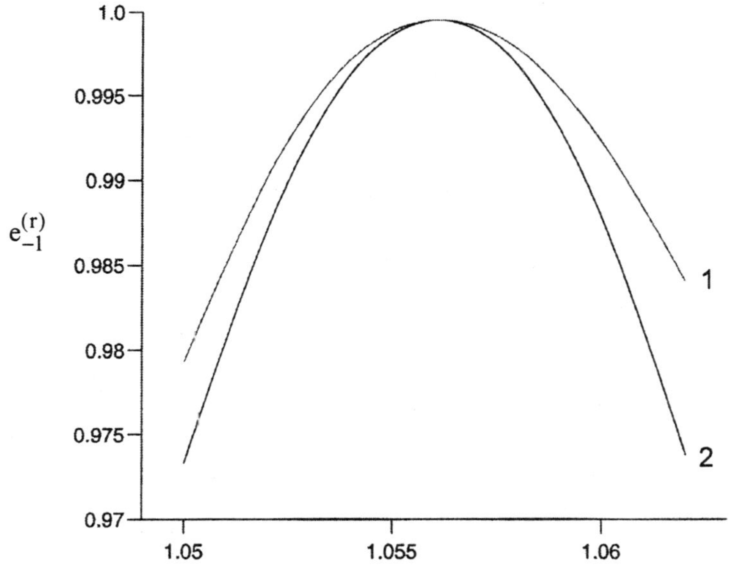

Fig. X.6. Spectral bandwidth of the perfectly blazed grating: (1) $\theta_2 = 72.2548^o$ and (2) $\theta_2 = 62.2548^o$.

However, it turns out that the tolerances with respect to a, c, t_2 are rather strict. They are wider when the grating is etched inside the low-index material (SiO_2) instead of being etched inside the high-index one, but then the modulated region must be deeper. Typically, for a grating etched in HfO_2, the relative errors on t_2, a and c have to be less than 2.5%, 10% and 5%, respectively, to allow for only a 1% reduction in efficiency. Such constraints have been fulfilled by the grating manufacturer Jobin-Yvon-Horiba and efficiencies as high as 99% have been measured, as shown in Fig. X.7.

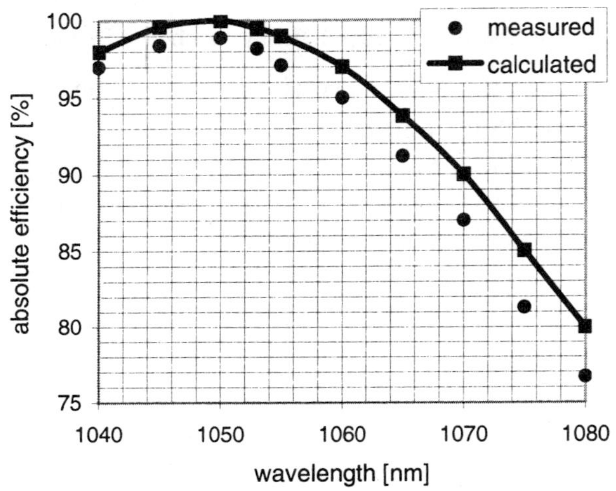

Fig. X.7. Comparison between predicted and measured efficiency of a perfectly blazed multilayer dielectric grating.

An interesting surprise came during the study. It may occur that the thicknesses t_h and t_l produced by the manufacturer of the multilayer stack depart by more than 10% of their specified values. This produces a change in wavelength and/or incidence at which the reflectivity reaches 100%. However, although at the desired blaze wavelength the reflectivity is lower than unity, if the exact thicknesses t_h and t_l are measured, it is then possible to redetermine some of the grating parameters (a, c, t_2, θ_L and d) to reach a quasiperfect blazing at the required value of λ_B. Also it was observed that an error in the value of c which may result from the ion-etching process could be corrected by a change in groove depth a, the higher groove depth compensating a too low groove width. These numerical discoveries were of great help in the manufacturing process.

Let us end this section with an interesting comment. The Heinsenberg uncertainty principle introduces the following relation:

$$\Delta\omega \cdot \tau = 2.5 \qquad \text{(X.2)}$$

between the pulse circular-frequency bandwidth $\Delta\omega$ and its duration τ. Thus a 100 fs pulse, like those produced in the CEA project, will have a 8-nm spectral bandwidth. Curves in Fig. X.6 show that the dominant harmonics of the pulse will entirely be diffracted with more than 99% efficiency. Thus the designed gratings not only avoid loosing a significant amount of energy, but they also warranty that the beam will be recompressed at its original shape, avoiding temporal wings.

X.3 Two-dimensional photonic crystals

As discussed in the beginning of the chapter, if one needs that the high reflectivity of the periodic structure is preserved within a larger (or the entire) angular interval, one of the possibilities is to use systems periodical in two or three dimensions. 2-D photonic crystals are better studied than the three-dimensional ones. The 2-D periodicity can be modeled with existing grating codes by using them for a stack of gratings, which form the photonic crystal. The same is valid for the 3-D structures, but one needs a code designed for crossed gratings, which implies greater difficulties. The main peculiarity of the 2-D photonic crystal case when compared with the classical grating is that, as a rule, it rarely consists of gratings with smooth or continuous profiles, more often of separate cylindrical objects like rods.

Most of the attention devoted to photonic crystal diffraction is directed toward structures having subwavelength periods. The grating equation implies that when $\lambda/d > 2$, the zeroth order is the only one to be diffracted whatever the incidence may be. Larger period structures can diffract light into non-specular orders, thus complicating the picture of propagating bands and band gaps. The simpler case of specular-only diffraction is common to solid-state physics, wherefrom the extensive theoretical investigation of photonic crystals has begun. However, in optics, non-specular diffraction has long time ago proven to be quite interesting and useful. We start our study with a short discussion of the principal property of 2-D photonic crystals to form band gaps, forbidden in transmission. Some peculiar applications of this property when only specular reflection and transmission orders propagate are illustrate further. Next, when the period is larger to allow for diffracted orders to propagate, with the wavelength lying inside the band gap for transmission, such systems can have almost 100% non-specular diffraction efficiency over the entire band gap in both TE and TM polarization [X.7]. Such behavior is called a *quasiperfect blazing* and can be quite useful for demultiplexing light in optical communication networks.

X.3.1 Existence of a propagation band

The system under consideration is presented in Fig. X.8. It has infinite length in x and z directions and is invariant with respect to z (e.g., periodically arranged cylindrical rods). In addition, it is periodical in y, but within a limited region between the sub- and superstrate. When this system is lighted under incidence θ_2, the field is given by the pseudo-Fourier series in x (I.29'). At the bottoms y_j of each of the rod rows, Eq. (III.13) applies:

$$V^{(j)} = T^{(j)} V^{j-1}.$$

Matrix $T^{(j)}$ designates the transmission matrix for one vertical period d_y, which links the vector $V^{(j-1)}$ at y_{j-1} to $V^{(j)}$ at y_j. Since all vertical periods are identical, it will be denoted by T.

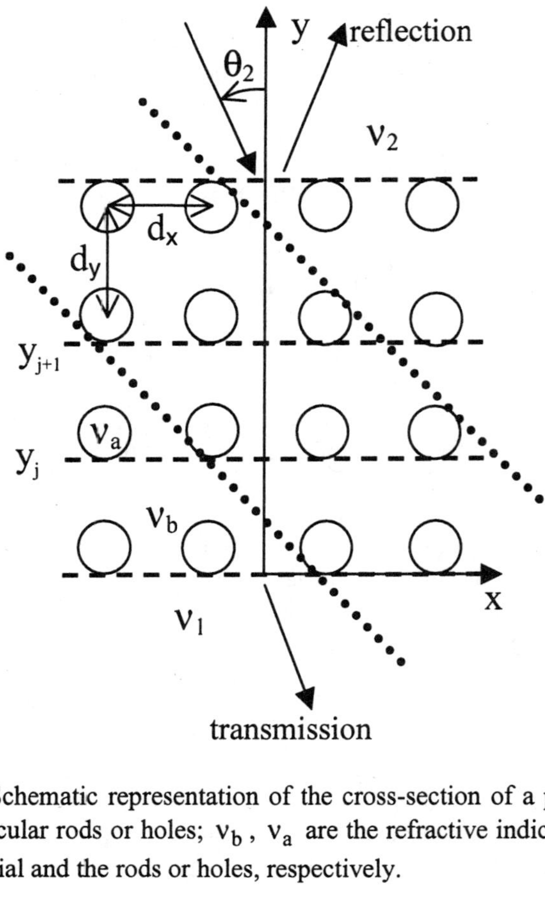

Fig. X.8. Schematic representation of the cross-section of a periodic array of circular rods or holes; ν_b, ν_a are the refractive indices of the bulk material and the rods or holes, respectively.

The transmission matrix can be diagonalized and written as:

$$T = QRQ^{-1} \tag{X.3}$$

where Q is obtained by putting the eigenvectors of T into columns and R is a diagonal matrix with elements equal to the corresponding eigenvalues ρ_m.

The transmission matrix through $\tilde{M}$ identical vertical periods is simply obtained by raising T to the $\tilde{M}^{th}$ power. Eq. (X.3) then leads to:

$$T^{\tilde{M}} = QR^{\tilde{M}}Q^{-1}. \tag{X.4}$$

As a result, the field in the superstrate is linked to the field in the substrate by:

$$V^{(M)} = T^{\tilde{M}}V^{(1)},$$

(note that $\tilde{M} = M - 2$), which leads to:

$$\tilde{V}^{(M)} = R^{\tilde{M}}\tilde{V}^{(1)}, \tag{X.5}$$

where $\tilde{V}^{(j)} = Q^{-1}V^{(j)}$. Recalling that ρ_m are the eigenvalues of T, we define constants Γ_m given by

$$\exp(i\Gamma_m) = \rho_m \quad \text{or} \quad \Gamma_m = -i\ln\rho_m. \tag{X.6}$$

Eq. (X.5) leads to:

$$\tilde{V}_m^{(M)} = \rho_m^{\tilde{M}}\tilde{V}_m^{(1)} = \exp(i\Gamma_m\tilde{M})\tilde{V}_m^{(1)}. \tag{X.7}$$

In view of studying the transmission of the 2-D photonic crystal, it is useful to let M tend toward infinity and to denote by Γ_m'' the imaginary part of Γ_m. Each eigenvalue with $\Gamma_m'' > 0$ gives a vector $\tilde{V}_m^{(M)}$, which corresponds to a field $V^{(M)}$ that is attenuated when $\tilde{M}$, and thus y, is increased. For a system infinitely thick in y (an unbounded 2-D photonic crystal), the corresponding field vector $\tilde{V}_m^{(\infty)}$ will be null, as the initial vector $\tilde{V}_m^{(1)}$ is bounded; thus, $V^{(\infty)} = Q\tilde{V}^{(\infty)}$ is null, too. When $\Gamma_m'' < 0$, the field is attenuated when y is decreased. Only solutions with $\Gamma_m'' = 0$ can propagate without decay. If no eigenvalue of T with $\Gamma_m'' = 0$ exists, all field components propagating downward through a sufficiently high number $\tilde{M}$ of crystal layer periods, will be almost null at the lower interface, due to the decay factor $\exp\left(-\Gamma''_m \tilde{M}\right) \approx 0$. Thus transmitted efficiencies of such a system will tend to zero, whatever the number of propagating diffracted orders in the superstrate may be. As a consequence, if all materials are lossless, the sum of reflected efficiencies will tend toward unity. The spectral (or angular) regions without transmission are called *forbidden bands* or *band gaps* in analogy with solid-state physics. Their complementary regions are called *propagation bands*. An important remark is due: Although light can pass without losses inside the propagation band, this does not mean that the transmission will always be 100%. Due to the reflection phenomena at the outermost interfaces, there is an oscillating behavior of transmission and reflection inside the propagation band, in the same manner as in the light reflection by a single dielectric layer or a Fabry-Pérot resonator.

The conclusion of this section is that the propagation band is the interval of wavelength λ in which at least one of the eigenvalues Γ_m has a null imaginary part. In order to find it, we need to compute as a function of λ the minimum with respect to m of the absolute values of the imaginary part of all eigenvalues Γ_m, denoted $\min_m |\mathrm{Im}(\Gamma_m)|$.

X.3.2 S-matrix approach and the photonic band gaps

The determination of Γ_m requires the knowledge of the transmission matrix T in Eq. (X.3) of each period in the vertical (y) direction. This can impose numerical problems due to the growing exponentially terms, as described in detail in Chapter II. To overcome these problems, it is necessary to use the S-matrix propagation algorithm, Chapter III. However, the reconstruction of the transmission matrix T from the S-matrix of the entire crystal layer leads to the same numerical difficulties as when the T-matrix is calculated directly by integration without using the S-matrix propagation algorithm. This fact has a simple explanation. By comparing the definitions of T and S matrices, Eqs. (III.14), (III.15) and (III..48), it is easy to find the expression for the block components of matrix T as a function of blocks of S:

$$\begin{aligned}
T_{11} &= S_{22}^{-1} \\
T_{12} &= -S_{22}^{-1} S_{21} \\
T_{21} &= S_{12} S_{22}^{-1} \\
T_{22} &= S_{11} - S_{12} S_{22}^{-1} S_{21}
\end{aligned} \tag{X.8}$$

In these expressions, the matrices S_{11} and S_{22}, which describe the transmission from the substrate to the cladding and in the opposite direction can become almost singular, due to the exponentially decaying terms for the field components which are evanescent inside the grating region. When inverted, S_{22} can (and it surely does) cause numerical problems in the calculations of matrix T and thus of Γ_m. To avoid this, it is necessary to find a possibility to calculate Γ_m without explicit determination of matrix T [X.8]. To this aim, let us rewrite Eq. (X.3) as:

$$TQ = QR \quad \Leftrightarrow \quad (T - \mathbb{1})Q = Q(R - \mathbb{1}).$$

This equation shows that $\rho_m - 1$ are eigenvalues of $T - \mathbb{1}$. Thus $1/(\rho_m - 1)$ are eigenvalues of $(T - \mathbb{1})^{-1}$:

$$(T - \mathbb{1})^{-1} Q = Q(R - \mathbb{1})^{-1}. \tag{X.9}$$

Let us define $\hat{\rho}_m$ as eigenvalues of $(T - \mathbb{1})^{-1}$:

$$\hat{\rho}_m = (\rho_m - 1)^{-1} \equiv [\exp(i\Gamma_m) - 1]^{-1}. \tag{X.10}$$

Thus, the large eigenvalues of T, which cause numerical problems, will result in eigenvalues of $(T - \mathbb{1})^{-1}$ tending to zero. Moreover, the eigenvalues of T which correspond to evanescent (decreasing) field components, and tend to zero, will produce no diverging eigenvalues of $(T - \mathbb{1})^{-1}$. This is the first advantage of working with $(T - \mathbb{1})^{-1}$ instead of T^{-1}. The other advantage is that the new matrix $(T - \mathbb{1})^{-1}$ presents no numerical difficulties to be calculated from the S-matrix. And indeed, using Eqs. (X.8), one can obtain $(T - \mathbb{1})^{-1}$ as a function of the blocks of matrix S:

$$(T - \mathbb{1})^{-1} = \left(\begin{array}{c|c} \mathcal{Z}\left(S_{11}S_{21}^{-1}S_{22} - S_{12} - S_{21}^{-1}S_{22}\right) & \mathcal{Z} \\ \hline -S_{21}^{-1}(\mathbb{1} - S_{22})\mathcal{Z}S_{12}(\mathbb{1} - S_{22})^{-1} & S_{21}^{-1}(\mathbb{1} - S_{22})\mathcal{Z} \end{array} \right) \tag{X.11}$$

with

$$\mathcal{Z} = \left(S_{11}S_{21}^{-1} - S_{21}^{-1} - S_{11}S_{21}^{-1}S_{22} + S_{12} + S_{21}^{-1}S_{22}\right)^{-1}. \tag{X.12}$$

The block S_{21} contains the reflection coefficients of waves at the same ordinate $y = 0$ and thus its inversion poses no numerical problems. The same is valid for

$\mathbb{1} - S_{22}$, because the quasisingularity of S_{22} is neutralized by the unit matrix $\mathbb{1}$. Once the matrix $(T - \mathbb{1})^{-1}$ is calculated, Eq. (X.10) serves to obtain Γ_m.

X.3.3 Examples of propagation bands and band gaps

Let us assume that Fig. X.8 presents a square array ($d_x = d_y = 1\,\mu m$) of cylindrical dielectric rods with refractive index $\nu_a = 2.9833$ hanging in air ($\nu_b = \nu_2 = \nu_1 = 1$) with rod radius r equal to $0.2\,\mu m$. The device is lighted under normal incidence by a TE or TM plane polarized wave. Fig. X.9 shows the obtained spectral dependence of $\min_m |\text{Im}(\Gamma_m)|$. As can be seen, a propagation band in both TE and TM polarization exists for $\lambda \in [1.75, 2.18\,\mu m]$, while the TE and TM band gaps partially overlap in the $[2.25, 2.32\,\mu m]$ wavelength range. In this range, the condition $\lambda > 2d_x$ holds, so that only the zero diffracted orders propagate, in reflection and transmission.

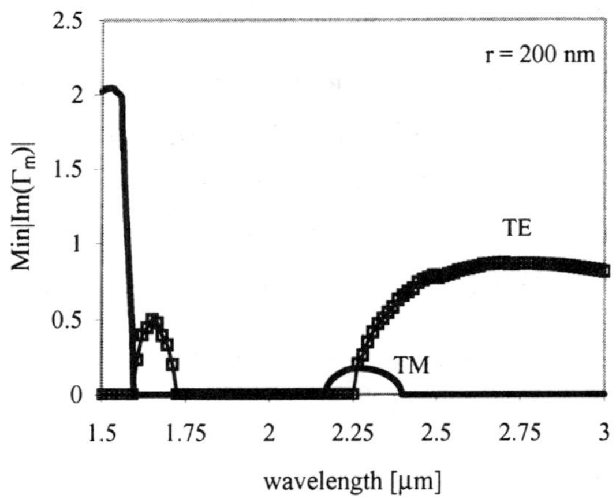

Fig. X.9. Example of spectral dependence of $\min_m |\text{Im}(\Gamma_m)|$ (after [X.7]).

X.3.4 Applications for light guiding

In view of the square symmetry of the structure in Fig. X.8, if it exhibits some properties for θ_2 close to normal incidence, these properties will be the same for θ_2 close to 90, 180 and 270°. In particular, if the band gap extends from 0 to 45°, any angle of incidence will fall in the band gap. Structures having higher symmetry could even be better suited for this purpose. If the cylinders are situated in a hexagonal lattice (Fig. X.10), the requirement for the angular width of the band gap reduces to extend from 0 to 30°.

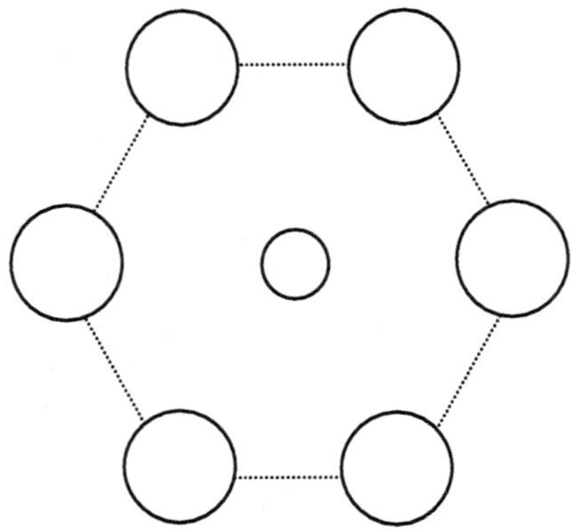

Fig. X.10. Cross-section of the unit cell of a photonic crystal with a hexagonal structure made of cylinders having two different diameters.

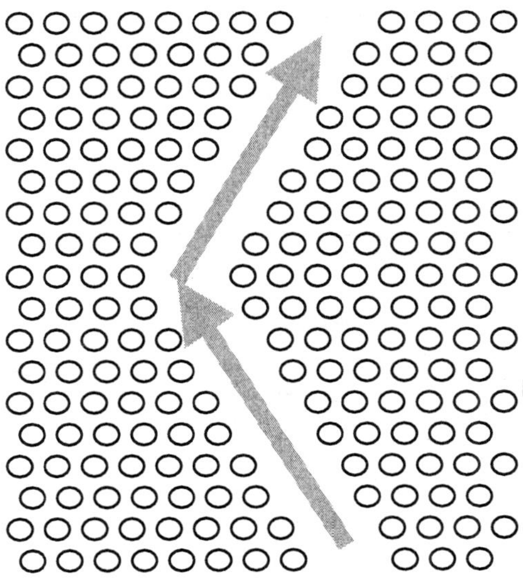

Fig. X.11. Light guiding in a cavity formed by photonic crystal walls consisting of cylindrical objects.

The property of totally reflecting the incident light whatever the direction can be of great importance for light guiding and manipulation. Light confinement and guiding in a single dimension is ensured by using planar waveguides. Channel waveguides and optical fibers confine light in two dimensions, but they suffer of two important limitations: dispersion and bending losses. High bending angles damage the guiding properties and lead to radiation losses. A waveguide constructed with photonic crystal walls can ensure bending, Fig. X.11, without losses even at 90°, as predicted numerically [X.9].

Of course, since the guiding effect is due to the total reflection on the cavity walls, the phase-matching between the waves reflected on opposite walls will lead to formation of a discrete modal spectrum, which will be characterized by some dispersion properties, as is the case for the classical planar or channel dielectric waveguides. Also remains the question about the manufacturing and the price of such photonic waveguides.

Applying this concept inside an optical fiber can reduce the dispersion. Holes are made inside the glass bulk before the fiber is drawn. These holes are preserved during the fiber formation, and their diameter is reduced proportionally with the reduction of the fiber diameter, the result is shown schematically in Fig. X.12. A numerical optimization of the positions and dimensions of the holes can ensure important reduction of dispersion and can maintain the single-mode regime over a large wavelength interval [X.10].

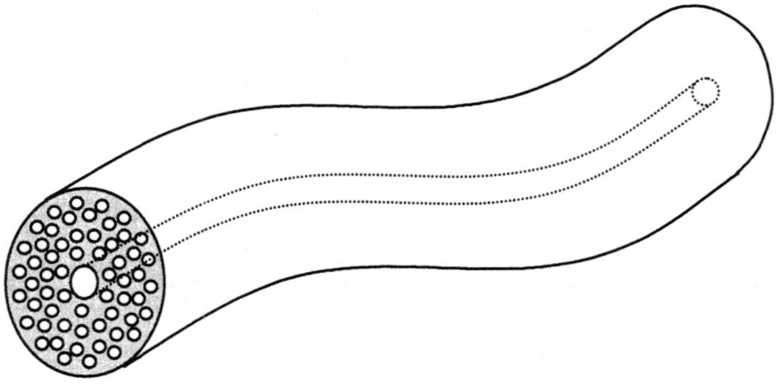

Fig. X.12. Schematic representation of a portion of optical fiber – photonic crystal hybrid structure. Small cylindrical holes run along the fiber length. A central hole (shown with a dashed line inside the fiber) serves as an energy propagator, which ensured low dispersion, low absorption losses and high damage threshold.

X.3.5 Negative refraction

Existence of a band gap with sharp boundaries imposes special properties on the dispersion properties of the matter. In particular, an artificial medium like the photonic crystal can exhibit exotic optical characteristics at wavelengths lying close to the boundaries of the band gap. The study of these properties goes beyond the frame of this book, but the following example is worth mentioning. At specific wavelengths two- and three-dimensional photonic crystals can act as homogeneous media having optical index smaller than unity, or even negative one. Rigorous numerical study [X.11] has demonstrated this property by investigating the electromagnetic field distribution when a limited incident beams falls on a photonic crystal slab.

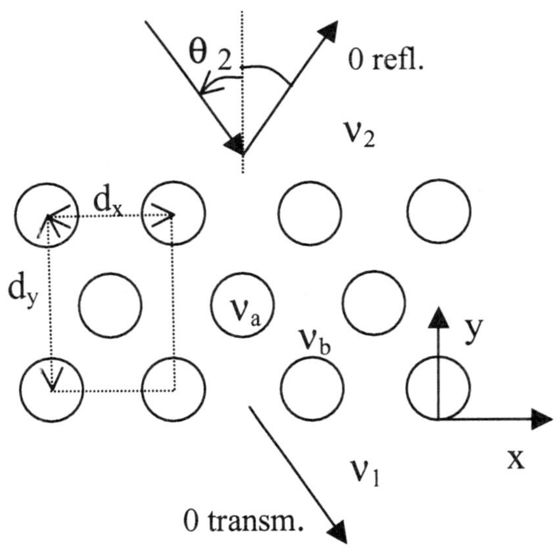

Fig. X.13. Schematic representation of one-and-a-half-layers of a 2-D photonic crystal made of circular rods with an equilateral triangular unit cell. $d_y = d_x\sqrt{3}$.

The structure has the geometry presented in Fig. X.8 in which the slab is cut along the dotted lines. The elementary crystal cell represents an equilateral triangle constructed of circular cylinders, Fig. X.13. For a specific set of parameters, $d_x = 1.27$ μm, $\nu_1 = \nu_2 = \nu_b = 1$, $\nu_a = 3$ and a number of vertical cells M = 7, Fig. X.14 presents a map of the modulus of the total electromagnetic field, when a TE polarized incident beam with a gaussian profile, with waist width of 10 μm and λ = 2.545 μm, falls on the structure under a principal direction $\theta_2 = 40°$. As one can observe, the energy flow inside the crystal has an opposite direction when compared with the incident, the reflected, and the transmitted beam, which can be interpreted as if the photonic crystal had a negative refractive index, at least concerning the Snell's law.

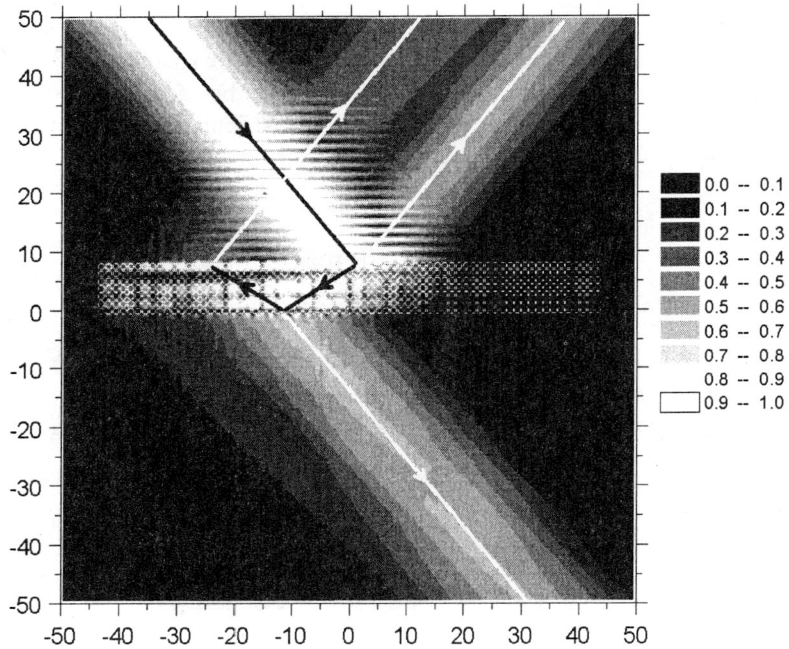

Fig. X.14. Negative refraction by a slab of 2-D photonic crystal (after [X.12])

X.3.6 Almost perfect blazing in natural light

In order to obtain non-specular propagating orders, we can choose the same crystal as in Fig. X.13, which presents a cut at 45^0 of the slab, as shown by the dotted lines in Fig. X.8. If the incidence in Fig. X.13 is at 45° and the period d_x is $1.414\ \mu m$, the -1st order can propagate if $\lambda < 2.414\ \mu m$. The rod radius is $0.35\ \mu m$ and the refractive indices are the same as in Fig. X.9 [X.7]. Fig. X.15 shows that a greater overlap over the TE and TM forbidden bands occurs, namely for $\lambda \in [1.97, 2.33\ \mu m]$.

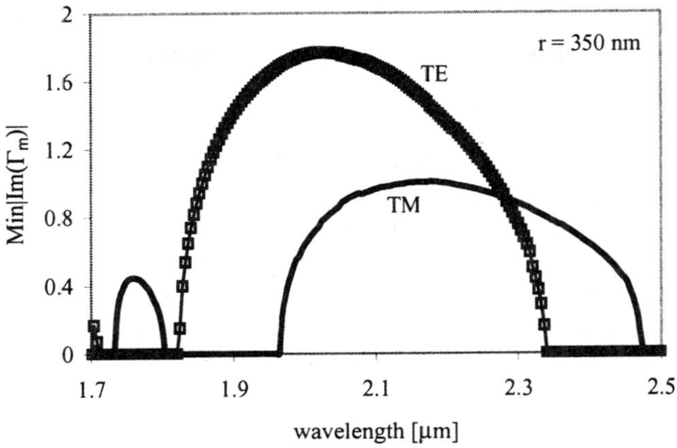

Fig. X.15. Forbidden bands for the crystal with $d = 1.414\,\mu m$, $r = 0.35\,\mu m$ and $\theta_2 = 45^o$ (after [X.7]).

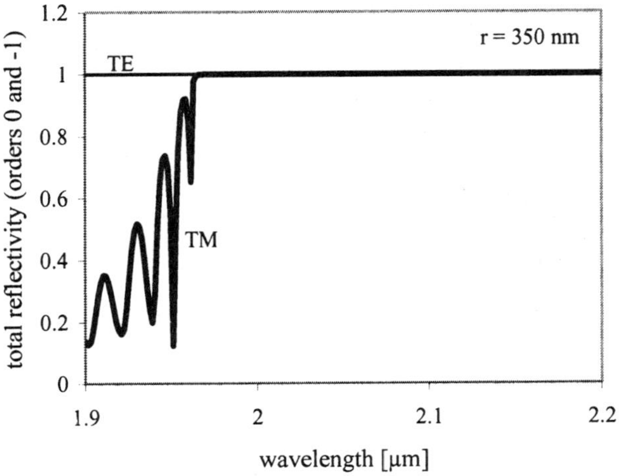

Fig. X.16. Spectral dependence of the total reflected energy (zeroth and minus first diffracted order) for a crystal made of 21 rod rows (after [X.7]).

For the choice of parameters described in Fig. X.15, Fig. X.16 shows the total energy as a function of wavelength, computed for a photonic crystal made with 21 rod rows (i.e., 10 vertical periods plus one row). As observed within the band gap, almost 99.9% of the incident light is reflected in the -1st and 0th orders. This does not warranty, however, that most energy goes in the dispersive order of interest. Indeed, Fig. X.17 shows that the energy is shared between the two propagating orders, the efficiency curves recalling those found for sinusoidal gratings with infinite conductivity [X.13]. But rod gratings offer a greater number of degrees of freedom than surface-relief ones.

By reducing the rod radius to 0.15 instead of 0.35 μm, the curves in Fig. X.17 are transformed into those in Fig. X.18, which present for both TE and TM polarization a 150 nm wide plateau. Going to 60° incidence increases the plateau width, as shown in Fig. X.19. This almost perfect blazing in unpolarized light with a plateau-like spectral dependence can have a great potentiality in optical demultiplexing. By reducing the period d_x to 0.96167 μm, the rod radius to 0.14 μm and taking for ν_a the refractive index of lithium niobate, it is possible to translate the center of the plateau observed in Fig. X.18 and Fig. X.19 from 2.15 μm to 1.5 μm, [X.7], the wavelength used in optical communications.

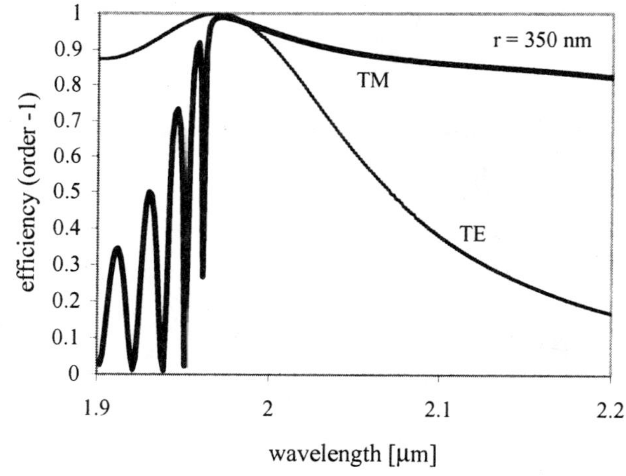

Fig. X.17. Spectral dependence of the -1st reflected order for the device in Fig. X.16 (after [X.7]).

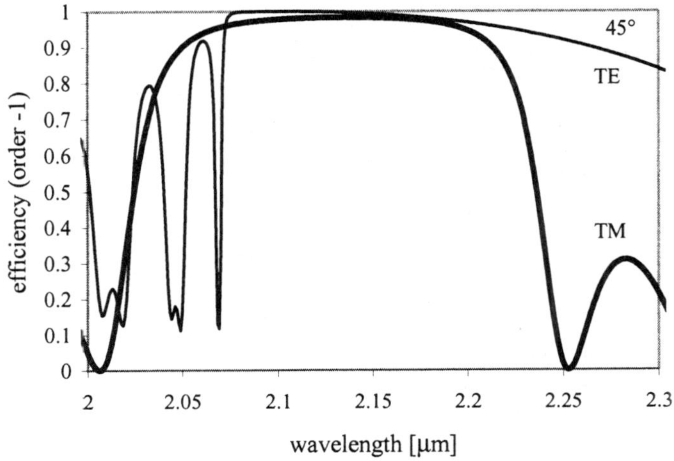

Fig. X.18. Spectral dependence of the -1st reflected order efficiency for a crystal made with 21 rows of 0.15 μm radius rods. Incidence is 45° (after [X.7]).

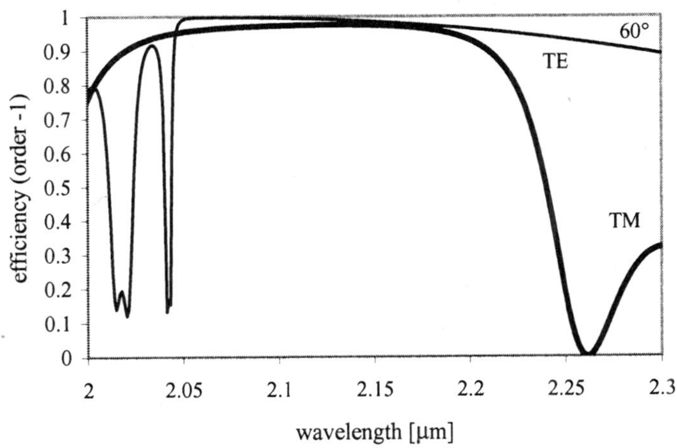

Fig. X.19. Same as in Fig. X.18, but the angle of incidence is 60° (after [X.7]).

X.4 Theory of three-dimensional photonic crystals

A 3-D photonic crystal is an infinite medium in which $\varepsilon(x, y, z)$ is periodic with period d_x, d_y and d_z along the x, y and z-axis, respectively, although the three directions of periodicity are not necessarily orthogonal. It is obtained by repeating a given unit cell in each direction. Since the 3-D periodicity has been investigated for a long time in solid-state physics for determining the electronic properties of semiconductor crystals, many authors [X.14-X.18] dealing with the photonic crystal problem compute the Bloch wave dispersion relation using an eigenvalue method which turns out to be the classical RCW method of Chapter VI. Unfortunately, as it has been shown [X.19], this method suffers from a poor convergence and its poor convergence is clearly linked with the discontinuity of permittivity. The 3-D problem does not allow for separation of polarization into TE and TM case, so that the poor convergence is manifested whatever the polarization of the incident light may be. Fast Fourier factorization allows writing the equations in a truncated Fourier space in such a way that the convergence problem disappears, but leads to a nonlinear eigenvalue problem.

X.4.1 Nonlinear eigenvalue approach

The periodicity of ε results in the pseudoperiodicity of the field which can be searched in the form of a triple series of harmonics $\vec{E}_{m,n,p}$:

$$\vec{E}(\vec{r}) = \sum_{m,n,p=-\infty}^{+\infty} \vec{E}_{m,n,p} = \sum_{m,n,p=-\infty}^{+\infty} \vec{A}_{m,n,p} \exp\left(i\vec{k}_{m,n,p} \cdot \vec{r}\right) \qquad (X.13)$$

where $\vec{A}_{m,n,p}$ are constants vectors and

$$\vec{k}_{m,n,p} = \vec{k}_0 + m\frac{2\pi}{d_x}\vec{x} + n\frac{2\pi}{d_y}\vec{y} + p\frac{2\pi}{d_z}\vec{z} \qquad (X.14)$$

is the direction of propagation of the (m,n,p) field component, $\vec{k}_0$ is the basic vector, chosen to have the smallest modulus among all $\vec{k}_{m,n,p}$ (this corresponds to the first Brillouin zone in solid-state physics) and $\vec{x}$, $\vec{y}$ and $\vec{z}$ are the unit vectors of the Cartesian axes. Similar expansion holds for $\vec{H}(\vec{r})$. The general propagation Eq. (V.22) holds, of course. The difference with 1-D and 2-D case is that, since the crystal is infinite, Eq. (X.13) is valid everywhere, so that $\frac{d}{dy} \to ik_{m,n,p,y}$. Moreover, $\varepsilon(x, y, z)$ is represented by a *triple* Fourier series with coefficients $\varepsilon_{m,n,p}$ which are y-independent. As soon as the shape of the basic object in the elementary cell is given by the equation $f(x, y, z) = 0$, the components of vector $\vec{N}$ can be calculated through Eq. (V.23). If the material is isotropic, then matrix Q_ε is given by Eq. (IX.11) in which *all elements are y-independent*. As a result, matrix M is y-independent too, and Eq. (V.22) reduces to an eigenvalue problem. These eigenvalues correspond to specific discrete links between $\vec{k}_0$ and λ (or ω), which determine the dispersive relations of the 3-D crystal and the formation of propagation and forbidden bands.

It is worth noting that before reducing Eq. (V.22) to a classical eigenvalue problem, transformations similar to those done for crossed gratings have to be conducted at a higher degree of complexity. If the periods d_x, d_y and d_z are about the same order of magnitude, the triple Fourier series of the fields in Eq. (X.13) will be truncated to the same number 2N+1 with respect to all subscripts m, n and p, i.e., from –N to +N. All the Fourier components of E_x, for example, will be placed inside a block $[E_x]$, with $(2N+1)^3$ elements $E_{x,r}$ where subscript r is derived from m, n and p in a manner, similar to Eq. (IX.13) for a crossed grating:

$$r = 1 + (m + N) + (2N + 1)(n + N) + (2N + 1)^2 (p + N) . \qquad \text{(X.15)}$$

Similar transformations have to be done for the elements $\varepsilon_{m'-m'',n'-n'',p'-p''}$ which appear in the Toeplitz block matrices allowing the calculation of Q_ε, and then of matrix M. It means that we obtain a square matrix $\mathcal{E}$ with elements $\mathcal{E}_{r',r''} = \varepsilon_{m'-m'',n'-n'',p'-p''}$, where subscript r' and r'' are defined, respectively, from (m', n', p') and (m'', n'', p'') through Eq. (X.15). When all these mono-subscript notations are used, the vector $\begin{pmatrix} [E_x] \\ [E_z] \\ [H_x] \\ [H_z] \end{pmatrix}$ is solution of an equation of the form:

$$ik_{r,y}\begin{pmatrix}E_{r,x}\\E_{r,z}\\H_{r,x}\\H_{r,z}\end{pmatrix}=M_{r,u}\begin{pmatrix}E_{u,x}\\E_{u,z}\\H_{u,x}\\H_{u,z}\end{pmatrix}, \qquad (X.16)$$

where the blocks $M_{r,u}$ depend on ω, as shown in Eq. (IX.12). The determination of the dispersion relation implies finding the value(s) of ω corresponding to a given value of the y-component of the initial vector $\vec{k}_0$, from which $\vec{k}_r$ is derived through Eqs. (X.14) and (X.15). Since, in Eqs. (IX.12) ω appears in both denominator and numerator, Eq. (X.16) leads to a nonlinear eigenvalue problem [X.20]. This problem turns out to be more difficult to resolve than the classical eigenvalue problem, at least from the numerical point of view. Moreover, as soon as the medium has 3 directions of periodicity, there is no reason to privilege one of them, namely the y-direction, as done in Eq. (X.16). This is not purely an academic question, as in some cases a solution with $k_{r,y}=0$ could exist, and then Eq. (X.16) is not suitable to be used. This is why 3-D periodicity requires special treatment that takes into account from the very beginning the fact that the y-independence of matrix M is known [X.21]. This results in a classical eigenvalue problem, which has been analyzed numerically with success.

X.4.2 Special approach for a 3-D periodicity

We go back to the time-harmonic Maxwell equations (I.8) and (I.9) and take advantage of the expression of $\vec{E}(\vec{r})$ given by Eq. (X.13) and of similar expression for $\vec{H}$, $\vec{B}$ and $\vec{D}$. For any field harmonic $\vec{E}_{m,n,p}$,

$$\mathrm{curl}\vec{E}_{m,n,p}=i\vec{k}_{m,n,p}\times\vec{E}_{m,n,p}. \qquad (X.17)$$

This equation can be written in a matrix form:

$$\mathrm{curl}\vec{E}_{m,n,p}=K_{m,n,p}\vec{E}_{m,n,p} \qquad (X.18)$$

where matrix $K_{m,n,p}$ is given by:

$$K_{m,n,p} = \begin{pmatrix} 0 & -k_{m,n,p,z} & k_{m,n,p,y} \\ k_{m,n,p,z} & 0 & -k_{m,n,p,x} \\ -k_{m,n,p,y} & k_{m,n,p,x} & 0 \end{pmatrix}. \tag{X.19}$$

Using the monosubscript block column vector notation introduced in Section X.4.1, we then construct a vector $\left[\vec{E}\right]$ made of 3 blocks $\left[E_x\right]$, $\left[E_y\right]$ and $\left[E_z\right]$ having $(2N+1)^3$ elements for each of them, each element defined by subscript r given by Eq. (X.15). Eq. (X.18) reads

$$\text{curl}\left[\vec{E}\right] = K\left[\vec{E}\right], \tag{X.20}$$

where the K matrix is now a matrix made with the nine blocks which appear in Eq. (X.19), each block having $(2N+1)^3$ elements labeled by subscript r.

From the equation $\left[\vec{D}\right] = Q_\varepsilon\left[\vec{E}\right]$ (Eq. (V.12)), where Q_ε is given by Eq. (IX.11), we obtain:

$$\text{curl}\left[\vec{E}\right] = KQ_\varepsilon^{-1}\left[\vec{D}\right]. \tag{X.21}$$

Taking advantage that the y-dependence of the field is known, the equation $\text{div}\vec{D} = 0$ allows eliminating the D_y component, if $k_{m,n,p,y} \neq 0$:

$$D_{m,n,p,y} = -\frac{k_{m,n,p,x}}{k_{m,n,p,y}} D_{m,n,p,x} - \frac{k_{m,n,p,z}}{k_{m,n,p,y}} D_{m,n,p,z}.$$

If, accidentally, one of $k_{n,m,p,y} = 0$, it is necessary to eliminate another component of $\vec{D}$ instead of D_y.

We then introduce a block matrix P with block size 3×2 given by:

$$P = \begin{pmatrix} 1 & 0 \\ -\dfrac{k_x}{k_y} & -\dfrac{k_z}{k_y} \\ 0 & 1 \end{pmatrix} \tag{X.22}$$

such that:

$$\begin{pmatrix} [D_x] \\ [D_y] \\ [D_z] \end{pmatrix} = P \begin{pmatrix} [D_x] \\ [D_z] \end{pmatrix}.$$

Defining a block column vector $\left[\vec{\tilde{D}}\right]$ by:

$$\left[\vec{\tilde{D}}\right] = \begin{pmatrix} [D_x] \\ [D_z] \end{pmatrix},$$

Eq. (X.21) reads:

$$\text{curl}\left[\vec{E}\right] = KQ_\varepsilon^{-1}P\left[\vec{\tilde{D}}\right];$$

thus Maxwell Eq. (I.8) gives:

$$KQ_\varepsilon^{-1}P\left[\vec{\tilde{D}}\right] = i\omega\mu_0\left[\vec{H}\right] \tag{X.23}$$

The second Maxwell equation (I.9) leads to:

$$\text{curl}\left[\vec{H}\right] \equiv K\left[\vec{H}\right] = -i\omega\left[\vec{D}\right]. \tag{X.24}$$

However, since we do not need D_y, we define a new matrix $\tilde{K}$, having the size of 2×3 blocks, derived from Eq. (X.19) by suppressing the central row:

$$\tilde{K}_{m,n,p} = \begin{pmatrix} 0 & -k_{m,n,p,z} & k_{m,n,p,y} \\ -k_{m,n,p,y} & k_{m,n,p,x} & 0 \end{pmatrix}$$

Eq. (X.24) then results into:

$$\tilde{K}\left[\vec{H}\right] = -i\omega\left[\vec{\tilde{D}}\right]. \tag{X.25}$$

Multiplying both members of Eq. (X.23) by $\tilde{K}$ and using Eq. (X.25) leads to:

$$\tilde{K}KQ_\varepsilon^{-1}P\left[\vec{\tilde{D}}\right] = \omega^2\mu_0\left[\vec{\tilde{D}}\right].$$

By introducing a matrix $Q = Q_\varepsilon/\varepsilon_0$, we finally obtain:

$$\tilde{K}KQ^{-1}P\left[\vec{\tilde{D}}\right]=\frac{\omega^2}{c^2}\left[\vec{\tilde{D}}\right] \tag{X.26}$$

where c is the speed of light in vacuum.

Eq. (X.26) has the form of a classical eigenvalue-eigenvector equation. Matrix $\tilde{K}KQ^{-1}P$ is known and, after truncations of the triple Fourier series in Eq. (X.13), it becomes a square matrix with $2\times(2N+1)^3$ dimension only. The search for the eigenvalues $\rho=\frac{\omega^2}{c^2}$ can be done using standard subroutines.

The main interest of the special approach is that, thanks to the elimination of D_y, it provides a smaller matrix than if we keep all the components of $\vec{D}$. Indeed, Eqs. (X.23) can be rewritten as:

$$KQ_\varepsilon^{-1}\left[\vec{D}\right]=i\omega\mu_0\left[\vec{H}\right],$$

and Eq. (X.24) leads to:

$$KQ_\varepsilon^{-1}\left[\vec{D}\right]=\omega^2\mu_0K^{-1}\left[\vec{D}\right]$$

or

$$K^2Q^{-1}\left[\vec{D}\right]=\frac{\omega^2}{c^2}\left[\vec{D}\right]. \tag{X.27}$$

This is an equivalent eigenvalue problem to be resolved. However, matrix K^2Q^{-1} is a square matrix with $3\times(2N+1)^2$ dimension. Moreover, since the components of $\vec{D}$ are not independent (due to $\mathrm{div}\vec{D}=0$), the set (X.27) has one third of the equations and unknowns which are not independent. As a result matrix K^2Q^{-1} is singular. One third of its eigenvalues is null. Working with Eq. (X.26) instead of Eq. (X.27) avoids searching null solutions and saves memory storage in computers. Anyway, the results found via both methods are in perfect agreement, the subroutine used for searching eigenvalues working perfectly, even when some eigenvalues are null.

X.4.3 Numerical example – a cubic photonic crystal

Let us apply the approach to a 3-D system with an elementary cubic cell ($d_x=d_y=d_z=d$) containing a single cube with size a centered in the cell center O. The discontinuity surfaces of ε are the cube walls. Vector $\vec{N}$, normal to the cube walls, is discontinuous on the cube edges and this cannot be avoided. The simplest prolongation of $\vec{N}$ in the entire elementary cell is to divide the cell into six pyramidal regions, which have the supports on the cell walls and their tops are in O. Inside each pyramid, $\vec{N}$ is taken perpendicular to the corresponding cube wall. A simple rule gives the components of $\vec{N}$ for each x, y and z, namely:

$$\begin{aligned} &N_x = 1, N_y = 0, N_z = 0 \quad \text{for} \quad |x| < |y|, |z| \\ &N_x = 0, N_y = 1, N_z = 0 \quad \text{for} \quad |y| < |x|, |z| . \\ &N_x = 0, N_y = 0, N_z = 1 \quad \text{for} \quad |z| < |x|, |y| \end{aligned} \tag{X.28}$$

This choice leads to a discontinuity of $\vec{N}$ on the pyramids' sidewalls. As a result, the regions where *both* ε and N are discontinuous are the cube *edges* only.

Two approaches have been proposed to numerically study the 3-D structure, using different representations of matrix Q in Eq. (X.27). They are called, respectively, E and H methods, and represent matrix Q in the form:

$$Q = \frac{1}{\varepsilon_0} \begin{pmatrix} [\![\varepsilon]\!] & 0 & 0 \\ 0 & [\![\varepsilon]\!] & 0 \\ 0 & 0 & [\![\varepsilon]\!] \end{pmatrix} \tag{X.29}$$

for the E method, and

$$Q = \frac{1}{\varepsilon_0} \begin{pmatrix} \left[\!\left[\frac{1}{\varepsilon}\right]\!\right]^{-1} & 0 & 0 \\ 0 & \left[\!\left[\frac{1}{\varepsilon}\right]\!\right]^{-1} & 0 \\ 0 & 0 & \left[\!\left[\frac{1}{\varepsilon}\right]\!\right]^{-1} \end{pmatrix} \tag{X.30}$$

for the H method [X.19], whereas the FFF approach requires using the more general form, given in Eq. (IX.11).

Fig. X.20 represents the normalized frequency of the first and second band at the Y point of the dispersion diagram, i.e., for $\vec{k}_0 = (0, \pi/d, 0)$, for a specific value of a/d and ε [X.21], comparing the convergence of the three approaches. The horizontal lines show the reference values obtained by using a completely different method based on a transfer matrix approach. The special approach developed in the previous section and based on the FFF method produces two different eigenvalues rapidly converging toward a common value, while the E and the H methods converge to the same value, but much more slowly.

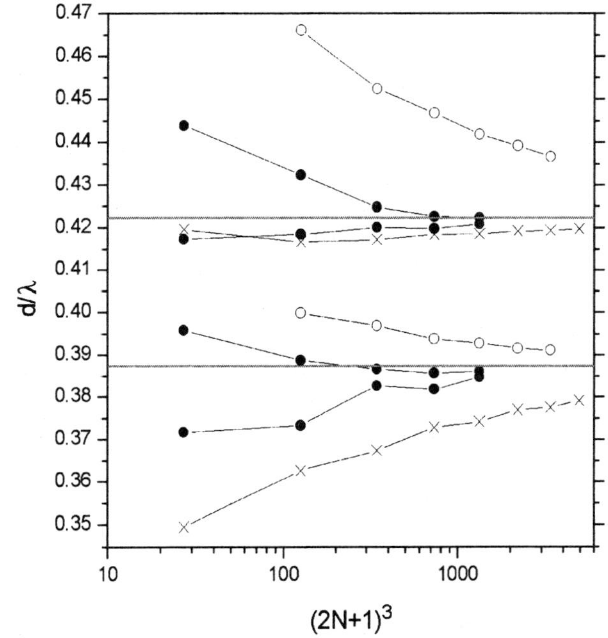

Fig. X.20. Normalized frequency of the first and second band at the Y point as a function of the total number of plane waves. The crystal is a cubic lattice of cubes with permittivity 10 lying in vacuum. The filling ratio is 0.125 (a/d = 0.5). The horizontal lines indicate the values obtained with a different method based on the transfer matrix approach. Filled circles, FFF method; empty circles, E method; and crosses, H method (after [X.21]).

X.5 Photonic crystals in biology: colors without pigments

Surveys of the natural world show an impressive array of bright colors in both the mineral and biological world. Pigments produce most colors, but in many cases they result from optical effects in structures [X.22]. Such structures may be found in minerals and give magnificent effects, as it occurs in opals [X.23], tigereyes, star sapphires and other gems. They are also found in the living world, e.g., on birds, insects [X.24], and marine animals like arthropods [X.25] and sea mice [X.26]. Among them, colors of butterflies have attracted a great amount of interest recently [X.27-X.31]. High-resolution microscopy has shown complex periodic structures [X.27-X.29] common to all colored wings, scales and cuticles, and the question is whether or not the diffraction of light on these structures can explain their iridescent colors.

Up to now, in view of the electromagnetic analysis, the theoretical approaches [X.27, X.30] have approximated the structure using a stack of thin films. Such a simple approach does not take into account the detailed microstructure of the colored wrings, as it can be seen in Fig. 6 of [X.28], for example. The electromagnetic theory of multilayered lamellar gratings can provide a much more accurate analysis [X.31]. The authors represent the complicated structure of the wings of Morpho rhetenor butterflies, with a tree-like geometry, by the model illustrated in Fig. X.21. In this model, the wing structure is a stack of slices and each slice is a lamellar grating, made of rectangular rods with refractive index ν. According to measurement reported in [X.27], the optical index ν of the tree-like structure is equal to $1.56 + i0.06$ in the whole visible region. The width of the rods varies with the number of the slice. The modal theory of gratings [X.32, X.33] gives the field reflected and transmitted by the grating for any optogeometrical condition on the incident field with arbitrary wavevector orientation, i.e., in conical diffraction mounting.

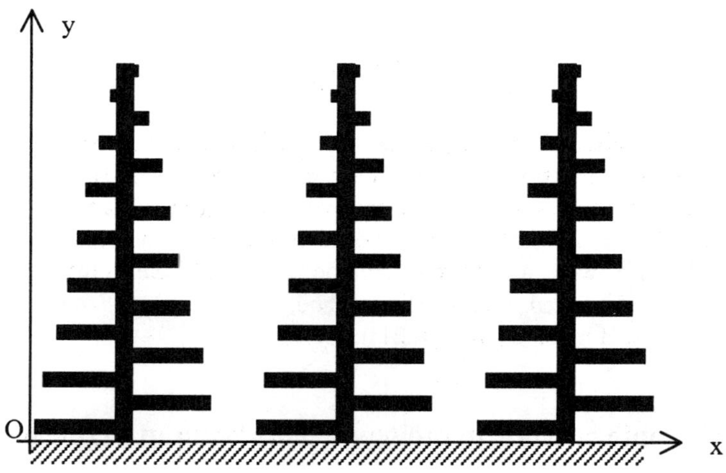

Fig. X.21. Schematic representation of a tree-like structure of the wing scale of Morpho rhetenor butterflies (after [X.31]).

However, the human eye does not recognize colors by their simple spectral position when the illumination has a complex spectral structure. An approach was used in [X.31] for converting reflection coefficients of the structure into colors by techniques commonly used in colors science and graphic imaging. This approach uses the standard normalized illuminant DGS of the Commission Internationale de l'Eclairage (CIE), which provides the energy distribution $e(\lambda)$ of the sky daylight. From this function and the reflectivity $R(\lambda)$, three parameters called CIE tristimulus values were computed. These parameters take into account the variation of sensitivity of the human eye with wavelength. These three parameters were transformed into three new ones (R, G, B), more convenient for the imaging process, which take into account the characteristics of monitors used in computer graphics. Their normalized values ($R, G, B \in [0,1]$) are then entered into a file called a *portable pix map* (PPM) file which gives a value for each pixel. These PPM files can easily be displayed on a computer monitor screen using classical image editors, and printed. Finally, the approach enables replacing a complicated set of data (e.g., the spectrum $R(\lambda)$) by a color point in a color map. A two-dimensional color map allows for two free parameters, e.g., the two angles determining the direction of diffraction. Such a representation gives a global insight on the reflecting properties of an object.

Applied to the model in Fig.X.21, the electromagnetic theory of stacks of lamellar gratings has provided the total reflected energy as function of wavelength, for both TE and TM polarization and normal incidence. The reflectivity exhibits a pronounced peak in the blue region of the spectrum, which is consistent with the experimental data reported in Fig. 3 of [X.27]. The full approach to present the color appearance of the wing scale, as presented in the previous paragraph, was then implemented. Results gave the expected blue appearance [X.31].

Immersing the scale inside an organic liquid like isopropyl alcohol produces drastic modifications in the reflected colors [X.29], due to a reduction of the refractive index contrast. The numerical simulations show that the global appearance of the scale becomes green, as reported in [X.30]. Other results presented in [X.31] also show that the electric field penetrates deeper inside the structured layers at longer wavelengths. As a result, longer wavelengths are much more sensitive to the bottom region of the scale, which contains absorbing pigments or defaults. This explains why the measured absorption of the wing is larger than the one predicted numerically, because the theory does not take into account these pigments.

In conclusion, the electromagnetic theory of diffraction by a stack of gratings is able to predict and explain some spectacular effects observed in the living world.

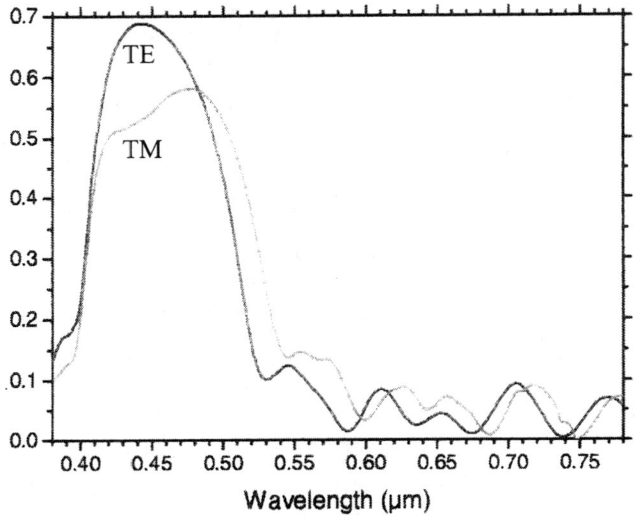

Fig. X.22. Entire reflection of the modeled wing (after [X.31]).

References

X.1. G. A. Mourou, C. P. J. Barty, and M. D. Perry, "Ultrahigh-intensity lasers: physics of the extreme on a tabletop," Physics Today, no.1, 51, 22-28, 1998

X.2. D. Strickland and G. Mourou, "Compression of amplified chirped optical pulses," Opt. Commun. 56, 219-221, 1985

X.3. O. E. Martinez, "Design of high-power ultrashort pulse amplifiers by expansion and recompression," IEEE J. Quantum Electronics 23, 1385-1387, 1987

X.4a. M. D. Perry, R. D. Boyd, J. A. Britten, D. Decker, B. W. Shore, C. Shannon, and E. Shults, "High efficiency multilayer dielectric diffraction gratings," Opt. Lett. 20, 940-942, 1995

X.4b. M. D. Perry, R. D. Boyd, J. A. Britten, D. Decker, B. W. Shore, C. Shannon, E. Shults, and L. Li"High efficiency multilayer dielectric diffraction gratings: erratum," Opt. Lett. 20, 1513, 1995

X.5. B. W. Shore, B. C. Stuart, M. D. Feit, A. M. Rubenchik, and M. D. Perry, "Laser induced damage in multilayer dielectric gratings due to ultrashort laser pulses," SPIE 2633, 714-723, 1977

X.6a. M. Nevière, J. Flamand, and F. Bonnemason, "TMR – Access to large scale facilities," Report on European Community RTD contract: "Gratings for Ultrabright Lasers," ERBFMGECT#980096

X.6b. B. Wattellier, "Amélioration des performances des chaînes lasers solides utilisant l'amplification à dérive de fréquence : nouveaux réseaux de diffraction à haute tenue au flux et mise en forme programmable de faisceaux lasers par modulation de la phase spatiale," PhD Thesis, Ecole Polytechnique, Paris, 2001

X.7. E. Popov, B. Bozhkov, and M. Nevière, "Almost perfect blazing by photonic crystal rod gratings," Appl. Opt. 40, 2417-2422, 2001

X.8. B. Gralak, "Etude théorique et numérique des propriétés des structures à bandes interdites photoniques," PhD Thesis, Université d'Aix-Marseille III, France, 2001

X.9. G. Tayeb and D. Maystre, "Rigorous theoretical study of finite size two-dimensional photonic crystals doped by microcavities," J. Opt. Soc. Am. A 14, 3323-3332, 1997

X.10a. H. Kubota, K. Suzuki, S. Kawanishi, M. Nakazawa, M. Tanaka, and M. Fijita, "Low-loss, 2km-long photonic crystal fiber with zero GVD in near IR suitable for picosecond pulse propagation at the 800 nm band," in *OSA Trends in Optics and Photonics*, v. 56, paper CPD-3 (Optical Society of America, Washington DC, 2001)

X.10b. D. Mogilevtsev, T. A. Birks, and P. St. J. Russel, "Group-velocity dispersion in photonic crystal fibers," Opt. Lett. 23, 1661-1664, 1998

X.10c. T. Birks, J. C. Knight, and P. St. J. Russell, "Endlessly single-mode photonic crystal fiber," Opt. Lett. 22, 961-963, 1997

X.11. S. Enoch, G. Tayeb, and D. Maystre, "Numerical evidence of ultra-refractive optics in photonic crystals," Opt. Commun. 161, 171-176, 1999

X.12. B. Gralak, S. Enoch, and G. Tayeb, "Anomalous refractive properties of photonic crystals," J. Opt. Soc. Am. A 17, 1012-1020, 2000

X.13. D. Maystre, M. Nevière, and R. Petit, "Experimental Verifications and Applications of the Theory," in *Electromagnetic Theory of Gratings*, R. Petit ed. (Springer-Verlag, Berlin, 1980), p.185

X.14. S. John, "Strong localization of photons in certain disordered dielectric superlattices," Phys. Rev. Lett. 58, 2486-2489, 1987

X.15. K. Ho, C. Chan, and C. Soukoulis, "Existence of a photonic gap in periodic dielectric structures," Phys. Rev. Lett. 65, 3152-3155, 1990

X.16. P. Villeneuve, S. Fan, and J. Joannopoulos, "Microcavities in photonic crystals," in *Microcavities and Photonic Bandgaps*, eds. J. Rarity and C. Weisbuch, (Kluwer, The Academic, Netherlands, 1996), pp. 133-151

X.17. Z. Zhang and S. Satpathy, "Electromagnetic wave propagation in periodic structures: Bloch wave solution of Maxwell's equations," Phys. Rev. Lett. 65, 2650-2653,1990

X.18. R. Meade, K. Brommer, A. Rappe, and J. Joannopoulos, "Electromagnetic Bloch waves at the surface of a photonic crystal," Phys. Rev. B 44, 10961-10964, 1991

X.19. H. S. Sözüer, J. W. Haus, and R. Inguva, "Photonic bands: convergence problems with the plane-wave method," Phys. Rev. B 45, 13962-13972, 1992

X.20. W. H. Press, S. A. Teukolsky, W. T. Vetterling, and B. P. Flannery, *Numerical Recipes in Fortran: the Art of Scientific Computing*, Second Ed. (Cambridge University Press, Cambridge, New York, Victoria, 1992), p. 455

X.21. S. Enoch, E. Popov, and M. Nevière, "3-D photonic crystals dispersion relation: improved convergence using fast Fourier factorization (FFF) method," Proc. SPIE 4438, 183-190, 2001

X.22a. Lord Rayleigh, "On the optical character of some brillant animal colours," Phil. Mag. 6th ser., 37, 98-111, 1919

X.22b. H. Onslow, "On a periodic structure in many insect scales and the cause of their iridescent colours," Phil. Trans. B, 211, 1-74, 1921

X.22c. Lord Rayleigh (junior), "Studies of iridescent colour and the structure producing it. IV. Iridescent beetles," Proc. Roy. Soc. A 103, 233-9, 1923

X.23. J. V. Sanders, "Diffraction of ligth by opals," Acta Crst. A24, 427-434, 1968

X.24a. R. M. Strong, "The metallic colours of feathers from the neck of the domestic pigeon," Biol. Bull. Woods Hole 3, 85-7, 1902

X.24b. A. Michelson, "On the metallic colouring in birds and insects," Phil. Mag. 6th ser., 21, 554-67, 1911

X.24c. W. Biedermann, "Farbe und Zeichnung der Insecten," *Handbuch der Vergleichenden Physiologie*, Teil II, pp. 1657-1994 (Verlag von Gustav Fischer, Jena ,1914)

X.24d. A. G. Mayer, "On the colour and colour pattern of moths and butterflies," Bull. Mus. Comp. Zool. Harv. 30, 169-259, 1897

X.24e. H. Kalmus, "Physiology and Ecology of cuticle colour in insects," Nature 148, 428-431, 1941

X.25. A. C. Neville, *Biology of arthropod cuticle* (McGaw-Hill, New York, 1975)

X.26a. A. R. Parker, R. C. McPhedran, D. R. McKenzie, L. C. Botten, and N. A. Nicorovici, "Photonic engineering Aphrodite iridescence," Nature 409, 36-37, 2001

X.26b. R. C. McPhedran, N. A. Nicorovici, D. R. McKenzie, L. C. Botten, A. R. Parker, and G. W. Rouse, "The sea mouse and the photonic crystal," Aust. J. Chem. 54, 241-244, 2001

X.27. P. Vukusic, J. R. Sambles, C. R. Lawrence, and R. J. Wootton, "Quantified interference and diffraction in single Morpho butterfly scales," Proceedings: Biological Sciences, The Royal Society of London 266, 1403-1411, 1999

X.28. P. Vukusic and J. R. Sambles, "Shedding light on butterfly wings," Proceeding of SPIE 4438, 85-95, 2001

X.29. S. Berthier, *Les couleurs des papillons ou l'impérative beauté* (Springer-Verlag France, 2000)

X.30. H. Tada, S. Mann, I. Miaoulis, and P. Wrong, "Effects of a butterfly scale microstructure on the iridescent colour observed at different angles," Opt. Express 5, 87, 1999

X.31. B. Gralak, G. Tayeb, and S. Enoch, "Morpho butterflies wings colour modeled with lamellar grating theory," Opt. Express 9, 568-578, 2002

X.32. S. E. Sandström, G. Tayeb, and R. Petit, "Lossy multistep lamellar gratings in conical diffraction mountings: an exact eigenfunction solution," J. of Electromagn. Waves and Applications 7, 631-649, 1993

X.33. L. Li, "A modal analysis of lamellar diffraction gratings in conical mountings," J. Modern Optics 40, 553-573, 1993

Chapter XI

X-Ray Gratings

X-ray spectroscopy is a domain of greatest interest. The emission and diffraction of X-rays is a powerful tool to analyze the structure of matter and its composition. In astrophysics, X-rays inform us on the evolution of stars, galaxies and the universe. X-ray analysis is done by recording spectral data, which requires using periodic structures. In the X-ray domain, these structures are crystals and gratings, depending on wavelength. Crystals work in hard X-rays, while gratings are used in X-UV and soft X-ray ranges.

There are two main types of gratings that operate in the X-ray domain, namely bare metallic gratings and multilayer gratings. The latter ones may be produced by two different ways. First, one can evaporate a multilayer on top of a bare metallic or dielectric profile, and obtain a multilayer coated grating. Second, one can etch a grating inside a flat multilayer and obtain a Bragg grating. In some cases, such Bragg gratings are used to focus beams, in addition to diffracting X-rays. To that end, the groove spacing slowly varies across the grating surface and the grating turns into a Bragg-Fresnel linear zone plate.

All kinds of X-ray gratings have a point in common. They operate at extremely low λ/d ratios, typically between 10^{-2} and 10^{-4}. This means that theoretically tens to hundreds of thousands of diffracted orders propagate. Thus for many years, such gratings seemed to be out of reach of the electromagnetic theories with computers available at that time, and it appeared quite natural to analyze them with the scalar theory. We will see that it may lead to erroneous results.

XI.1 Bare metallic gratings

XI.1.1 Scalar diffraction approach

In the asymptotic regime when $\frac{\lambda}{d} << 1$, the phenomenon of diffraction by a metallic triangular grating can be tackled with the scalar theory of diffraction [XI.1]. Such an approach starts from Fresnel-Kirckhoff diffraction formula, uses the Fresnel reflectivity on the large facets of the profile, neglects multiple reflection at edges and assumes Fraunhofer diffraction regime. It then derives an analytical expression of the intensity diffracted in direction $\theta_{2,n}$, as a function of the wavelength λ, the period d, the angle of incidence θ_2, the blaze angle φ and the diffraction angle $\theta_{2,n}$. From this expression, the directions of the diffracted orders are derived and perfectly coincide to the ones given by Fraunhofer equation (I.39). The theory is also able to provide a relation between the groove depth a and the blaze wavelength λ_B required in order to get a maximum -1st order efficiency. Denoting by a_s the optimal value of a predicted by the scalar theory, the result for a triangular profile is:

$$a_s = \frac{\lambda_B}{2\cos\frac{A.D.}{2}}, \tag{XI.1}$$

where A.D. is the angle between incident and -1st order direction. Extended to a sinusoidal profile, the scalar theory gives:

$$a_s = \frac{\lambda_B}{3.4\cos\frac{A.D.}{2}}, \tag{XI.2}$$

while for a lamellar profile, the result is:

$$a_s = \frac{\lambda_B}{4\cos\frac{A.D.}{2}}, \tag{XI.3}$$

These simple results may be useful to start an optimization process.

Things become more questionable when one wants to estimate the corresponding optimized efficiencies. From the intensity dependence on λ, d, θ_2, φ and $\theta_{2,n}$, there exists at least three different ways of calculating the grating efficiency, which lead to results differing by a factor 2. For instance, Madden and Strong suggested to use the incidence angle on the large facets of an echelette grating, in view of calculating the Fresnel reflectivity [XI.1]. On the other hand, M. C. Hutley [I.2] advocated to take the bisector between incident and the -1st order direction, which preserves the reciprocity. The arguments, presented by the different authors are not completely clear, and the scalar approach remains unsatisfactory.

XI.2 Electromagnetic approach

The classical differential theory presented in Chapter II seems *a priori* to lead to huge differential sets to integrate, linked with the high number of propagating orders. However, for wavelengths below 100 nm, the reflectance of all known materials drops to very low levels, not exceeding some parts in one thousand. In the soft X-ray range, the refractive indices of all materials, metal and dielectrics, writen as $\nu = 1 - \tilde{n} + i\kappa$ are such that $\tilde{n} \sim 10^{-2}$ to 10^{-5}, and $\kappa \sim 10^{-3}$ to 10^{-6}. Thus, their permitivity is very close to ε_0, and such materials only appear as weak perturbations of vacuum. Thus, the coupling between the various equations in the set (II.2), made by the nonzero Fourier components of $\varepsilon(x, y)$, is very weak. It means that if a grating is blazed in the -1st order, only a few neighboring orders will propagate a non-negligible amount of energy. As a result, a severe truncation of the set (II. 2) or (II. 2') is able to produce accurate results for the desired -1st order efficiency. Numerical experiments have shown that 13 or 15 Fourier coefficients of the field (from –7 to +7) are enough to resolve most of X-ray grating problems, even when, theoretically, thousands of orders are propagating. Most of them carry less than 10^{-6} percentage of energy.

The direct consequence of the values of the refractive index is a poor reflectance under normal incidence. Below 30 nm, grazing incidence is used in order to get significant reflected powers. Near normal incidence and even in grazing incidence, Eqs. (XI.1)-(XI.3) predict optimal groove depths not too far from the blaze wavelength λ_B, which results in modulation ratios a/d typically about one or a few percents. As a result, the integration domain is very small. No numerical contamination occurs, in general, during the integration process, and the classical differential theory turns out to be an efficient tool for optimizing XUV and X-ray gratings [XI.2-XI.4], requiring short computation times.

XI.3 Limitation of the scalar approach

Having at our disposal an efficient tool to analyze X-ray gratings with the electromagnetic theory, the limits of the scalar approach can be checked, concerning both the validity of the blazing conditions, Eqs. (XI.1)-(XI.3), and the predicted efficiencies.

XI.3.1 Validity of the blazing condition

As long as gratings are used under near normal incidence, XUV gratings lead to a simple analysis [XI.5]. They behave in a scalar manner, which results in efficiency curves free of anomalies and polarization effects. If infinite conductivity is assumed, efficiency curves can all be derived for a given type of profile from an universal efficiency curve whatever the groove depth, groove spacing or incidence may be [XI.5]. Figure XI.1 shows the -1st order universal efficiency curve of a sinusoidal grating used in the scalar domain. The efficiency can be represented as a function of λ/a ratio, only. It culminates at about 34% when $\lambda/a = 3.4$ [XI.6]. Lamellar grating efficiency will culminate at 40% when $\lambda/a = 4$, while a triangular profile can produce 100% efficiency for $\lambda/a = 2$. Thus, in the asymptotic domain, when the period is much larger than the wavelength, triangular profiles have much better potentialities when compared with other groove shapes. The blaze wavelengths are consistent with Eqs. (XI.1)-(XI.3) near normal incidence and when $\lambda/d \ll 1$.

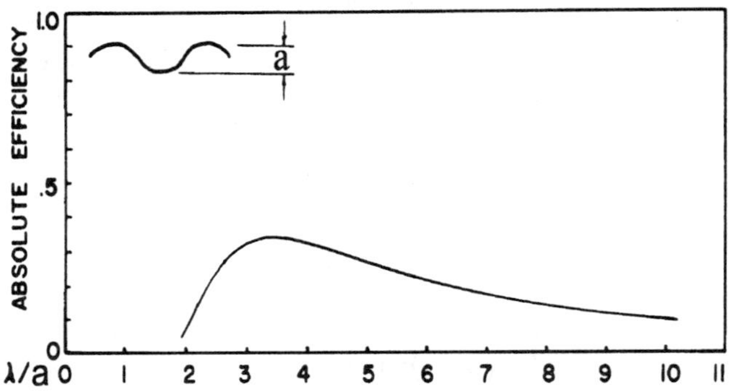

Fig. XI.1. The universal efficiency curve of a sinusoidal grating used in the scalar domain (after [XI.5]).

When the real conductivity has to be considered, it merely accounts for the metal reflectance (calculed under normal incidence). It must be pointed out, however, that the limits of the simple behavior not only depend on the λ/d and a/d ratios. As established in [XI.6], the angle of incidence plays an important role in the deterioration of the performance of the scalar behavior as this angle increases. The Seya-Namioka mount seems to be a limit beyond which a nonscalar behavior is often observed [XI.7]. As a simple rule of a thumb, in the asymptotic limit $\lambda/d << 1$, the scalar theory is safe to be used for incidence $\theta_2 < 40^o$, and may still give useful information as far as 60°.

Under $\lambda = 40$ nm, the reflectivity of all metals is less than 10% and grazing incidence is often used. In order to clarify the limits of the scalar approach, we chose a sinusoidal profile for which the optimal groove depth is given by Eq. (XI.2), if we rely on scalar theory. We choose a 1200 groove/mm grating used under five different angles of incidence between 71° and 88° and various selected wavelengths in the $2-40$ nm range. For each wavelength and incidence, the -1st order efficiency is calculated as a function of the groove depth in order to determine the real optimal value a_r of the groove depth a. Figure XI.2 plots the relative discrepancy $\frac{a_s - a_r}{a_r}$ between the scalar and electromagnetic theory predictions. The discrepancy is less than 10% when the incidence remains below 80°, but may be greater than 20% at steeper grazing incidence. It is worth having in mind that 25% error in groove depth can be allowed without loosing more than 10% in efficiency, as can be established by rigorous computations. These conclusions hold for both TM and TE polarization.

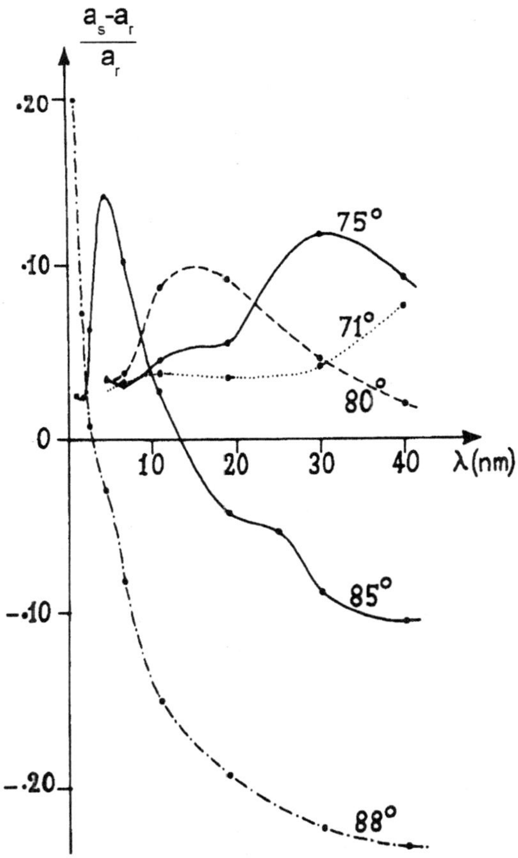

Fig. XI.2. Discrepancy in the optimal groove depth (after [XI.2]).

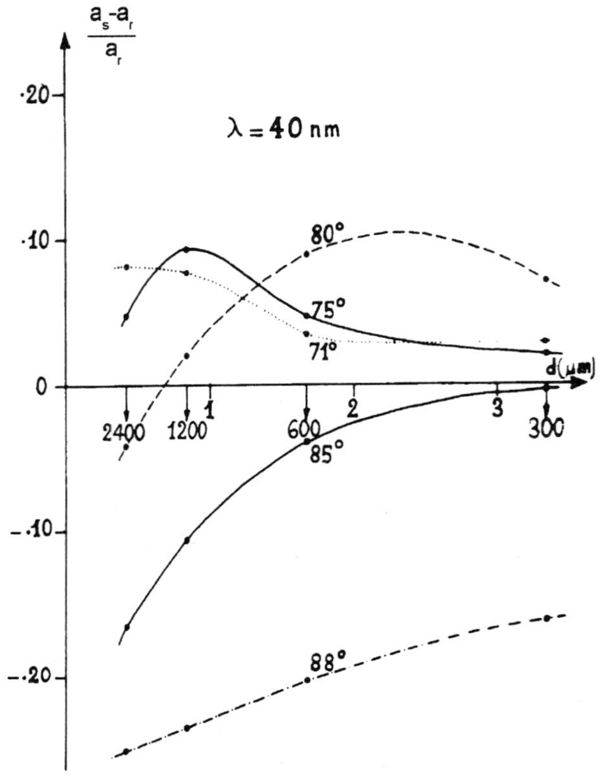

Fig. XI.3. Influence of the groove spacing (after [XI.2]).

In Fig.XI.3, we stick to $\lambda = 40$ nm and investigate the influence of the groove frequency in the range between 300 and 2400 grooves/mm, for the same values of incidence as in Fig.XI.2. The conclusions are similar. As expected, the scalar behavior generally improves when the groove spacing is increased. Nevertheless the relative discrepancy remains below 10% for all groove spacings provided $\theta_2 < 80^o$.

XI.3.2 Validity of the predicted efficiencies

Despite the conclusion made in the previous section, however, we cannot conclude that gratings used under 71° to 80° incidence angles behave in a purely scalar fashion. Due to the fact that TE and TM reflectivity of metallic flat surfaces are somehow different, similar differences are found in the TE and TM -1st order efficiency curves [XI.7]. On top of that, differences come from groove depth and groove spacing, so that the efficiency curves can no longer be predicted with simple rules from an universal curve, as it occurs near normal incidence. However, the superiority of triangular profiles over the other ones remains, in general. Of course, at very steep grazing incidence, the TE and TM reflectivity both approach unity and the polarization effects are expected to disappear. However, the predictions of the scalar theory are questionable anyway. A detailed study of this question [XI.8] was undertaken during the preparation of the X-ray multimirror (XMM) project.

XI.3.2.1 The XMM grating requirements

The X-ray multimirror mission spacecraft launched by the European Space Agency in January 2000 consists of a high throughput X-ray telescope aimed to work in the 0.5 – 5 nm wavelength range. This range includes lines of the most abundant elements (O, N, C, etc.) in the universe. In order to investigate their proportions, a reflection grating spectrometer was designed to achieve high energy resolution and high efficiency in that spectral region. Since very limited experimental results were existing in the literature for reflection gratings below 5 nm wavelength, various grating samples were asked to the grating manufacturers in order to determine what performance could be expected. In parallel, theoretical studies were done with both the scalar and the electromagnetic theory in order to see how close the real gratings approach the theoretical limit.

The reflection grating spectrometer consists of 600 identical replicas of the same shallow triangular gold grating used under grazing incidence. The groove frequency was $525\ \text{mm}^{-1}$, the blaze angle φ was 0.75°, apex angle 90° and the grazing angle θ defined by $\theta = 90° - \theta_2$ was varied between 0 and 3°. The preliminary studies of the XMM mission involved both investigating the original sample and studying the replication process.

XI.3.2.2 Theoretically predicted efficiencies for the XMM gratings

With the above-specified grating parameters, and using the refractive indices of gold given in [XI.9], the 0th, -1st, -2nd and -3rd order efficiencies were calculated as functions of the grazing angle θ, for $\lambda = 0.834\ \text{nm}$, assuming a perfect groove geometry. The results are shown in Fig. XI.4. The results from the scalar theory of diffraction (SD) were obtained following the lines given by Madden and Strong [XI.1]. Strong discrepancies are found, with often a factor 2 between the two predictions at a given angle θ. The question which remained to be answered was what approach gives the most reliable results. Although a spectacular agreement between electromagnetic theory and measurements was already obtained [XI.10] at $\lambda = 6.7$ nm , measurements of all produced samples were made to answer the question in the 0.5 – 5 nm spectral range.

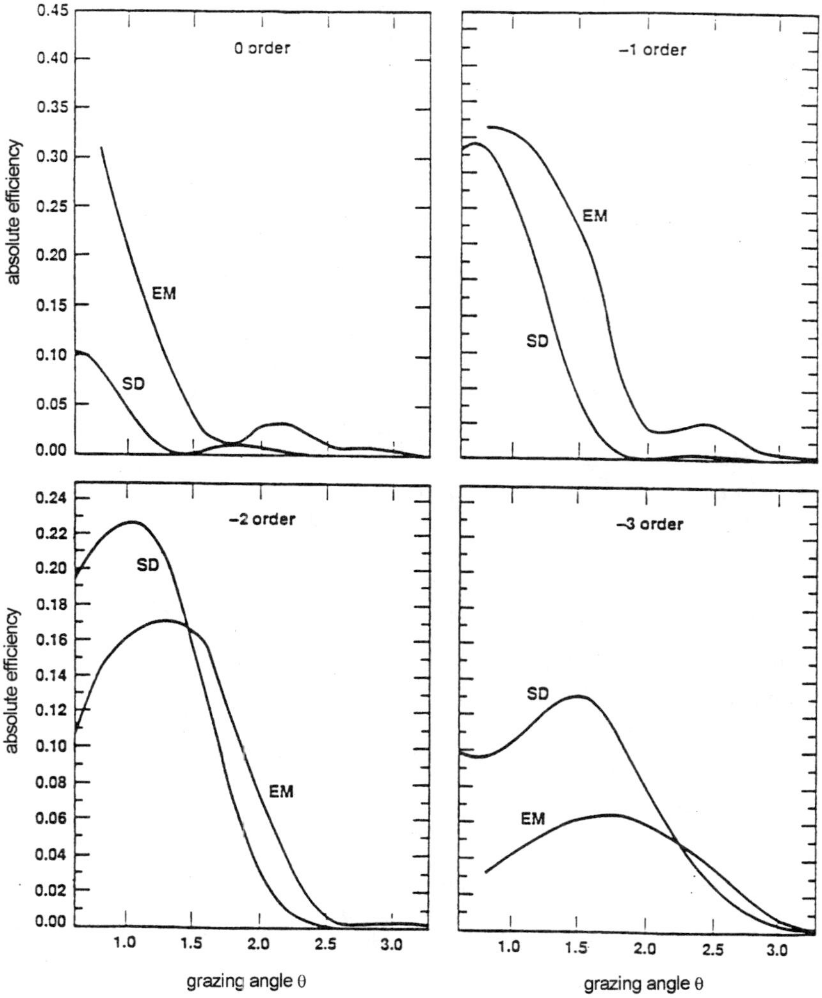

Fig. XI.4. Theoretical efficiencies predicted by the scalar theory of diffraction (SD) and the electromagnetic theory (EM), as functions of the grazing angle θ (in degrees) (after [XI.8]).

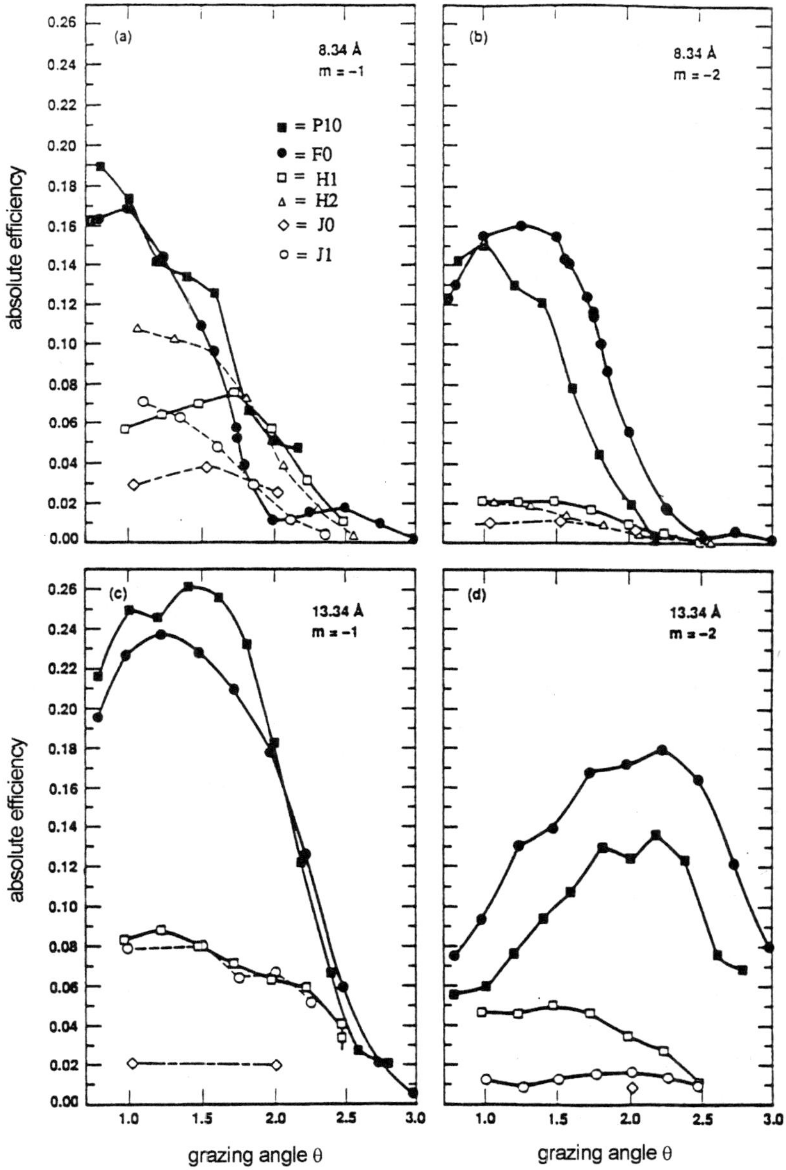

Fig. XI.5. Measured grating sample efficiencies (after [XI.8]).

XI.3.2.3 XMM grating-sample measurements

Figure XI.5 shows the measured -1st and -2nd order efficiencies as functions of the grazing angle θ at wavelengths 0.834 and 1.334 nm, for various grating samples named J1, F0, P10, etc., provided by several grating manufacturers. Efficiencies as high or close to 20% can be obtained, which is quite surprising at these short wavelengths. All gratings were supposed to closely approach the ideal grating profile described in Section XI.3.2.1. But due to the difficulty in producing profiles with 0.75^{0} blaze angle, a great dispersion of the results is observed, reaching a ratio 10 in the efficiencies between two different samples at the same incidence and wavelength. Two samples, P10 and F0, looked much better than all others, the F0 grating looking the best one, if we consider both -2nd and -1st order at 0.834 and 1.334 nm wavelength. The question is how closely it approaches the theoretical limit.

XI.3.2.4 Comparison of measured and computed efficiencies of the F0-XMM grating sample

Figure XI.6 compares the measured -1st and -2nd order efficiencies of the F0 sample with the theoretical values given by the differential theory. At $\lambda = 0.834$ nm, the theoretical results are taken from Fig. XI.4, while the measured ones are taken from Fig. XI.5. At $\lambda = 1.334$ nm, new theoretical results from the electromagnetic theory are plotted and compared with the measurements extracted from Fig. XI.5. A spectacular argument is observed. The peak values have a few percent difference, while the theoretical and experimental curves as functions of θ closely follow each other. Of course, the discrepancy is much higher if we compare the measurements with the results provided by scalar theory extracted from Fig. XI.4

XI.3.2.5 Surface metrology of the various samples

In order to understand the superiority of the F0 grating over the other competitors, surface metrology was done with stylus tracing. Figure XI.7 shows Talystep and Dektak tracings of the groove profiles for five different samples. It is shown that in most cases, the profile turns out to be far from the perfect triangle, except in one case, the F0 sample, which looks almost ideal. Hence the explanation of the agreement between the measured and the predicted efficiencies. This quasiperfect profile was produced with a new technology which consists of ion-etching on a highly polished blank.

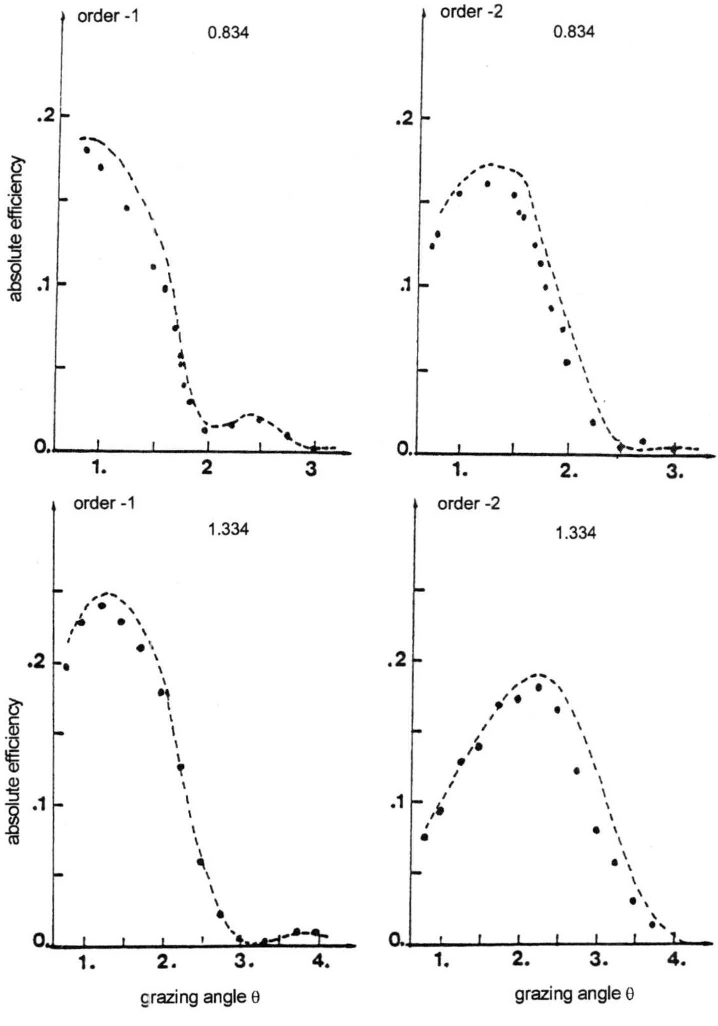

Fig. XI.6. Comparison of the measured -1st and -2nd order efficiencies (dots) of the F0 grating as functions of θ with the theoretical efficiencies (dashed lines) predicted by the electromagnetic theory, at two different wavelengths (after [XI.8]).

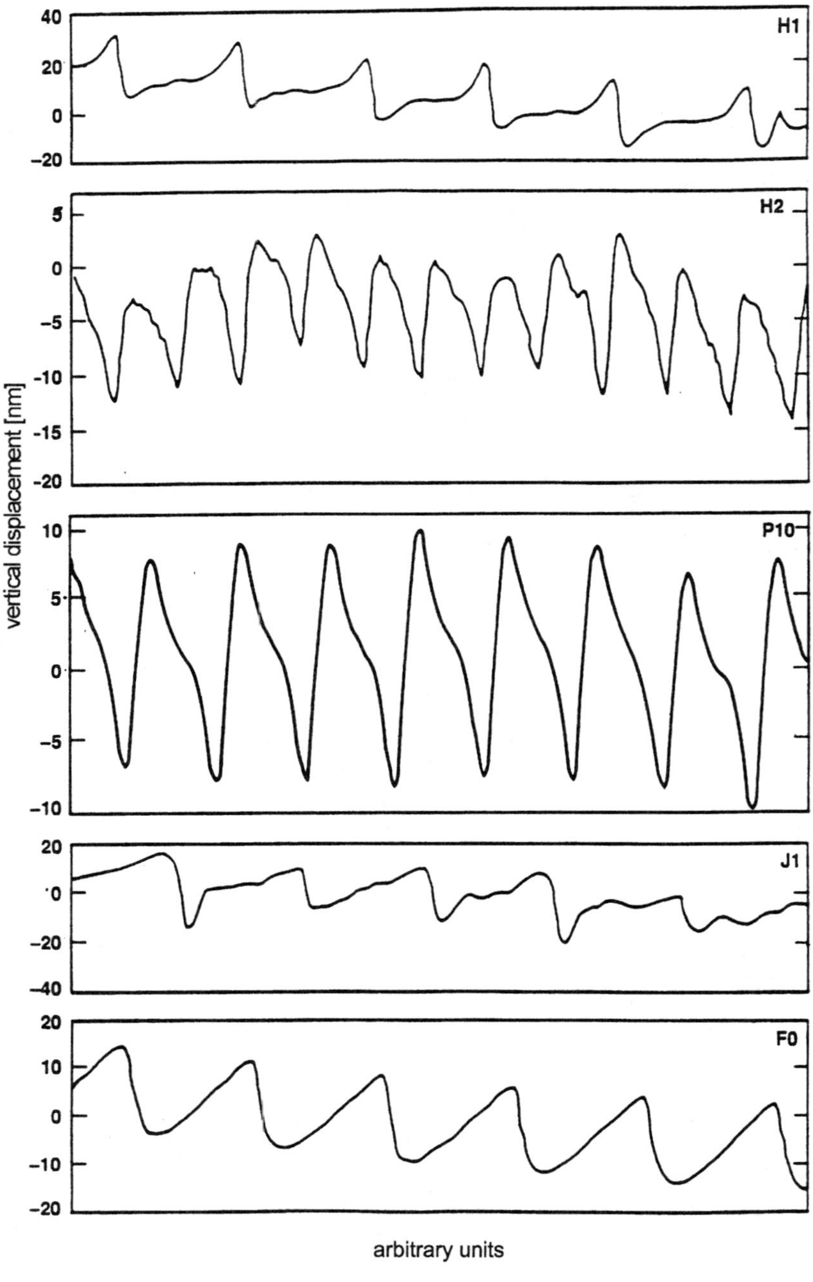

Fig. XI.7. Stylus tracing of groove profiles for 5 grating samples, as indicated. Horizontal scales are in arbitrary units (after [XI.8]).

XI.3.2.6 Study of the P10 sample

Fig.XI.7 also reveals that the P10 sample, which presented in Fig. XI.5 competitive efficiencies, has a good surface quality, except that the large facet of the groove is not straight, but a little wavy. Talystep measurements were done to get an accurate description of the wavy facet. The computer code was then modified to analyze a wavy-facet grating. Theoretical results then fairly reproduced the measurements shown in Fig. XI.5 for that sample.

XI.3.2.7 Study of the influence of the replication process

In view of producing 600 copies of the master grating, the replication process was analyzed in detail [XI.11]. The process converts a grating profile into its symmetric with respect to the grating plane. Thus a grating like the F0 sample, which presents somehow rounded valleys (Fig.XI.7), will produce a first generation replica (F1) with rounded tops. Measurements were done to compare the F0 and F1 sample efficiencies. On the other hand, rounded top profiles were included in the computer code, and the corresponding efficiencies computed.

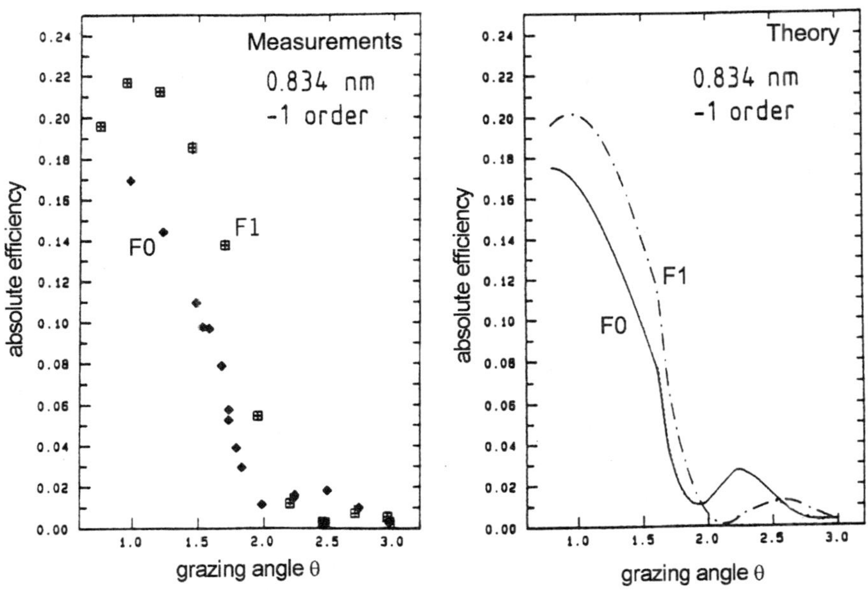

Fig. XI.8. Theoretical and measured -1st order efficiencies as functions of θ for the F0 samples and its replica F1 (after [XI.11]).

Figure XI.8 shows both the measured and the computed efficiencies at 0.834 nm wavelength. In both figures, the rounded-top grating has a higher maximum than the master (the contrary was observed at $\lambda = 1.334$ nm). Once again, the theory fully accounts for the observed effect. The comparisons between the measurements and the theory done in preparation of the XMM mission fully convinced the scientific community that it is often necessary to use Maxwell equations to optimize gratings, even in the X-ray domain. The observed agreement is a proof that the macroscopic concept of refractive index is still useful at $\lambda = 0.834$ nm, a wavelength not so far from the size of atoms. We will see that the concept still works much below this wavelength.

XI.4 Multilayer gratings for X-ray diffraction

The use of gratings under grazing incidence is not a strong handicap in making spectra. However, in many situations, gratings are not only asked to diffract light, but also to focus or collimate it. In the X-ray domain, the result is high aberrations, which are difficult to correct in the sense that they require making grooves on a sophisticated curved surface. After the pioneering papers of Barbee and Spiller [XI.12], a solution was found to increase mirror reflectance under near normal incidence with the use of a stack of layers of alternating dielectric and absorbing materials deposited on the mirror surface. The optimization of their thicknesses from numerical methods and classical thin film theory [XI.13] has permitted interesting reflectance in the soft X-ray range. The use of such multilayers to increase grating efficiencies is thus evident. In addition, with the substantial increase of grating efficiencies near normal incidence [XI.14], it became possible to use them not only for diffraction, but also for beam shaping [XI.15], like Fresnel plates and concave gratings serve in traditional optics to substantially reduce the number of optical components.

The two kinds of multilayer gratings described at the beginning of the chapter, namely multilayer coated gratings and Bragg gratings require completely different analysis.

XI.5 Multilayer coated gratings

A typical multilayer coated grating is represented in Fig. XI.9a and the detail of one period is shown in Fig. XI.9b. All the profiles are identical and obtained from each other by a vertical translation. In the X-ray domain, the thickness of each metallic or dielectric layer is typically about 2 nm. As a result, the different profiles interpenetrate, which means that the bottom of the jth profile is lower than the top of the $(j-1)^{th}$ one. Contrary to what was assumed in an early work [XI.16], no homogeneous layer can be found between neighboring profiles. It is thus impossible to compute the T-matrix of a profile formed by a single interface, from which we should be able deduce the global T and S matrix of the modulated stack. It is however possible to analyze this complicated device with the classical differential theory [XI.17], taking into account the profile interpenetration.

XI.5.1 Theory

For the differential theory, the only differences between the stack of profiles in Fig. XI.9 and a single interface grating is that the function $k^2(x,y)$ has a much more complicated form, and that the total modulated region may be much thicker than the groove depth of one interface.

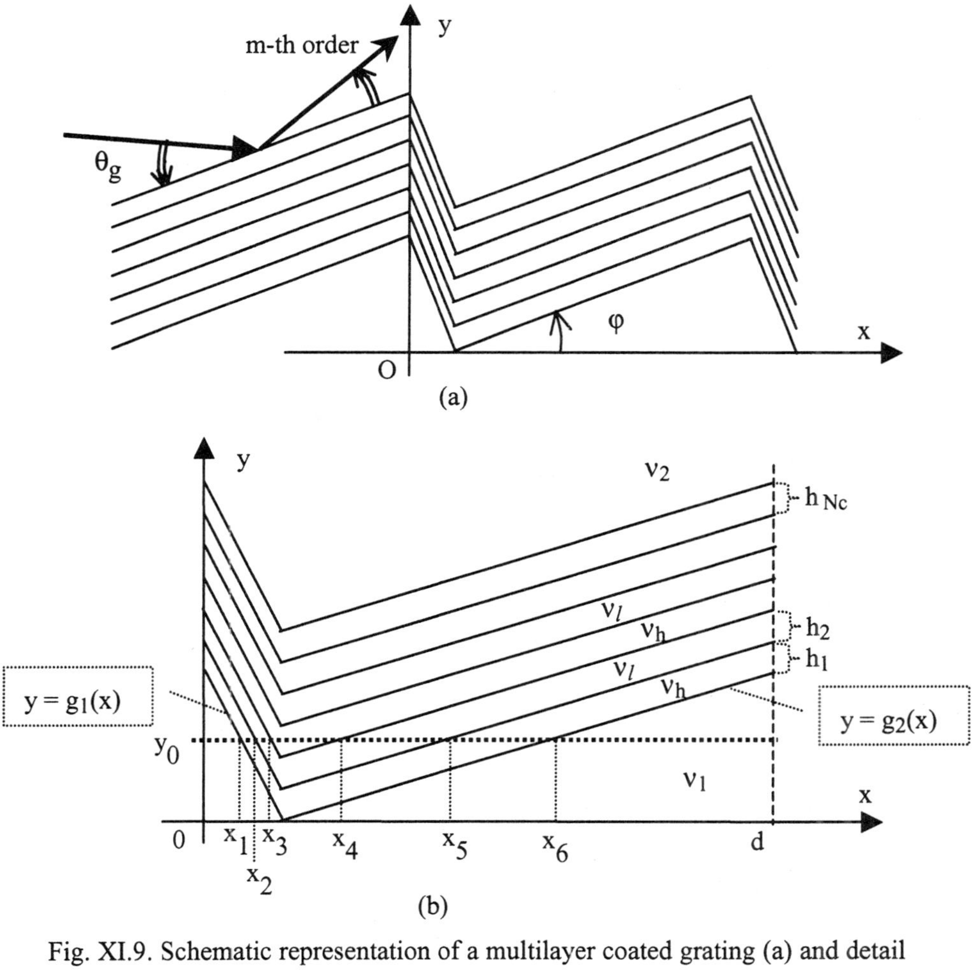

Fig. XI.9. Schematic representation of a multilayer coated grating (a) and detail of one period (b).

Instead of being a simple step function as illustrated in Fig. (II.1), at the given ordinate y_0 illustrated in Fig. XI.9b, $k^2(x, y_0)$ has the multistep shape shown in Fig. XI.10, in which the refractive indices ν_l and ν_h of dielectric and metallic layers are assumed to be real. The only difficulty is the determination of the Fourier components $(k^2)_n$, $\forall y$. To that end, we first establish formulae that give the Fourier coefficients of a periodic multistep function having an arbitrary number of discontinuity points. In principle, this could be done numerically using the FFT algorithm, but we prefer the following analytical approach.

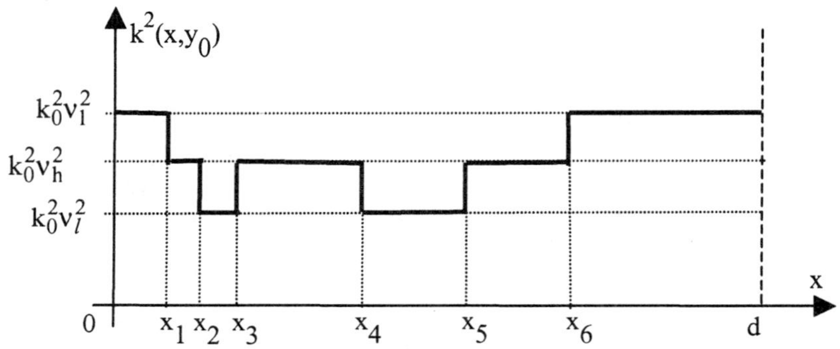

Fig. XI.10. Shape of the multistep function $k^2(x, y_0) = k_0^2 \nu^2$ over one period, when ν_1, ν_l and ν_h are real numbers.

XI.5.1.1 Fourier coefficients of an arbitrary step function

Consider an arbitrary complex step function $y = f(x)$ with no particular symmetry property, periodic with period d with respect to x, whose real part $\mathrm{Re}[f(x)]$ is represented on a single period in Fig. XI.10. This function has N_p $(N_p \geq 2)$ discontinuity points x_p, $p \in (1, N_p)$ and takes values y_p at $x = x_p - \hat{\delta}$, where $\hat{\delta}$ is arbitrarily small.

The derivative of $f(x)$ in the sense of distributions for $x \in (0, d)$ is

$$f'(x) = 0 + \sum_{p=1}^{N_p} \sigma_p \delta(x - x_p), \qquad \text{(XI.4)}$$

where 0 comes from the derivative in the sense of functions and

$$\sigma_p = y_{p+1} - y_p, \qquad p \in [1, N_p - 1],$$
$$\sigma_{N_p} = y_1 - y_{N_p},$$

and $\delta(x - x_p)$ is the Dirac function. It must be pointed out that when f(x) is considered on the entire x-axis, the Dirac function $\delta(x)$ is replaced by the Dirac comb with period d.

Let us introduce the Fourier coefficients f_n of function $f(x)$:

$$f(x) = \sum_n f_n \exp(inKx),$$

where $K = 2\pi/d$, which gives

$$f'(x) = \sum_n inKf_n \exp(inKx) \qquad \text{(XI.5)}$$

Since the Fourier coefficients of a Dirac comb are constant and all equal to $1/d$, for $x \in [0, d]$ we have

$$\delta(x) = \frac{1}{d}\sum_n \exp(inKx). \qquad \text{(XI.6)}$$

Thus the use of Eqs. (XI.4)-(XI.6) leads to

$$f_n = \frac{1}{2i\pi n}\sum_{p=1}^{N_p} \sigma_p \exp(-inKx_p), \qquad n \neq 0$$

or

$$f_n = -\frac{1}{2\pi n}\sum_{k=1}^{N_p} \sigma_p[\sin(nKx_p) + i\cos(nKx_p)]. \qquad \text{(XI.7)}$$

For $n = 0$,

$$f_0 = \frac{1}{d}\int_0^d f(x)dx = y_1(x_1 + d - x_{N_p}) + y_2(x_2 - x_1)$$
$$+ y_3(x_3 - x_2) + \ldots y_{N_p - 1}(x_{N_p - 1} - x_{N_p - 2}) + y_{N_p}(x_{N_p} - x_{N_p - 1}),$$

$$f_0 = \frac{1}{d}\left[y_1 d + \sum_{p=1}^{N_p - 1} x_p(y_p - y_{p+1}) + x_{N_p}(y_{N_p} - y_1) \right] \Rightarrow$$

$$f_0 = \frac{1}{d}\left(y_1 d - \sum_{p=1}^{N_p} \sigma_p x_p \right). \qquad \text{(XI.8)}$$

The problem which remains to solve is now to find the abscissae and ordinates of all discontinuities of function $k^2(x,y)$ inside the modulated area. If h_i are the vertical thicknesses of the various layers illustrated in Fig. XI.9b, the modulated area now lies between $y=0$ and $y_{Max} = a + \sum_{i=1}^{N_C} h_i$, where N_C is the number of layers. We thus have to find a way of determining x_p and y_p $\forall y_0 \in [0, y_{Max}]$.

XI.5.1.2 Determination of σ_p, x_p, and y_l

The values of some of the jumps of k^2 will depend on the value of the refractive index of the upper layer, and thus on the parity of the number of layers N_C. To be most general, we introduce two refractive indices $\tilde{\nu}_h$ and $\tilde{\nu}_l$ defined by

$$\tilde{\nu}_h = \begin{cases} \nu_h & \text{if } N_C = 2q+1, \\ \nu_l & \text{if } N_C = 2q \end{cases}, \text{ q integer}$$

$$\tilde{\nu}_l = \begin{cases} \nu_l & \text{if } \; N_C = 2q+1 \\ \nu_h & \text{if } \; N_C = 2q \end{cases}.$$

Although multilayers in the X-ray domain generally are a superposition of identical bilayers, some special applications [IX.18, IX.19] require varying thicknesses through the stack in order to enlarge the high efficiency plateau. We thus develop the theory for arbitrary thicknesses h_j, which complicates the process.

In order to determine the jumps and discontinuity points of function $k^2(x, y_0)$ whatever y_0 may be inside the modulated region, let us assume that the dielectric stack thickness is higher than the groove depth:

$$\sum_{i=1}^{N_C} h_i > a \,.$$

This situation will be the most common in the X-ray range, although the alternative one is not more difficult to handle [XI.20]. We then have three different ranges for ordinate y_0.

(i) $y_0 < a$:

Since the thicknesses $h_1, h_2, \ldots\ldots, h_{N_C}$ of the different layers may take arbitrary values, in order to calculate the number of discontinuities, we first compute

$$h_{es}(n) = \sum_{j=1}^{n} h_j$$

until we get $h_{es}(n) > y_0$. Then we define the number $N_p = n$, which turns out to be the number of profiles having a minimum lower than y_0. Since the maximum of all of them is greater than a and thus greater than y_0, the number of discontinuity points, here called N_{S2}, is equal to $2N_p$. In order to calculate the abscissas of the discontinuity points, we introduce abscissas $X_{1,j}$ of the intersection point of the first segment of the j-th profile with the line $y = y_0$, and $X_{2,j}$ for the intersection of the second segment of the same profile with the same line. This definition gives:

$$\begin{aligned}
X_{1,1} &= g_1(y), \\
X_{2,1} &= g_2(y), \\
X_{1,j} &= g_1[y - h_{es}(j-1)], \qquad \forall j \in [2, N_p], \\
X_{2,j} &= g_2[y - h_{es}(j-1)], \qquad \forall j \in [2, N_p].
\end{aligned}$$

In fact, if we look at Fig. XI.9b, we see that the values x_p introduced in Section XI.5.1 are related to $X_{1,j}$ and $X_{2,j}$ by:

$$x_p = \begin{cases} X_{1,p} & p \in [1, N_p] \\ X_{2,2N-p+1} & p \in [N_p + 1, 2N_p] \end{cases}.$$

The values of the jumps σ_p are

$$\begin{aligned}
\sigma_1 &= k_0^2(\tilde{v}_h^2 - v_2^2), \\
\sigma_{2N} &= -\sigma_1, \\
\sigma_p &= (-1)^p k_0^2(\tilde{v}_l^2 - \tilde{v}_h^2), \\
y_1 &= k_0^2 v_1^2.
\end{aligned}$$

The Fourier coefficients $(k^2)_n$ of the function $k^2(x, y_0)$ are thus obtained through Eqs. (XI.7) and (XI.8) in which N_p is replaced by $N_{S2} = 2N_p$.

(ii) $\mathbf{a < y_0 < \sum_{i=1}^{N_C} h_i}$ **:**

We first calculate

$$h_{es}(n) = \sum_{j=1}^{n} h_j$$

until we get $h_{es}(n) > y_0$ and define $N_p = n$. We then calculate

$$a_{es}(n') = a + \sum_{j=1}^{n'} h_j$$

until we get $a_{es}(n') > y_0$ and define $N' = n'$. N' is the number of profiles having maxima lower than y_0. Introducing $N_S = N_p - N'$, the number of discontinuities N_{S2} is equal to $2N_S$, i.e., to $2(N_p - N')$. Following the same methods as in the previous case, we obtain

$$X_{1,i} = g_1[y - h_{es}(N' + i - 1)], \qquad \forall i \in [1, N_S],$$
$$X_{2,i} = g_2[y - h_{es}(N' + i - 1)], \qquad \forall i \in [1, N_S],$$
$$x_p = \begin{cases} X_{1,p} & p \in [1, N_S] \\ X_{2,2N_S - p + 1} & p \in [N_S + 1, 2N_S] \end{cases}$$

$$\sigma_p = (-1)^{(p+N')} k_0^2 (\tilde{v}_l^2 - \tilde{v}_h^2), \qquad \forall p \in [1, 2N_S],$$
$$y_1 = \begin{cases} k_0^2 \tilde{v}_h^2 & N' = 2q + 1, \\ k_0^2 \tilde{v}_l^2 & N' = 2q \end{cases} \quad , \text{q integer.}$$

(iii) $\mathbf{\sum_{i=1}^{N_C} h_i < y_0 < y_{Max} = a + \sum_{i=1}^{N_C} h_i}$ **.**

In that case we define N_p as being equal to $N_C + 1$. We calculate

$$a_{es}(n') = a + \sum_{i=1}^{n'} h_i$$

until we get $a_{es}(n') > y_0$. Then we define $N' = n'$, which leads to

$$N_{S2} = 2N_S = 2(N_p - N') = 2(N_C + 1 - N').$$

For second time, we calculate all the possible values of

$$h_{es}(n) = \sum_{i=1}^{n} h_i ,$$

for $n \in (1, N_C)$. The discontinuity abscissas are given by

$$X_{1,i} = g_1[y - h_{es}(N' + i - 1)], \qquad \forall i \in [1, N_S],$$
$$X_{2,i} = g_2[y - h_{es}(N' + i - 1)], \qquad \forall i \in [1, N_S],$$

$$x_p = \begin{cases} X_{1,p} & p \in [1, N_S] \\ X_{2,2N_S-p+1} & p \in [N_S + 1, 2N_S] \end{cases}.$$

Concerning the jumps, we obtain

$$\sigma_p = (-1)^{p+N'} k_0^2(\tilde{v}_l^2 - \tilde{v}_h^2), \qquad \forall p \in [1, N_S - 1], \forall p \in [N_S + 2, 2N_S],$$
$$\sigma_{N_S} = k_0^2(v_2^2 - \tilde{v}_h^2),$$
$$\sigma_{N_S+1} = -\sigma_{N_S},$$

while

$$y_1 = \begin{cases} k_0^2 \tilde{v}_h^2 & N' = 2q + 1, \\ k_0^2 \tilde{v}_l^2 & N' = 2q \end{cases}, \text{ q integer.}$$

XI.5.1.3 Discussion

The previous section provides the values of the jumps and abscissae of the discontinuity points of $k^2(x, y_0)$ $\forall y_0$ inside the modulated area. Eqs. (XI.7) and (XI.8) then provide the Fourier components $(k^2)_n$ $\forall y_0$, and the classical differential theory is used as in Chapter II, except that the modulated region ranges between 0 and $y_{Max} = a + \sum_{i=1}^{N_C} h_i$. In general, in the X-ray domain, reliable results are obtained without the use of the S-matrix propagation algorithm, due to the fact that all refractive indices are close to unity, and groove depths are small. However, for thick stacks including hundreds of layers, it could be necessary to include the S-matrix propagation algorithm. It is worth noting that, in that case the modulated region has to be cut into M-2 horizontal slices, which results in truncating several interfaces. It is not possible to compute the S matrices of the various physical interfaces separately, due to their interpenetration.

The FFF method could be included for TM polarization. However, in X-ray domain, the jumps of permittivity between neighboring media are so small that no problem occurs with the classical formulation. Polarization effects are generally small, so that TM results are close to TE ones and the speed of convergence of the Fourier series is about the same for both polarizations.

The numerical difficulties are much more serious when the X-ray multilayer grating is used in a very high order and is then called an "echelle." New needs in X-ray fluorescence spectroscopy require the achievement of ultrahigh ($\approx 10^5$) resolution and collection of important flux, which implies the use of gratings in high (i.e., from 10th to 60th) orders, at not too small ($\approx 5^o$) grazing angles of incidence. The latter requirement precludes using bare gold gratings whose reflectivity would be too small in the 4 – 25 nm range of wavelength. Taking into account both the high-resolution and the high-flux requirements, monochromators designers are led to use a high-blaze-angle ($\approx 10^o$) echelette grating overcoated by a metal-dielectric (e.g., tungsten-carbon) multilayer. The electromagnetic analysis of a deep X-ray multilayer coated grating used in a high order has been conducted with the S-matrix propagation algorithm and with an asymmetric truncation of the propagation equation set [XI.20].

XI.5.2 Example of design: a grating working at 1.33 nm

The design of a multilayer coated grating must optimize both the reflectivity of the multilayer and the blazing potentialities of the grating [XI.14]. Let us consider a flat stack of N_C metal-dielectric layers. The thickness D of each bilayer is called the multilayer period. We call Γ the ratio between the thickness of the absorbing layer (metal) and the multilayer period (D). When a flat multilayer is lighted under a grazing angle θ_g, defined in Fig. XI.9a, Bragg's law gives the peak wavelength λ at which the reflectivity is maximum:

$$\lambda = \frac{2D}{n} \sin\theta_g, \tag{XI.9}$$

where n is an integer which will be chosen equal to unity. When the multilayer is evaporated on an echelette (or echelle) grating, the Bragg condition (Eq.

(XI.9)) must be combined with the blaze condition of the mth order of the grating. This requires that the directions of the $-m^{th}$ order and the incident wave be symmetrical with respect to the normal on the large facet of the grating (Fig. XI.9a). This condition issued from scalar optics has been checked with numerical computations [XI.17-XI.20] and turns out to work pretty well in the X-ray domain, at which the wavelength-to-groove-spacing ratio is generally less than $1/100$. The result is the following condition:

$$\lambda = (2d \sin \varphi \sin \theta_g)/m, \qquad \text{(XI.10)}$$

where φ is the blaze (groove) angle.

As shown in [XI.14], Eqs. (XI.9) and (XI.10) result in:

$$\frac{D}{n} = \frac{d \sin \varphi}{m} \qquad \text{(XI.11)}$$

and it is worth noting that this blazing condition is independent of incidence.

In order to design a grating to work around 1.33 nm wavelength, we choose a stack of 61 layers made of 30 rhodium-carbon identical bilayers, plus an extra rhodium layer at the top with same thickness as the other rhodium layers. We choose D = 3.3 nm and $\Gamma = 0.33$. The stack is evaporated over a 0.36 μm period gold echelette grating. Bragg's law (Eq. (XI.9)) gives $\theta_g \approx$ 12.2° if we choose $n = 1$. Then Eq. (XI.11) gives $\varphi \simeq 0.5^{o}$, and it can be verified that, as assumed in Section XI.5.1.2, the stack thickness is higher than the groove depth. From Henke's tables [XI.21], the refractive indices of rhodium, carbon and gold are derived for any wavelength. Fig. XI.11 shows the -1st order efficiency e_{-1} as a function of wavelength when the grating is illuminated under 78.3^{o} incidence, corresponding to $\theta_g = 12.2^{o}$. A blaze peak as high as 0.28 is obtained, which is quite spectacular in this spectral region, as compared to a bare gold grating with efficiency that culminates at $\sim 10^{-3}$. Moreover, the blaze peak concentrates about 99.9% of the total reflected energy and is almost equal to the reflectivity of the flat multilayer. The blaze peak position ($\lambda = 1.37$ nm) is slightly shifted from the predicted one, (1.33 nm) due to the fact that we have taken $\varphi = 0.5^{o}$ while the exact value should be $\varphi = 0.5252^{o}$. But since the blazing condition (XI.11) is θ_g-independent, the blaze peak can be recentered without reduction using a small increase (0.2^{o}) of incidence, e.g. $\theta_g \simeq 12^{o}$.

An interesting question is whether the high blaze peak remains when this grating is used in a higher order in view of increasing the resolution. If we use the grating in the echelle regime, e.g. in the 10th order $(m = -10)$, Eq. (XI.11) implies multiplying $\sin\varphi$ by 10, on the other hand θ_g is unchanged ($\theta_g = 12.2^{o}$). If $\varphi = 5^{o}$, the angle of incidence thus turns to be equal to 82.8^{o} . The dotted curve in Fig. XI.11 shows the -10th order efficiency. The blaze peak remains visible. A gain by a factor 10 in the resolution is only paid by a reduction by a factor 2 of the efficiency. The solid curve plots the sum of all nonnegligible propagating-order efficiencies and one can observe that in the echelle regime, the blaze peak carries about 75% of all the diffracted energy. Such a concentration does not occur if nontriangular profiles are used.

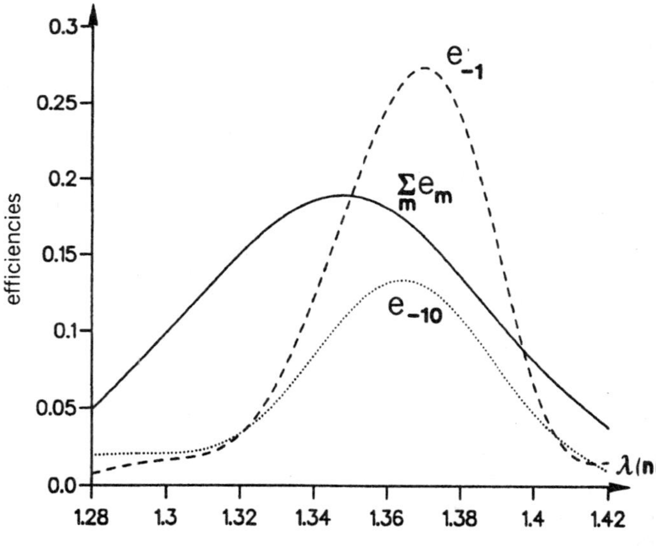

Fig. XI.11. Blazing performance in the usual (dashed curve) and echelle regime (solid and dotted curves) of a multilayer coated echelette gratings (after [XI.20]).

XI.6. X-ray Bragg gratings

When a grating is etched inside a flat metal-dielectric multilayer, we obtain an X-ray Bragg grating depicted in Fig. XI.12. The analysis of such a device falls in the frame of Chapter III. The modulated region can be decomposed into several elementary gratings as shown in Fig. XI.13. For each of them, the T-matrix is computed by the differential formalism. The T-matrix of the unmodulated stack is obtained analytically (see Section III.4 or [XI.22]).

In the X-ray domain, the T-matrix propagation algorithm is generally able to provide the diffracted efficiencies [XI.22]. If the stack is too thick, the S-matrix propagation algorithm gives the results anyway.

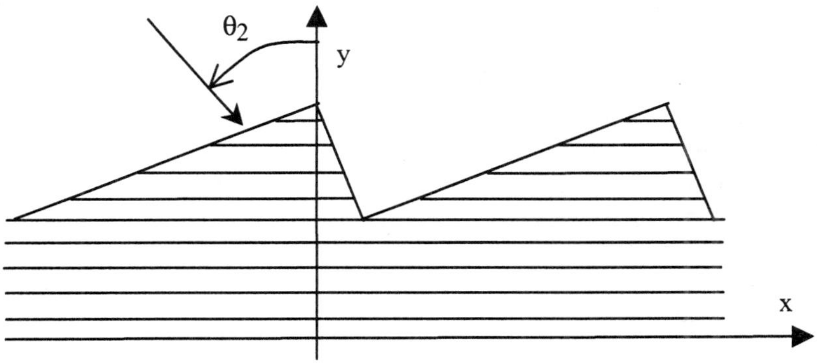

Fig. XI.12. Schematic representation of an X-ray Bragg grating.

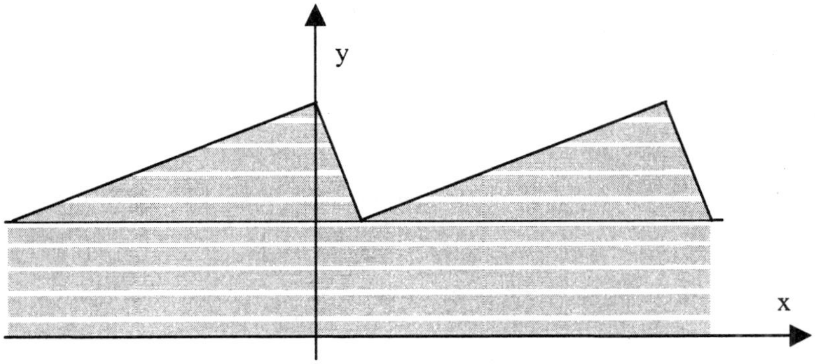

Fig. XI.13. Decomposition of the Bragg grating in terms of elementary gratings.

XI.6.1 Bragg grating design and its performance

We consider a flat stack of N_C metal-dielectric layers in which we etch a triangular profile with vertical small facets. Bragg's law still relates the peak wavelength (λ) to the multilayer period (D) and grazing angle (θ_g). But if we try again to apply the blazing condition (XI.11) to concentrate the diffracted energy into the $-m^{th}$ order, this time we obtain a very shallow grating with efficiencies lower than 10^{-13}. In fact only the upper bilayer will be etched. Let us show that on an example.

We choose a 61 layer stack made of 30 C-W identical bilayers, plus an extra W layer at the top with same thickness as the other W layers. We choose $D = 3.3\ \text{nm}$, $\Gamma = 0.33$; the stack is evaporated on a gold substrate. As in Section XI.5.2, we derive $\theta_g = 12.2^o$ from Eq. (XI.9), and $\varphi = 0.5252°$ from Eq. (XI.11). As a consequence, the groove depth a, equal to $d \tan\varphi$, only reaches 3.3 nm, if the grating period is kept equal to 0.36 μm. As previously announced, only the top bilayer is modulated and the stack mainly behaves like a multilayer mirror, concentrating almost all reflected energy inside the zeroth order. This is the main difference with the previous section in which all layers in the stack follow the gold grating groove shape. The conclusion is that the usual blazing condition (XI.10) issued from scalar optics does not apply to Bragg gratings. There are two ways of increasing the groove depth so that all layers be etched.

(1) Keeping the groove spacing unchanged (0.36 μm) and increasing the blaze angle up to 15.539°, that is to violate Eq. (XI.11);
(2) Keeping the blaze angle unchanged (0.5252°) and increasing the groove spacing to 10.92 μm. This will respect Eq. (XI.11), but will reduce the grating resolution.

In case (1), a maximum -1st order efficiency of 5.35% was found for $\theta_g = 11.30^o$. This is approximately 32% of the multilayer reflectivity. The behavior is more interesting in case 2, for which efficiency of 0.132 was found in the first internal (-1st) order. Calculations have shown that all spectral orders successively (sequently) present a peak efficiency when θ_g is varied in the vicinity of the value 12.2° derived from Eq. (XI.9). The -1st order has the highest peak, which then concentrates about 77% of the multilayer reflectivity.

This interesting energetic performance is, however, of limited interest due to the poor spectral resolution of a coarse grating. However, a good compromise can be found between cases 1 and 2 by increasing both blaze angle and groove spacing in order to etch the entire multilayer, and we can arrive at reasonable values of grating parameters, making grating manufacturing possible and keeping dispersion high enough.

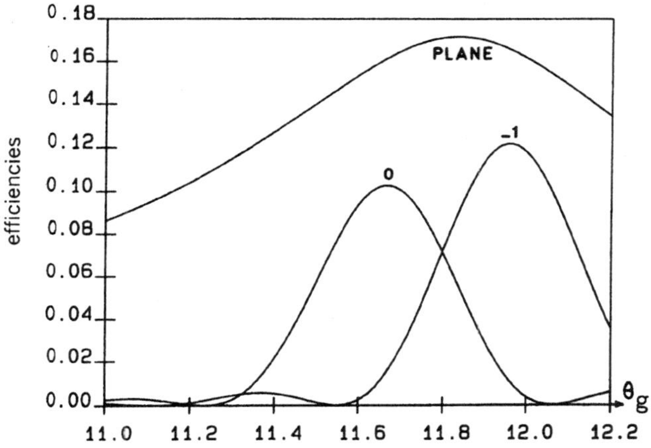

Fig. XI.14. Blaze peaks of a 4-μm groove spacing Bragg grating at $\lambda = 1.33$ nm and reflectivity of the flat stack (after [XI.22]).

Fig. XI.14 shows the reflectivity of the tungsten-carbon plane multilayer and the 0th and −1st order efficiencies of a 4-μm-period, 1.433° blaze angle, 88.567° apex angle triangular grating at wavelength 1.33 nm, as functions of θ_g in degrees. The −1-order efficiency reaches 71% of the reflectivity of the multilayer and culminates higher than the zeroth-order efficiency.

XI.6.2 Comparison with measurements in the hard X-ray domain

In order to design a Bragg grating for hard X-rays (working at $\lambda = 0.154$ nm), we choose a multilayer made of 60 silicon-tungsten bilayers deposited onto a crystalline silicon substrate. The uppermost layer is W. The Si layers have a 2 nm thickness, and the W layers have a 1.1 nm thickness. At $\lambda = 0.154$ nm, such a multilayer has a reflectivity which reaches 0.813 under 88.545° incidence ($\theta_g = 1.455^o$). A grating with a 4 μm period is etched inside the whole stack, and results in a symmetric lamellar profile. Taking $0.99999244 + i1.7\times10^{-7}$ and $0.99999543 + i4\times10^{-6}$ for the refractive indices of Si and W, respectively, the computed grating efficiencies as functions of the incidence θ_2 are shown in Fig. XI.15 for the central orders. Even at this short wavelength, the peak efficiencies approaches 40%.

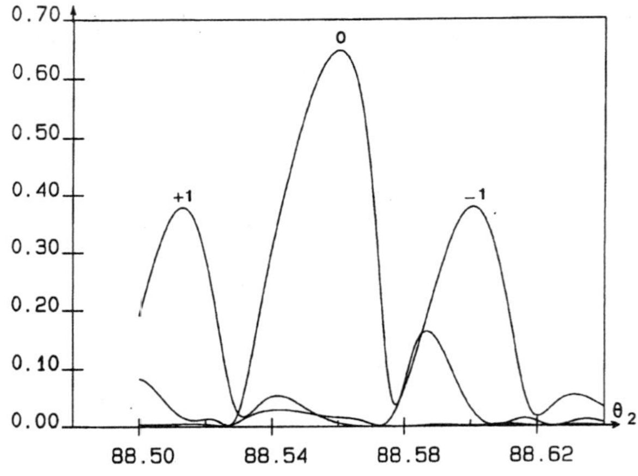

Fig. XI.15. Absolute efficiencies in the 0 and ±1 orders of a lamellar multilayer grating as functions of incidence at 0.154 nm wavelength (after [XI.22]).

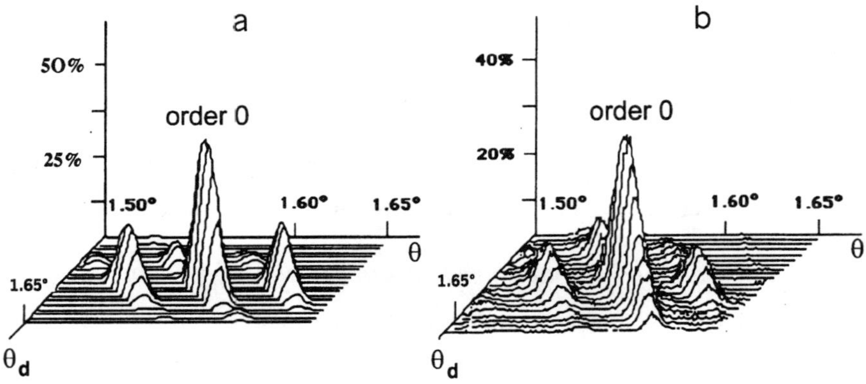

Fig. XI.16. Comparison between theoretical (a) and measured (b) efficiencies at a 0.154 nm wavelength (after [XI.23]).

The surprisingly high efficiencies obtained by the differential theory turns out to be in perfect agreement with those predicted independently by the modal theory [XI.23]. A stack of 80 carbon-tungsten bilayers with D=3 nm was produced [XI.23] and etched to make a symmetrical lamellar grating with a 4 μm period and 0.23 μm groove depth. Figure XI.16 shows a comparison [XI.23] between theoretical results and measurements, when the grating was lighted by a monochromatic beam of CuK_α radiation ($\lambda = 0.154$ nm). The efficiencies were measured in the two-dimensional scan mode [XI.23], i.e., as a function of both the grazing incident angle $\theta = \pi/2 - \theta_2$ and the grazing diffracted angle θ_d. The results prove that the macroscopic electromagnetic theory of gratings remains a reliable tool to optimize grating efficiencies, even at wavelengths close to the size of atoms at which the concept of refractive index could become questionable.

References

XI.1. R. P. Madden and J. Strong, *Diffraction Gratings,* Appendix P, in series "Concepts in Classical Optics" (Freeman and Co, San Francisco, USA, 1958)

XI.2. M. Nevière and J. Flamand, "Electromagnetic theory as it applies to X-ray and X-UV gratings," Nucl. Instr. and Meth. 172, 273-279, 1980

XI.3. M. Nevière, J. Flamand, and J. M. Lerner, "Optimization of gratings for soft X-ray monochromators," Nucl. Instr. and Meth., 195, 183-189, 1982

XI.4. M. Nevière, P. Vincent, and D. Maystre, "X-ray efficiencies of gratings," Appl. Opt. 17, 843-845, 1978

XI.5. E. G. Loewen and M. Nevière, "Simple selection rules for VUV and XUV diffraction gratings," Appl. Opt. 17, 1087-1092, 1978

XI.6. E. G. Loewen, M. Nevière, and D. Maystre, "On an asymptotic theory of diffraction gratings used in the scalar domain," J. Opt. Soc. Am. 68, 496-502, 1978

XI.7. E. G. Loewen and M. Nevière, "Grating efficiency theory as it applies to blazed and holographic gratings," Appl. Opt. 16, 2711-2721, 1977

XI.8. A. J. F. den Boggende, P. A. J. De Korte, P. H. Videler, A. C. Brinkman, S. M. Kahn, W. W. Craig, C. J. Hailey, and M. Nevière, "Efficiency of X-ray reflection gratings," in *X-ray Instrumentation in Astronomy II*, L. Golub ed., SPIE 982, 283-298, 1988

XI.9a. A. P. Lukirskii, E. P. Savinov, O. A. Ershov, and Y. U. F. Shepelev, "Reflection coefficients of radiation in the wavelength range from 23.6 to 113 Å for a number of elements and substances and the determination of the refractive index and absorption coefficient," Opt. Spectr. 16, 168-172, 1964

XI.9b. O. A. Ershov, I. A. Brytov, and A. P. Lukirskii, "Reflection of X rays from certain substances in the region from 7 to 44 Å," Opt. Spectr. 22, 66-69, 1967

XI.10. M. Berland, P. Dhez, M. Nevière, and J. Flamand, "X-ray ultraviolet grating measurement at LURE : comparison with electromagnetic theory predictions," in *Reflecting Optics for Synchrotron Radiation*, SPIE 315, 155-159, 1981

XI.11. A. J. F. den Boggende, M. Nevière, P. H. Videler, P. A. J. De Korte, A. C. Brinkman, and S. M. Kahn, "The X-ray efficiency of replicas of a master blazed reflection grating," in *X-ray instrumentation in Medecine and Biology, Plasma Physics, Astrophysics and Synchrotron Radiation,* R. Benattar ed., SPIE 1140, 533-541, 1989

XI.12a. T. W. Barbee, Jr., "Sputtered layered synthetic microstructure (LSM) dispersion elements," in *Low-Energy X-Ray Diagnostics* 1981, D. T. Atwood and B. L. Henke, eds., Vol. 75 of the AIP Conference Proceedings (American Institute of Physics, Washington, D. C., 1981), p. 131

XI.12b. E. Spiller, "Reflective multilayer coatings for the far UV region," Appl. Opt. 15, 2333-2338, 1976

XI.13a. T. W. Barbee, Jr., "Combined microstructure X-ray optics," Rev. Sci. Instrum. 60, 1588-1595, 1989

XI.13b. B. Vidal and P. Vincent, "Metallic multilayers for X-rays using classical thin-film theory," Appl. Opt. 23, 1794-1801, 1984

XI.14. C. Rife, W. R. Hunter, T. W. Barbee, Jr., and R. G. Cruddace, "Multilayer-coated blazed grating performance in the soft X-ray region," Appl. Opt. 28, 2984-2986, 1989

XI.15. V. V. Aristov, A.I. Erko, and V. V. Martynov, "Principles of Bragg-Fresnel multilayer optics," Rev. Phys. Appl. 23, 1623-1630, 1988

XI.16. B. Vidal, P. Vincent, P. Dhez, and M. Nevière, "Thin films and gratings: theories used to optimize the high reflectivity of mirrors and gratings for X-ray optics," in *Applications of Thin Film Multilayered Structures to Figured X-ray Optics*, G. F. Marshall, ed., Proc. SPIE 563, 142-149, 1985

XI.17. M. Nevière, "Multilayer coated gratings for X-ray diffraction: differential theory," J. Opt. Soc. Am. A8, 1468-1473, 1991

XI.18. A. J. F. den Boggende, M. P. Bruijn, J. Verhoeven, H. Zeijlemaker, E. J. Puik, and H. A. Padmore, "A broad-band multilayer coated blazed grating for X-ray wavelengths below 0.6 nm," in *Advanced X-Ray/EUV Radiation Sources and Applications*, J. P. Knauer and G. K. Shenoy eds., Proc. SPIE 1345, 189-197, 1990

XI.19. M. Nevière and A. J. F. den Boggende, "X-ray multilayer coated reflection gratings: theory and applications," in *Application and Theory of*

Periodic Structures, J. M. Lerner and W. R. McKinney, eds., SPIE 1547, ch. 7, 76-87, 1991

XI.20. M. Nevière and F. Montiel, "Soft X-ray multilayer coated echelle gratings: electromagnetic and phenomenological study," J. Opt. Soc. Am. A13, 811-818, 1996

XI.21. B. L. Henke, P. Lee, T. J. Tanaka, R. L. Shimabukuro, and B. K. Fugikawa, "Low energy X-ray interaction coefficients: photo absorption, scattering and reflection," in Vol. 27 of *Atomic and Nuclear Data Tables*, (American Institute of Physics, New York, 1982), pp. 1-144

XI.22. M. Nevière, "Bragg-Fresnel multilayer gratings: electromagnetic theory," J. Opt. Soc. Am. A 11, 1835-1845, 1994

XI.23. V. V. Martynov, B. Vidal, P. Vincent, M. Brunel, D. V. Roschupkin, Yu. Agafonov, A. Erko, and A. Yuakshin, "Comparison of modal and differential methods for multilayer gratings," Nucl. Instrum Methods Phys. Res. A 339, 617-625, 1994

Chapter XII

Transmission Gratings

Most diffraction gratings operate under reflection. Thus, transmission gratings have attracted a limited amount of interest, and the most well-known books on the topic (I.2, I.3, I.9) do not provide any set of efficiency curves that could help grating designers. Indeed, transmissions gratings are useful in various domains. In astrophysics, they are combined with a prism to make a dispersive element called a *Carpenter grating* or *grism*, which transforms a telescope into a spectrograph. In photolithography, chromium rod gratings (also called Ronchi gratings) are used as masks. The same gratings may be useful to make high-accuracy optical translation-rotation encoders. Holographic gratings deposited onto a flat waveguide are used in integrated optics to couple a laser beam into a slab. Blazed transmission gratings can be used as beam dividers and trapezoidal gratings as beam splitters [I.3]. The most useful applications will be considered in this chapter, with the exception of grating couplers, which will be analyzed in Chapter XIII.

XII.1 Blazing of transmission gratings

In view of their applications in astronomy, we first concentrate on echelette profiles.

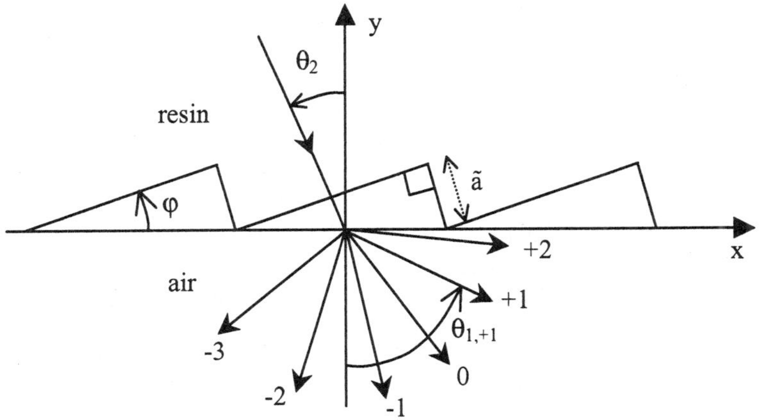

Fig. XII.1. Blazed transmission gratings and notations.

XII.1.1 Echelette profiles

The grating profile is represented in Fig. XII.1. It separates the superstrate, which is generally a resin and has refractive index ν_2, from air or vacuum ($\nu_1 = 1$). If we follow the reasoning of physical optics (scalar theory), the transmitted field will be maximum if it is phase matched to the incident field. Let us call $\tilde{a}$ the height of the small facet, which is equal to $d \sin\varphi$. In order to obtain the phase matching, we first choose the incident wavevector normal to the large facet ($\theta_2 = \varphi$). We then choose the grating period d so that the selected transmitted order propagates in the same direction as the incident wave. Finally we choose the small facet height $\tilde{a}$ so that the difference between optical paths in air and resist is equal to an integer number of wavelengths. The latter condition implies $(\nu_2 - 1)\tilde{a} = p\lambda_0$, where p is integer and λ_0 is the wavelength in vacuum. If we choose $p = 1$, we then derive the blaze angle φ by:

$$(\nu_2 - 1)\sin\varphi = \frac{\lambda_0}{d}. \tag{XII.1}$$

On the other hand, the Fraunhofer equation for transmitted orders gives:

$$\sin\theta_{1,m} = \nu_2 \sin\varphi + m\frac{\lambda_0}{d}.$$

If we choose $m = -1$ and $\theta_{1,m} = \varphi$, this equation identifies with Eq. (XII.1), ensuring both the phase matching and the non-deviation of the selected order. This represents the scalar blazing condition for transmission gratings and is called *the grism equation*, [XII.1]. Of course, reasoning from scalar optics are not rigorous from the electromagnetics point of view, and calculations may exhibit significant departures from this simple prediction, when the groove frequency of the grating is high. However, grism equation provides a nice starting point for an optimization process based on Maxwell equations.

To study the behavior of grisms, we first choose a regular resin that is commonly used by a well-known grating manufacturer, and interpolate its refractive index ν_2 from the values found in [XII.1]. We then choose six values of wavelength, called λ_c, in the visible (and near-infrared region) and four groove frequencies: 300, 600, 900 and 1200 mm^{-1}. Table XII.I shows both the values of ν_2 used in the following and the values of $\varphi = \arcsin[\lambda_0/d(\nu_2 - 1)]$, which fulfill the grism condition. Of course, when φ is too large, the -1st transmitted order no longer exists. Thus we have limited the values of φ to a maximum of ~ 66°.

Table XII.1. Blaze angles corresponding to wavelengths λ_c and four groove frequencies for a resin with refractive index ν_2, deduced from the grism equation (after XII.2]).

		φ [deg]			
λ_c (μm)	ν_2	300 mm^{-1}	600 mm^{-1}	900 mm^{-1}	1200 mm^{-1}
0.4	1.610	11.345	23.169	36.169	51.896
0.5	1.597	14.552	30.166	48.918	
0.6	1.590	17.763	37.602	66.242	
0.7	1.584	21.075	45.987		
0.8	1.580	24.443	55.851		
0.9	1.577	27.916			

To obtain a broad description of the behavior of transmission gratings, Figs. XII.2-XII.16 show the results for all the gratings described in Table XII.1, obtained with the electromagnetic theory in both TE and TM polarization. The efficiency in natural light is also plotted in the figures.

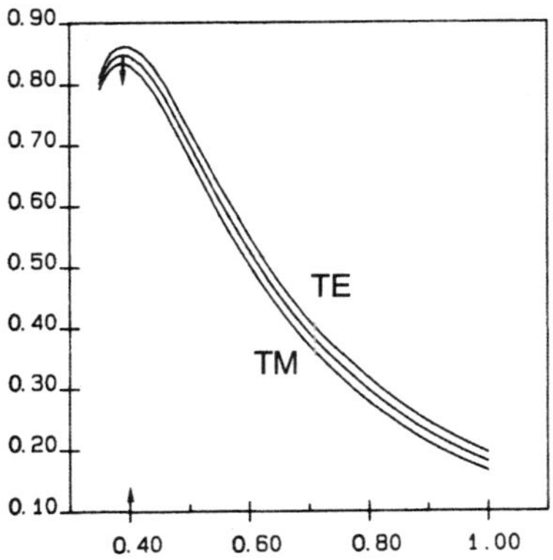

Fig. XII.2. -1st order efficiencies in TE polarization, TM polarization and for natural light as functions of wavelength λ_0 (in μm) for a 300-groove/mm transmission grating with $\varphi = 11.345^{\circ}$ ($\lambda_c = 0.4$ μm).

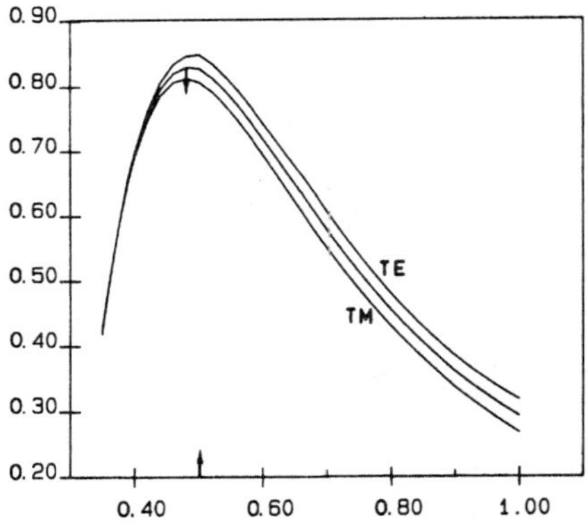

Fig. XII.3. Same as Fig. XII.2 but with $\varphi = 14.552^{\circ}$, $\lambda_c = 0.5\ \mu m$.

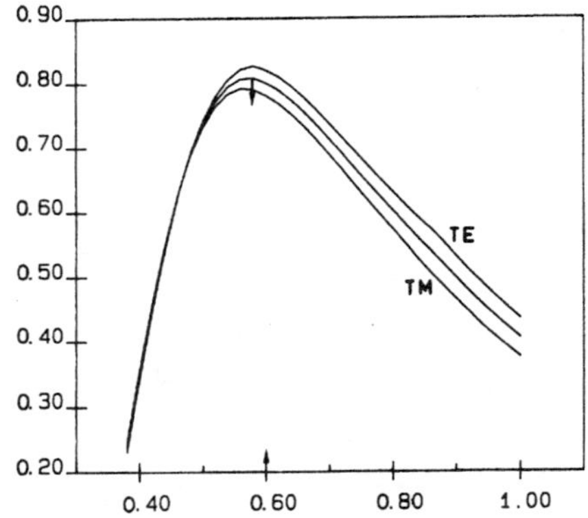

Fig. XII.4. $\varphi = 17.763^{\circ}$, $\lambda_c = 0.6\ \mu m$.

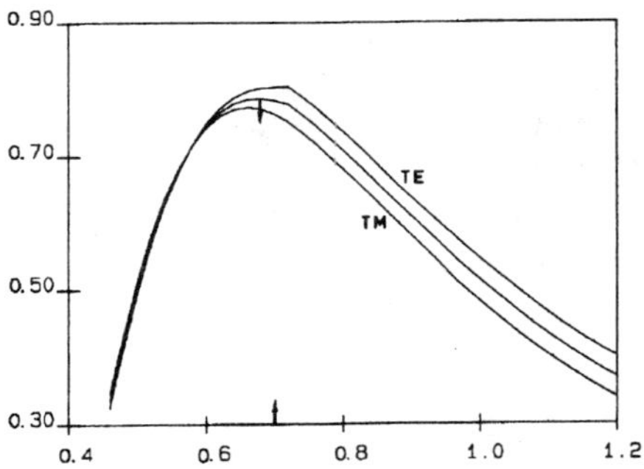

Fig. XII.5. $\varphi = 21.075^{\circ}$ $\lambda_c = 0.7\ \mu m$.

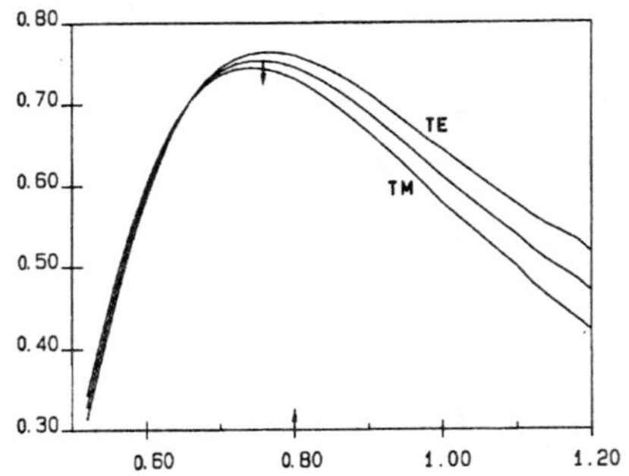

Fig. XII.6. $\varphi = 24.443^{\circ}$, $\lambda_c = 0.8\ \mu m$.

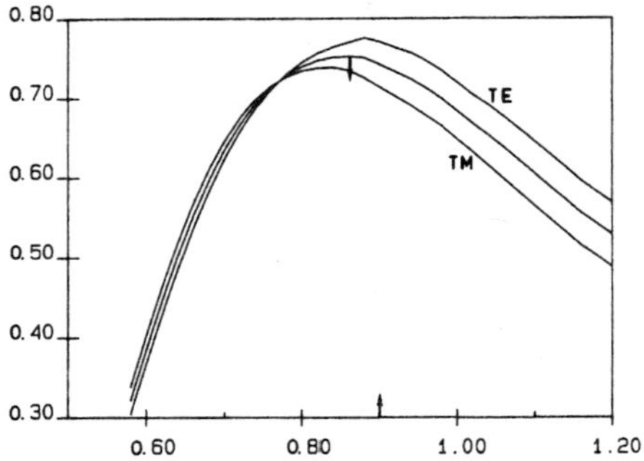

Fig. XII.7. Same as Fig. XII.2 but with $\varphi = 27.916^{\circ}$, $\lambda_c = 0.9\ \mu m$ (Figs. XII.2-XII.7, after [XII.2]).

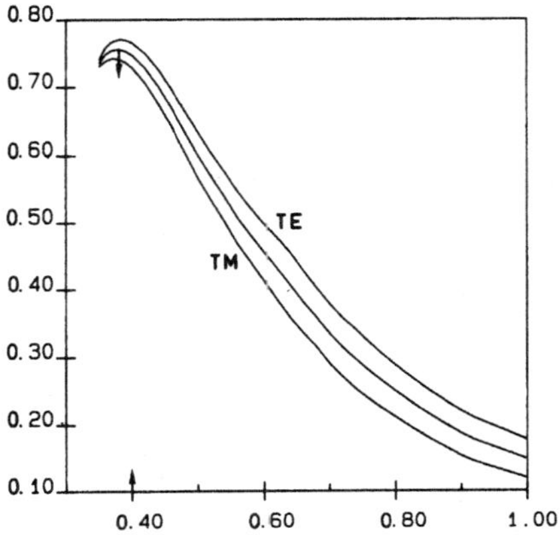

Fig. XII.8. -1st order efficiencies as functions of wavelength for a 600-groove/mm transmission grating with $\varphi = 23.169^{\circ}$ and $\lambda_c = 0.4\ \mu m$ (Figs.XII.8 – 12, after [XII.2]).

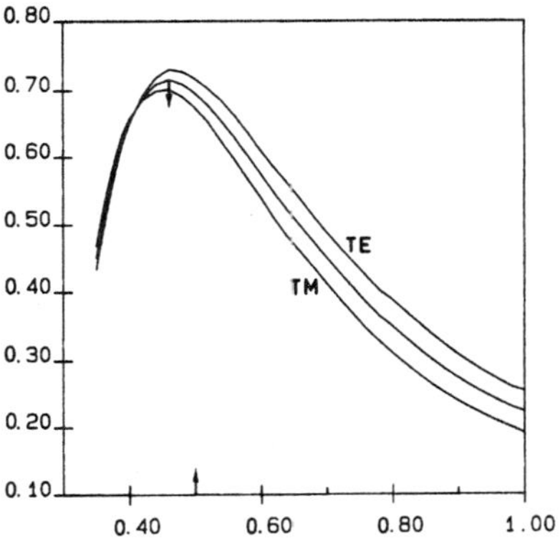

Fig. XII.9. $\varphi = 30.166^{\circ}$, $\lambda_c = 0.5\ \mu m$.

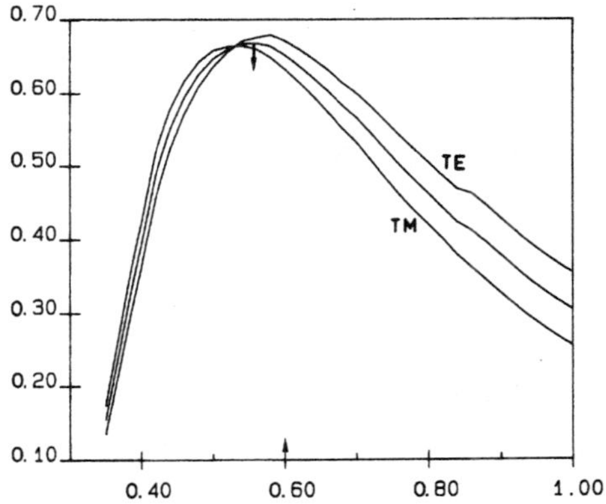

Fig. XII.10. $\varphi = 37.602^{\circ}$, $\lambda_c = 0.6\ \mu m$.

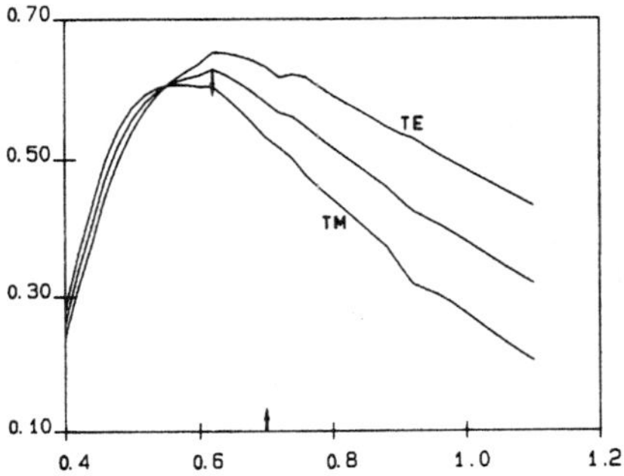

Fig. XII.11. $\varphi = 45.987^{\circ}$, $\lambda_c = 0.7\ \mu m$.

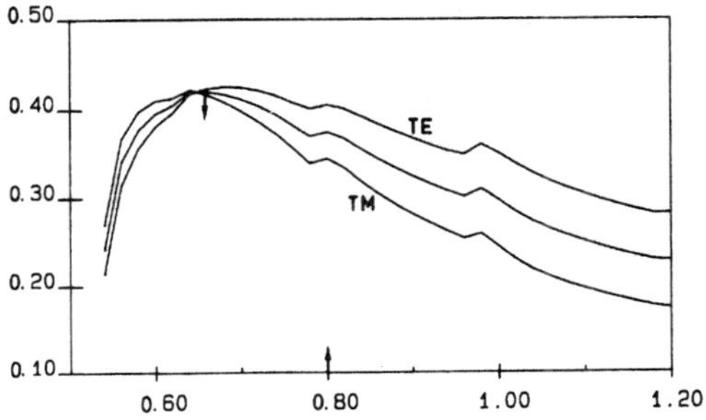

Fig. XII.12. $\varphi = 55.851^{\circ}$, $\lambda_c = 0.8\ \mu m$.

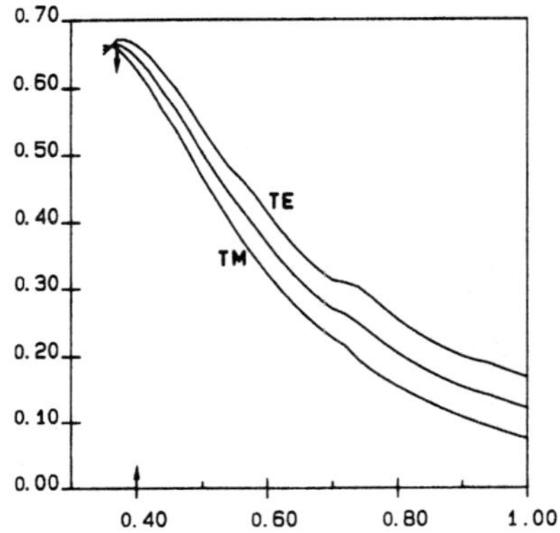

Fig. XII.13. Efficiencies as functions of wavelength for a 900-groove/mm transmission grating with $\varphi = 36.169^{\circ}$, $\lambda_c = 0.4\ \mu m$ (after [XII.2]).

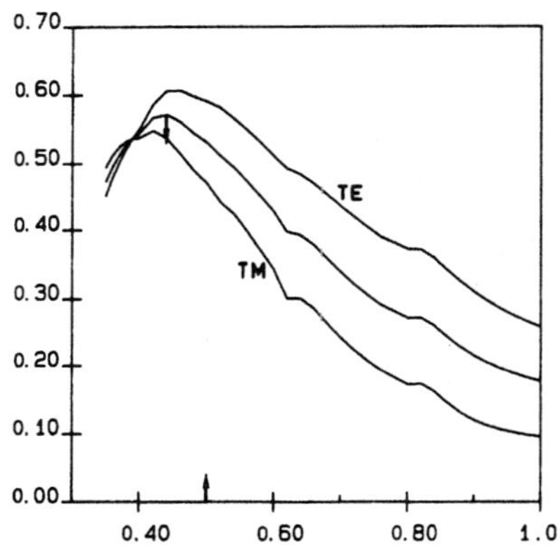

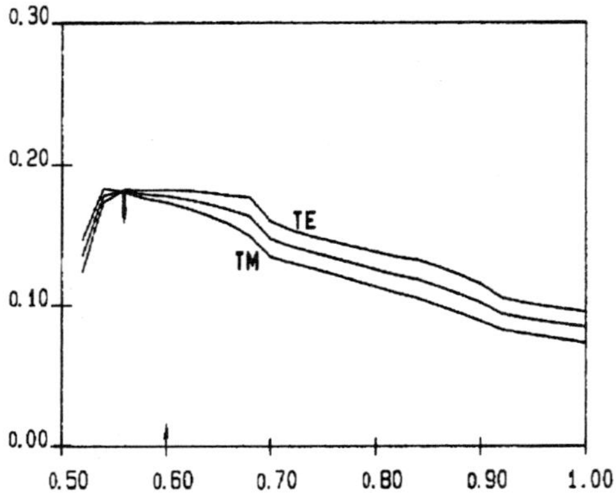

Fig.XII.14. $\varphi = 48.918^{\circ}$, $\lambda_c = 0.5\ \mu m$. Fig. XII.15. $\varphi = 66.242^{\circ}$ $\lambda_c = 0.6\ \mu m$.

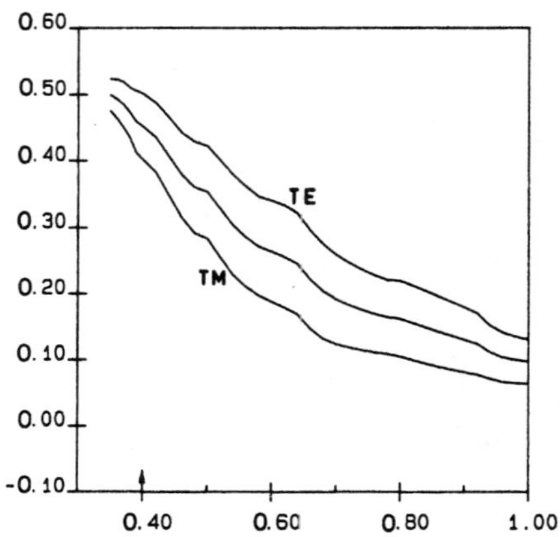

Fig. XII.16. Efficiencies as functions of wavelength for a 1200-groove/mm transmission grating with $\varphi = 51.896^{\circ}$, $\lambda_c = 0.4\ \mu m$ (after [XII.2]).

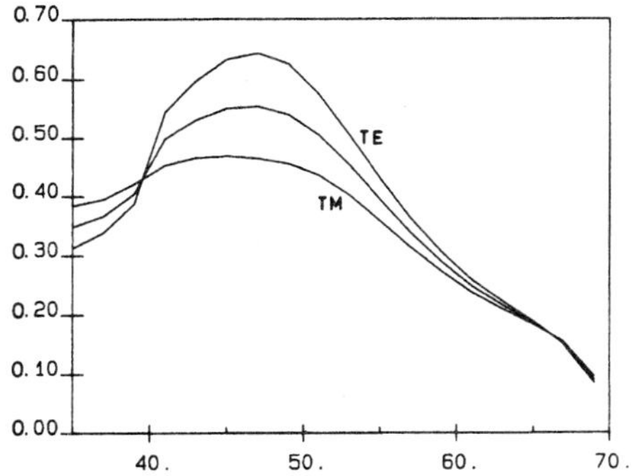

Fig. XII.17. Efficiencies as functions of angle of incidence θ (in degrees) for the grating of Fig. XII.12 ($\varphi = 55.851^o$, 600 groove/mm) at wavelength $\lambda_0 = 0.8\ \mu m$ (after [XII.2]).

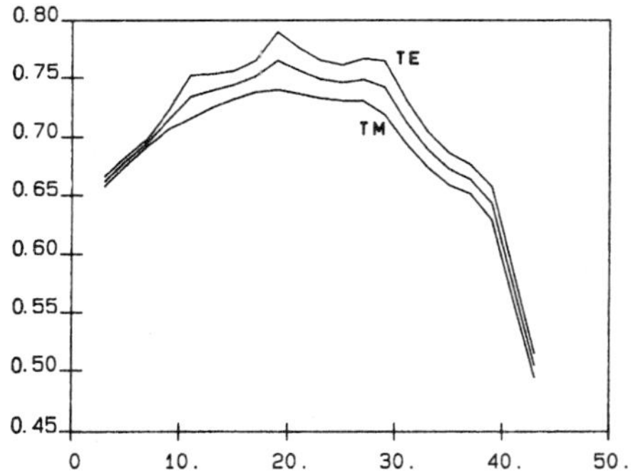

Fig. XII.18. Efficiencies as functions of angle of incidence θ for the grating of Fig. XII.8 ($\varphi = 23.169^o$, 600 groove/mm) at wavelength $\lambda_0 = 0.4\ \mu m$ (after [XII.2]).

The first feeling when looking at these curves is that, on the whole, they are much more regular than those obtained for reflection gratings. Anomalies are weak or nonexistent. TE and TM curves present one crossover only, and follow similar shapes. The TE blaze peak is always slightly higher than the TM peak. Their shift is always small or negligible, and polarization effects are low, except in a few cases. A comparison between Figs. XII.2, XII.8, XII.13 and XII.16, or between Figs. XII.3, XII.9 and XII.14, or between Figs. XII.4, XII.10 and XII.15 shows, as expected, that polarization effects increase with the groove frequency for a given λ_c. At the same time, the height of the natural light blaze peak reduces from 84% to 50% for the first series, and from 82% to 57% for the second series. The losses in efficiency become more severe when λ_c is shifted from 0.4 to 0.5 and to 0.6 μm, which assumes stronger increases of blaze angles and incidences. If the groove frequency is kept constant, a comparison between Figs. XII.2-XII.7, or between Figs. XII.8-XII.12, or between Figs. XII.13-XII.15 shows that polarization effects increase with λ_c, i.e., with the blaze angle and the incidence. For the first series (Figs. XII.2-XII.7) only a slow reduction of blaze peak (from 84% to 75%) is seen when λ_c

moves from 0.4 to 0.9 μm, but blaze peaks deteriorate much faster on other series when the blaze angle comes close to 50° or more.

Another interesting question to examine is whether the grism condition really predicts the correct blaze angle to get maximum efficiency at the chosen wavelength λ_c. To answer the question, we have drawn two arrows in Figs. XII.2-XII.15, one corresponding to the chosen wavelength λ_c and another corresponding to the real blaze wavelength λ_B (abscissa of the natural light efficiency maximum). It is seen that for low blaze angles, λ_B is almost identical to λ_c, and that the difference $\lambda_c - \lambda_B$ does not exceed 50 nm if $\varphi < 40^o$. On the other hand, in Fig. XII.12, for example, ($\lambda_c = 0.8\,\mu m$, $\varphi = 55.851^o$), the shift may approach 150 nm, and the efficiency at $\lambda = \lambda_c$ is only 38%. The question is whether a change in incidence ($\theta_2 \neq \varphi$) could improve the efficiency. Figure XII.17 answers the question; a tilt of the grating to reduce the incidence from 55° to 47° increases the natural light efficiency from 38 to 55%. Similar results have already been observed experimentally and reported in [XII.3] for an echelette grating used in higher orders. The conclusion is that, for high blaze angles, the grism condition alone is not sufficient to find the optimal use of a transmission grating. In fact, if we maintain constant the groove frequency, wavelength, and resist, a complete optimization of both blaze angle and incidence through the electromagnetic theory leads to an optimal efficiency of 56.4% reached for $\varphi = 51.375^o$ and $\theta_2 = 44.156^o$. This is far more than the 38% obtained through the grism condition, but not far from the 55% reached in Fig. XII.17 after a slight tilting of the grating of Fig. XII.12.

On the other hand, the efficiency curves as a function of incidence illustrated in Fig. XII.18 show that, for low and medium blaze angles, the grism condition indeed leads to an angle of incidence close to the optimum. A shift of the incidence from 23° to an optimum value of 19° increases the efficiency only from 74.9 to 76.5%, i.e., by a mere 1.6%.

XII.1.2 Sinusoidal profiles

One may wonder about the origin of the blaze effect of transmission gratings. Following the reasoning of scalar optics, generally pointed out by grating users and manufacturers, the blaze effects occurs "when the microprisms that are formed by each groove refract light at the same angle as the angle of diffraction of the ruling" [XII.1]. It is easy to see that this is equivalent to the grism equation. Thus, for sure, there is some physical insight in this explanation. But one might incorrectly think that the blaze effect would have to disappear if the microprisms disappear, e.g., if we replace the triangular profile by a sinusoidal or a lamellar one. To clear up this point, we chose the grating parameters of Fig. XII.9 (600 grooves/mm, angle of incidence $\theta_2 = 30.166^o$) and we computed the efficiencies at $0.5\ \mu m$ wavelength for different sinusoidal gratings, with grooves depth a varying from 0.1 to $0.9\ \mu m$.

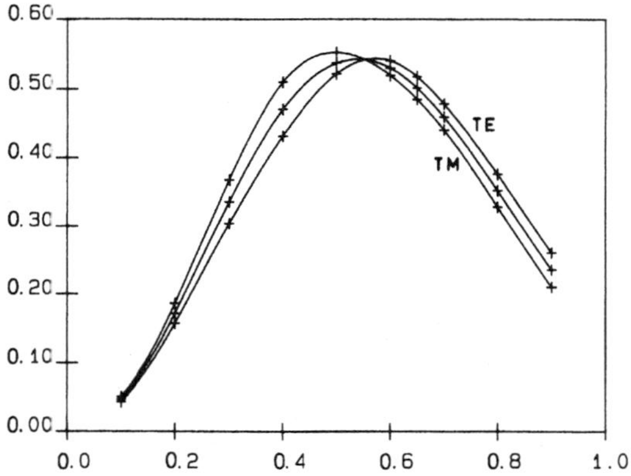

Fig. XII.19. Efficiencies as functions of groove depth a for a 600-groove/mm sinusoidal grating ($\theta_2 = 30.166^o, \lambda = 0.5\ \mu m$) (after XII.2]).

Fig. XII.19 shows that, for natural light, a blaze peak of 0.542 is obtained when $a \simeq 0.53\ \mu m$. Thus a blaze effect still exists, although it is somewhat weaker than the one obtained with the triangular profile that can reach 72%. The conclusion is that the blaze effect is not strictly related to the existence of microprisms, but asymmetric profiles have a greater potentiality to concentrate energy into one given order. For memory, we recall that, for reflection gratings, it has been established that triangular, sinusoidal, and lamellar profiles are equal competitors [XII.4] to concentrate energy in a given direction. The difference in behavior comes from the fact that reflection gratings are generally used in the resonance domain ($\lambda/d \sim 1$), where smaller number of diffraction orders propagates, whereas transmission gratings are used at lower λ/d ratios, where a scalar behavior begins to appear and the gratings supports multiple diffraction orders. In the scalar domain, also for reflection gratings, sinusoidal profiles give poorer results than triangular profiles.

The main impression which arises from Sections XII.1.1 and XII.1.2 is that current, medium modulated transmission gratings have poorer efficiency performances than their homologous reflection ones. However, modern technology (e.g., ion etching) now allows producing gratings with modulation ratios a/d which exceed unity. Such deep gratings have completely different blazing capabilities, as shown in the next section.

XII.1.3 Deep transmission gratings

Assuming a fixed wavelength λ_0 in vacuum such that $\lambda_0/d = 1.414$ and Littrow incidence, Fig. XII.20 shows the -1st order efficiency of a resin grating ($\nu_2 = 1.66$) as function of the modulation a/d for TE polarization. The three common grating profiles reach efficiencies between 90% and 100%, for modulations close to 2. It is worth noting that for triangular profiles, symmetrical ones gives 90% efficiencies, while for metallic gratings the blaze effect is obtained with asymmetrical profiles. Another important remark from the manufacturer's point of view is that the maximum value does not significantly depends on the groove shape, although it does influence the groove depth value at which the high blazing occurs. This is of great importance since it is difficult to obtain a deep grating with a carefully specified profile.

The spectral dependence of diffraction efficiencies of deep gratings has been studied in details in [XII.5]. Roughly speaking, the TM efficiency curves resembles the TE efficiency behavior of metallic gratings and vice versa. The blaze peak is higher for low values of the resin index (ν_2), at the cost of a deeper optimal modulation.

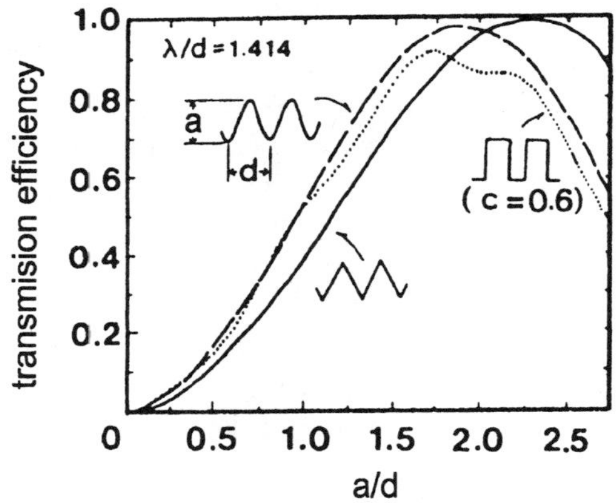

Fig. XII.20. Groove depth dependence of the -1st order transmission efficiency for different profiles lighted under Littrow incidence; for rectangular groove, the groove width-to-groove spacing ratio is 0.6 (after [XII.5]).

XII.1.4 An analytical solution for perfect blazing of triangular profiles

Since the work of Maréchal and Stroke [XII.6], an explicit solution for obtaining the perfect blazing of a perfectly conducting echelette grating is well known. It consist of lighting the grating with a wavevector normal to the large facet ($\theta_2 = \varphi$) in TM polarization, and -1st order Littrow mount, i.e., $\lambda = 2d\sin\theta_2$. Since $\theta_2 = \varphi$, Littrow condition then implies that $d\sin\varphi = \lambda/2$, which means that the height of the small facets $\tilde{a}$ is equal to $\lambda/2$. It can then easily be verified that the boundary conditions of electromagnetism are satisfied by adding to the incident wave a counter-propagating wave with the same amplitude and a phase-mismatch of π. This counter-propagating direction is that of the -1st order in Littrow mount, which, thus, carries 100% efficiency.

Using similar arguments, the authors of [XII.7] have proposed a solution for transmission gratings. They first choose a refractive index ν_1 given by:

$$\nu_1 = \frac{p+2q}{p}, \qquad \text{(XII.2)}$$

with p, q integers.

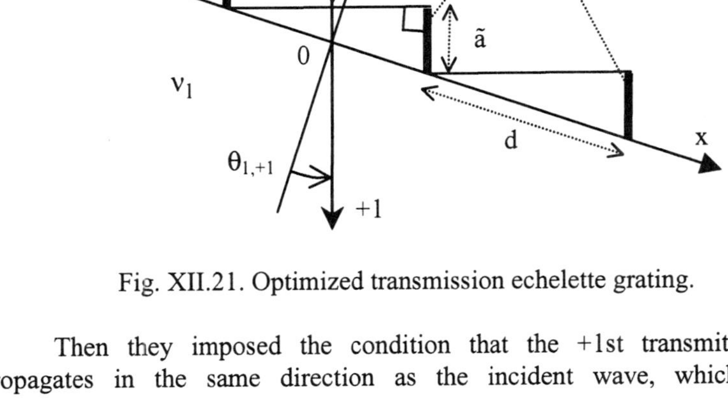

Fig. XII.21. Optimized transmission echelette grating.

Then they imposed the condition that the $+1$st transmitted order propagates in the same direction as the incident wave, which implies

$\tilde{a} = \frac{\lambda}{\nu_2 - 1}$. Putting an infinitely thin layer of perfectly conducting metal on the small facets ensures that the reflected orders reduce to the Fresnel reflected wave on the large facet, and that the transmitted orders reduce to the +1st order which carries all the non-reflected energy. If the grating is made with glass ($\nu_2 = 1.5$, which is obtained by taking $q = 1$ and $p = 4$ in Eq. (XII.2)), the +1st transmitted order thus carries 96% of the incident light. One may go to perfect blazing ($e_{+1}^{(t)} = 1$) by adding an antireflection coating on the large facets.

XII.2 Grism design

For certain applications, it is useful to have a dispersive system that provides in-line viewing at one central wavelength. This is realized in direct vision spectroscopes and is particularly useful in astronomical spectrographs. Moderns observatories work at the top of high mountains where due to the elevation, any change in the observing apparatus is difficult and time consuming. It is then particularly useful to be able to quickly transform a telescope into a spectrograph which provides the spectrum in the same chamber as the image chamber, *whatever the dispersion may be.* This is achieved by Carpenter gratings or grisms, obtained by replicating a resin echelette grating onto the hypotenuse face of a right-angle prism. On top of saving the telescope time, grisms make the mounting insensible to small vibrations, due to the stationary character of the deviation [XII.8].

XII.2.1 Optimization process

Although grisms were initially designed to be used with the incident beam normal to the entrance face of the prism, both experiments [XII.3] and calculations [XII.2] have shown the interest of slightly tilting them from that position. The detailed ray path is then illustrated in Fig. XII.22, which defines the various angles used in that section. The prism angle is denoted by φ_p. The refractive index of glass is chosen to match the one of the resin as closely as possible. The choice of the groove spacing d is linked with the required dispersion, but the groove parameters φ and d have to closely satisfy the grism equation (XII.1) which is the starting point of the optimization process. We recall that this equation also ensures a direct vision at a selected wavelength λ_B. The full optimization process is done through the following steps.

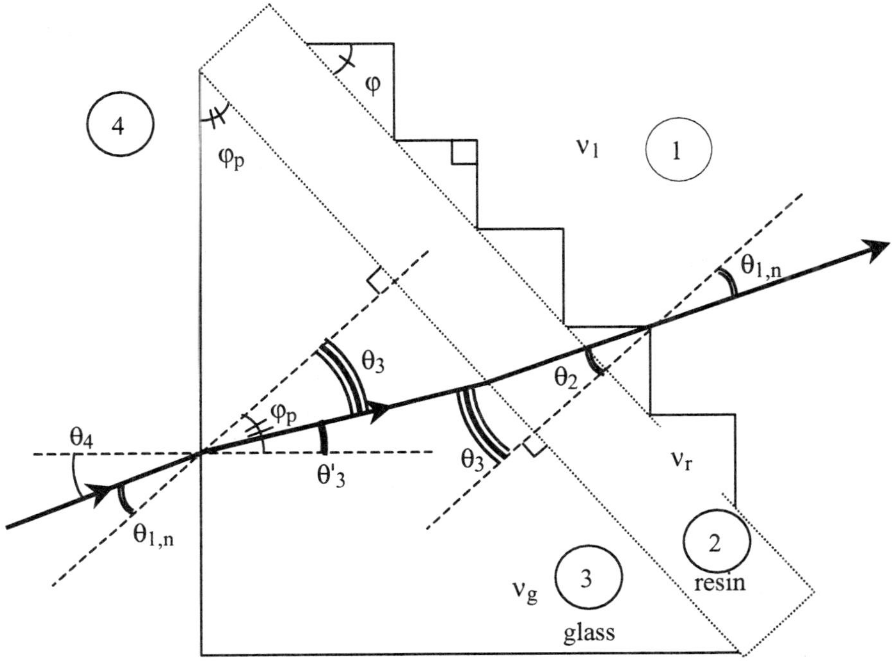

Fig. XII.22. Detailed ray path of a grism.

(i) The modulated interface is first studied separately. The transmitted efficiency $e_n^{(t)}$ in the selected order (e.g., $n = 1$) is computed at wavelength λ_B as a function of all free parameters, i.e., as a function of d, φ, apex angle φ_a of the triangular profile (chosen to be 90° in the figure) and incidence θ_2. This can be done with the classical differential theory (Chapter II) in most cases. The starting values of φ_a and θ_2 may be respectively chosen equal to 90° and φ. A code using the classical differential theory as subroutine looks for the maximum of the function $e_n^{(t)}(d, \varphi, \varphi_a, \theta_2)$ through an iterative process, using a step-by-step method that does not assume the derivability of the function. The process gives the optimized values of all free parameters, among which are φ and θ_2.

(ii) If we designate by ν_g and ν_r the relative indices of glass and resin, Snell's law reads:

$$\sin\theta_3' = \frac{\sin\theta_4}{\nu_g} \quad \text{(XII.3)}$$

$$\nu_r \sin\theta_2 = \nu_g \sin\theta_3 \quad \text{(XII.4)}$$

As it is made clear on the figure:

$$\theta_3 = \varphi_p - \theta_3' \quad \text{(XII.5)}$$

and

$$\varphi_p = \theta_4 + \theta_{1,n} \quad \text{(XII.6)}$$

thus we obtain:

$$\mathrm{tg}\varphi_p = \frac{n\lambda_B}{d(\cos\theta_{1,n} - \nu_g \cos\theta_3)} \quad \text{(XII.7)}$$

where $\theta_{1,n}$ is given by $\sin\theta_{1,n} = \nu_r \sin\theta_2 + n\frac{\lambda_B}{d}$ and θ_3 is derived through Eq. (XII.4) from the optimized value of θ_2. Eq. (XII.7) gives the correct prism angle φ_p which ensures direct vision at the chosen wavelength λ_B (in vacuum). From θ_3, Eq. (XII.5) gives θ_3'. The incidence θ_4 on the entrance face comes from Eq. (XII.3).

(iii) Knowing incidence (θ_4 and θ_3) and the refractive indices ν_g and ν_r, it is then possible to compute the Fresnel transmission coefficients for the two faces of the prism. Their product multiplied by the TE and TM efficiency of the grating gives the TE and TM efficiencies of the entire grism.

XII.2.2 Grisms for Palomar Observatory and for European Southern Observatory

As an example of a full grism performance, we show in Fig. XII.23 the efficiency behavior of a grism which was made for the Department of Astronomy at the Palomar Observatory.

The desired dispersion in the −1st order required a groove frequency close to 800 mm^{-1}. The central wavelength λ_B was 0.435 μm. The prism was made with UBK7 glass, and the resin was resin A of Bausch and Lomb grating manufacturer [XII.1]. Keeping $\varphi_a = 90°$, the optimization of d, φ and θ_2 led to the optimal values: $d = 1.30208\,\mu m$ (768 gr/mm), $\varphi = 36.25°$, and

$\theta_2 = 38.75^o$. From these values, we deduced $\varphi_p = 39.33^o$; $\theta_4 = -2.78^o$ and $\theta_{1,n} (\equiv \theta_{1,-1}) = 42.10^o$. The natural light transmitted efficiency of the entire grism reached 67.7% at the desired wavelength λ_B, and the TE and TM spectral dependence are shown in Fig. XII.23. As can be deduced from the curves presented in Section XII.1.1, when smaller dispersion is required, higher efficiencies can be obtained.

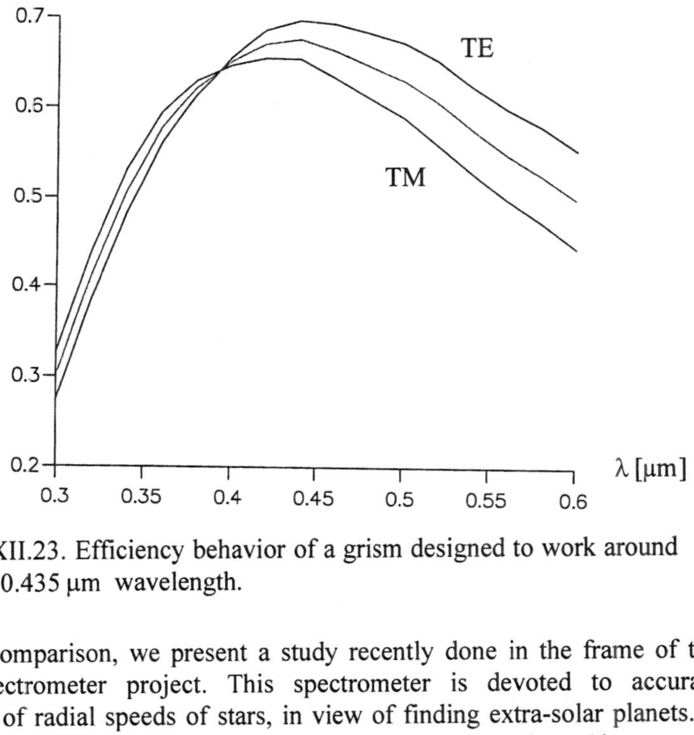

Fig. XII.23. Efficiency behavior of a grism designed to work around a 0.435 μm wavelength.

For comparison, we present a study recently done in the frame of the HARPS spectrometer project. This spectrometer is devoted to accurate mesurement of radial speeds of stars, in view of finding extra-solar planets. It will be mounted on the 3.60 m telescope of the European Southern Observatory. The central wavelength is $\lambda_B = 0.49\ \mu m$. The prism material is FK5 from Schott Inc., and the resin is again resin A of Bausch and Lomb Inc. Its angle φ_p is equal to 13.235°, while the groove frequency is 257.14 mm^{-1}. The optimization process of the remaining free parameters led to $\varphi = 12.235^o$ and $\theta_2 = 8.75^o$. This resulted in $\theta_4 = 5.74553^o$, which meant that the incident beam had to be slightly tilted toward the bottom of the prism. The predicted maximum natural light efficiency of the entire grism, which includes an antireflection coating on the entrance face of the prism, this time culminated at 84%. Following the design specification, a grism was made and measured under predicted working conditions. Fig. XII.24 shows the comparison between predicted and measured performances. It clearly shows how useful an electromagnetic study of gratings may be, specialy in a domain in which photons are expensive.

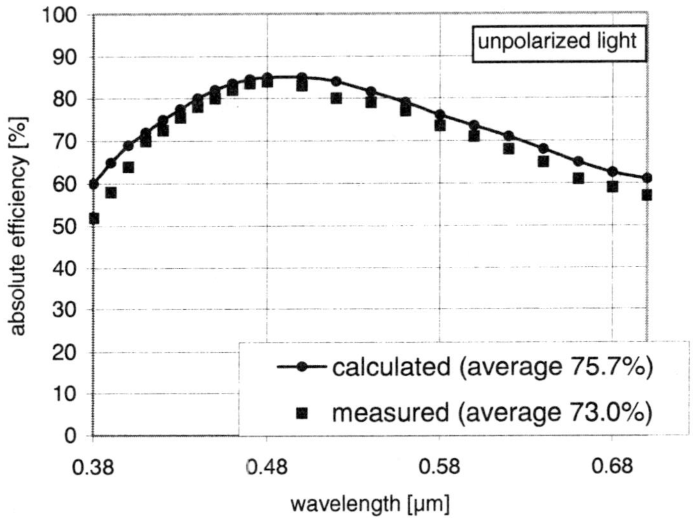

Fig. XII.24. Theoretical and measured efficiency of the HARPS grism. The "averaged" values in the legend are the values of the efficiencies averaged over the 0.38 – 0.69 μm spectral domain of interest.

XII.3 Ronchi gratings as photolithography masks

Ronchi gratings are gratings made with opaque rods. They are commonly made in chromium patterns by photolithographic methods. Assuming rectangular rod cross-section, the modal theory [I.16] can simply be adapted to that situation, and is able to accurately predict their efficiencies.

An interesting and simple situation is obtained when the grating in used in -1st order Littrow mount and the ratio λ/d is chosen in such a way that only two reflected and transmitted orders (the 0th and -1st ones) propagate, Fig.XII.25. This implies:

$$\lambda = 2d \sin \theta_2 \tag{XII.8}$$

or $\alpha_0 = \pi/d$ and

$$\frac{2}{3} < \frac{\lambda}{d} < 2 . \tag{XII.9}$$

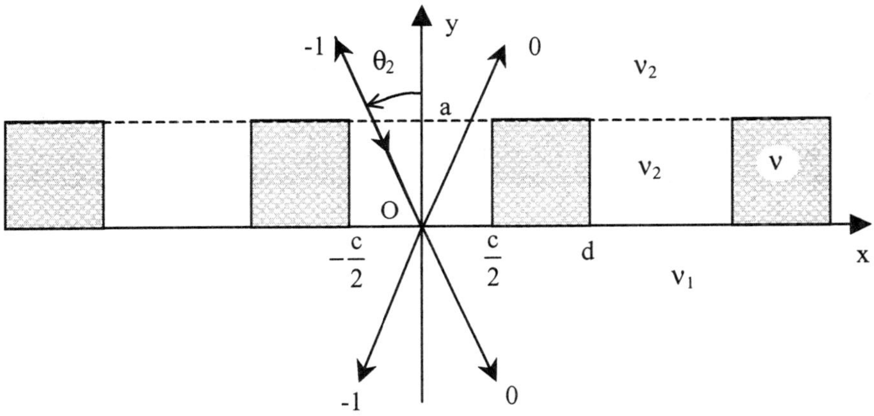

Fig. XII.25. Photolithographic setup.

When the rods are highly conducting, the field E inside the rectangular grooves is approximately given by the classical theory of cavities. For $x \in \left[-\frac{c}{2}, \frac{c}{2}\right]$ and $y \in [0, a]$ in Fig. XII.25, it is given [XII.9] by the modal series:

$$E(x, y) = \sum_{m=1}^{\infty} \left[a_m \, e^{i\rho_m (y+a)} + b_m \, e^{-i\rho_m (y-a)} \right] \sin\left[m \frac{\pi}{c} \left(x + \frac{c}{2} \right) \right] \tag{XII.10}$$

suggested by Eq. (I.6), where

$$\rho_m = \sqrt{(\nu_2 k_0)^2 - \left(m \frac{\pi}{c} \right)^2} , \tag{XII.11}$$

with $\mathrm{Im}(\rho_m) > 0$.

Let us discuss the propagative behavior of these modes assuming $\nu_2 = 1$. A mode will propagate toward $\pm y$ if $\mathrm{Re}(\rho_m) > 0$, which implies

$$m < k_0\, c/\pi\,. \qquad \text{(XII.12)}$$

As a result, if Eq. (XII.12) if violated for the lowest value of m, it will be violated for all modes. The conclusion is that if c is smaller than a limit value $c_l = \frac{\lambda}{2}$, all modes are evanescent. If the groove depth a is high enough, only the dominating mode remains at the bottom of the grooves, and we have [XII.10]:

$$E(x,0) \approx (a_1 + b_1)e^{i\rho_1 a}\cos\left(\frac{\pi}{c}x\right) \qquad \text{(XII.13)}$$

Of course, when we move from the first groove to the next one on its right, the pseudoperiodicity of the field implies that E given by Eq. (XII.13) is multiplied by $\exp(i\alpha_0 d) = -1$. The same factor applies to the neighboring groove on the left. As a result, the field at $y = 0$ is periodic with respect to x with period 2d and $E(x,0)$ is an *even function* of x. The same parity applies to the transmitted field which is derived from $E(x,0)$ through the boundary conditions, and thus the 0th and -1st order Rayleigh coefficients have equal modulus. We recall that, due to Littrow mount, these orders propagate in directions symmetric with respect to the grating normal.

Two different regimes of interest can then be distinguished, which both respect Eq. (XII.9):

(i) If $\mathbf{c} < \frac{\boldsymbol{\lambda}}{\mathbf{2}}$, all modes are evanescent; thus the transmitted field is expected to decrease exponentially when increasing a, as implied by Eq. (XII.13) in which $\rho_1 = i\sqrt{\left(\frac{\pi}{c}\right)^2 - k_0^2}$. The device behaves as a low-pass filter.

(ii) If $\frac{\boldsymbol{\lambda}}{\mathbf{2}} < \mathbf{c} < \boldsymbol{\lambda}$, the field at $y = 0$ can be approximated by Eq. (XII.13) in which ρ_1 is real ($\rho_1 = \sqrt{k_0^2 - \left(\frac{\pi}{c}\right)^2}$), provided a is high enough to allow neglecting the higher-order evanescent modes. The field at the bottom of the grooves is no longer negligible. So are the transmitted 0th and -1st order efficiencies, which are typically in the range 15 to 25% [XII.9]. The device then produces two transmitted beams with equal amplitudes, which propagate in two directions symmetric with respect to the grating normal.

This property is used in a process for a fairful optical reproduction on a support of a mask made with periodically distributed slits [XII.11]. Nowadays the structures constituting modern electronic elements have submicrometer dimensions. When duplicated by photolithography, a complex diffraction phenomenon occurs, which may produce duplicated images, and prevents reproduction on the support of a cast shadow which should correspond fairly well to the mask and should otherwise have perfectly contrasted contours. The invention cited in [XII.11] remedies these disadvantages. The two symmetric, equal amplitude transmitted orders interfere with 100% contrast whatever the distance between the mask and the support maybe. This process [XII.11], discovered at the end of the 1970s through numerical computations quite a long time before receiving a theoretical explanation [XII.10], has been reinvented by several companies [XII.12-XII.15] and published under the name of the *photomask self-interference method*. It is nowadays widely used in industry. Electromagnetic studies [XII.9] have shown that the mounting presents wide tolerances with respect to incidence, modulation and even polarization. As a result, it turns out to be useful to duplicate double-frequency gratings, non-periodic gratings such as Bragg-Fresnel linear zone plates, curved gratings, etc. [XII.16].

XII.4 High-accuracy translation-rotation encoders

The property of transmission gratings used in Littrow mount to produce only two propagating orders with equal intensity, discussed in the previous section, holds not only for Ronchi (rod) gratings, but also for deep lamellar dielectric (e.g., plastic) gratings and can be used to make translation-rotation encoders [XII.17] and [XII.18]. This can be achieved by using a second grating to recombine into an intereference pattern the beams produced by the first grating.

XII.4.1 Encoder used in normal incidence

A combination of two gratings can serve to measure interferometrically the relative displacement between two gratings. A transmission grating (called upper grating) used under normal incidence [XII.19, XII.20] produces several transmitted beams, Fig. XII.26. A detection is made by a second grating which causes two of the beams (e.g., the +1 and −1 orders) diffracted by the upper grating to interfere. If the upper grating moves at a distance Δx along an axis Ox normal to the grooves and situated in the mean plane of the upper grating, the +1 order experiences a phase shift $\tilde{\varphi}$, given by

$$\tilde{\varphi} = 2\pi(\Delta x/d)\,, \qquad \text{(XII.14)}$$

where d is the period of the upper grating, while the -1st order experiences a phase shift $-\tilde{\varphi}$. The phase difference between the two beams results in an interference pattern below the second grating, which depends on Δx. Equation (XII.14) shows that the displacement that will be measured by detection of the interference fringes produced below the second grating is independent of the wavelength λ of the incident beam but depends on the groove spacing d. Most displacement encoders produced until recently used gratings with periods ≥ 4 μm, which limited the resolution to $\sim 1\,\mu m$. Increasing requirements for positioning accuracy of precision machine tools in recent years (e.g., in semiconductor technology and in production of diffractive-optics components), as well as for information storage and reading heads in compact disk devices, have fueled a demand for encoders with resolutions in the submicrometer range. One might think *a priori* that the classical He-He laser interferometers [XII.21] could solve the problem. Despite their long-term wavelength stability [XII.22], high measuring accuracy requires operating in vacuum, a possibility which is precluded in many practical situations in which high-precision machine tools and most coordinate measuring devices are used.

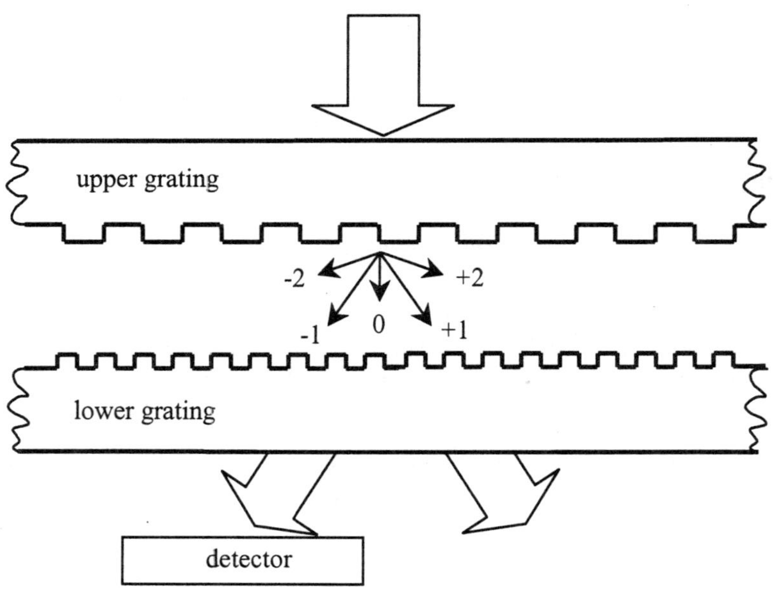

Fig. XII.26. Grating encoder used in normal incidence (after [XII.17]).

As soon as measurements in vacuum are precluded, the accuracy of the measurements in air strongly depends on the refractive-index fluctuations

[XII.23]. It implies strict control of air pressure, temperature and relative humidity, which always have local fluctuations. This deteriorates accuracy with increasing measurement length. Grating interferometers do not suffer from these inconveniences. In particular, they bring a solution that is independent of

(i) the wavelength, permitting the use of non-stabilized sources such as light-emitting diodes,
(ii) the environmental parameters (temperature, pressure, relative humidity, etc.), and
(iii) the spacing between the moving and the read-out gratings.

Moreover, the encoder described in the next section presents a large degree of tolerance with respect to the parallelism of the two gratings and to their groove shape, permitting the use of cheap photoresist or plastic gratings. It provides high visibility of the fringes without requiring equal efficiencies of the two transmitted orders diffracted by the upper grating.

XII.4.2 Encoder made of identical gratings

The translation-rotation encoder can be made of two identical transmission gratings (see Fig. XII.27), which can reduce the costs and, in addition, can lead to a substantial increase in the fringe contrast even without using optimized gratings. The groove shape of the gratings, which have equal periods, could be symmetric with respect to the normal on the mean plane of the grating, i.e., anyone of the profiles currently used (e.g., sinusoidal, lamellar, trapezoidal), but they also can consist of parallel metal rods. Each grating is adjacent to each of the moving parts whose displacement has to be measured, and the two gratings are approximately parallel. The upper grating is illuminated in a -1st order Littrow mounting, and the λ/d ratio is chosen in such a way that only two orders propagate in the substrate and superstrate. As a result, Eqs. (XII.8) and (XII.9) hold. Eq. (XII.9) implies that only the zeroth and the -1st orders will propagate below the upper grating and that they will fall upon the second grating under the same Littrow condition as given in Eq. (XII.8). The result is that, below the second grating, only four orders will propagate, respectively denoted (-1, 0), (-1, +1), (0, -1) and (0, 0), under diffraction angles $\pm\theta_2$. The orders in the pair (-1,+1) and (0,0) are parallel to each other, as well as the orders in the pair (-1, 0) and (0, -1).

In practice, the upper grating is illuminated with a beam having a finite width L, so that the upper grating produces two transmitted beams with the same widths L. These two beams are spatially separated below the upper grating after a vertical distance M_y, given by

$$M_y = L/(2\sin\theta_2),$$

thus the distance between the two gratings must be shorter than this value in order to create the desired interference pattern. After propagating along a vertical distance longer than M_y, the interference patterns of the (-1, 0) and (0,-1) orders on one side, and of the (-1,+1) and (0,0) orders on the other side, do not overlap and can then be captured by a detector. It is worth noting that no extraparasitic order (e.g., +1 or –2 order), which could lower the contrast, propagates. The incident beam must be linearly polarized (TE or TM).

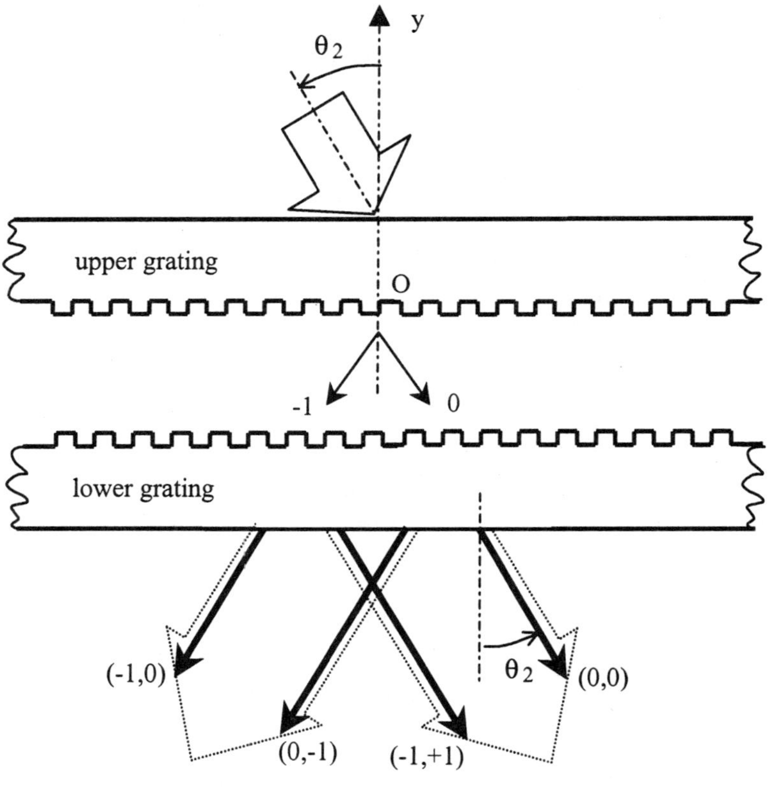

Fig. XII.27. Schematic representation of a grating encoder made of identical gratings (after [XII.17]).

XII.4.3 The interference pattern

XII.4.3.1 Theory

Under the illumination conditions specified above and assuming that the incident amplitude is equal to unity, the upper grating will produce two diffracted orders, with amplitudes t_0 and t_{-1}, which fall upon the second grating under angles of incidence $\pm\theta_2$. The -1 order falling upon the second grating then produces two transmitted orders with amplitudes $t_{-1,0}$ and $t_{-1,+1}$, whereas the zeroth order gives transmitted orders with amplitudes $t_{0,-1}$ and $t_{0,0}$. Inasmuch as the two profiles are identical and symmetrical with respect to the grating normal Oy, we find that

$$\begin{aligned} t_{-1,0} &= t_{-1}\, t_0, \\ t_{-1,+1} &= t_{-1}\, t_{-1}, \\ t_{0,0} &= t_0\, t_0, \\ t_{0,-1} &= t_0\, t_{-1} = t_{-1,0}. \end{aligned}$$

The result is that a detector that receives the interference pattern of the pair of (-1,0) and (0,-1) orders measures the intensity of the sum of two propagating waves with equal amplitudes and with opposite phases $\tilde{\varphi}$ and $-\tilde{\varphi}$ given by Eq. (XII.14) when the relative displacement between the two gratings is equal to Δx. Then, as it is well known [I.47], the intensity e_{-1} of the -1st diffraction order of the entire system of the two gratings, constituted of the pair (-1,0) and (0,-1) contributions, is given by

$$e_{-1} = 2\left|t_{0,-1}\right|^2 \left(1+\cos\tilde{\varphi}\right) \tag{XII.15}$$

and that it varies between the minimum, $e_{-1,\min} = \left|t_{-1,0} - t_{0,-1}\right|^2 = 0$, and the maximum, $e_{-1,\max} = 4\left|t_{0,-1}\right|^2$. The visibility V_{-1}, defined by

$$V_{-1} = \frac{e_{-1,\max} - e_{-1,\min}}{e_{-1,\max} + e_{-1,\min}}, \tag{XII.16}$$

will be optimal (equal to unity), independently of the relative values of the grating efficiencies in the zeroth and the -1st orders. Even for weakly modulated gratings, the contrast will be 100%. On the other hand, the luminosity of the interference pattern depends on the product $\left|t_0\, t_{-1}\right|^2$. Because the sum $\left|t_0\right|^2 + \left|t_{-1}\right|^2$ is bounded by the energy-balance criterion, the luminosity will be maximum when

$$\left|t_0\right|^2 = \left|t_{-1}\right|^2. \tag{XII.17}$$

This condition can be fulfilled by the use of deep gratings with optimized groove shape. However, electronic detection does not require the highest luminosity. Only good visibility is required, which can be obtained with identical cheap photoresist gratings.

Note that the interference pattern between the (-1,+1) and (0,0) orders is generally less favorable for use. Indeed, in general, $t_{-1,+1}$ is different from $t_{0,0}$, and the interference pattern's intensity $e_{0,min}$, formed by the pair (-1,+1) and (0,0) contributions, varies between its minimum $e_{0,min}$ and its maximum $e_{0,max}$ values:

$$e_{0,min} = \left|t_{0,0}\right|^2 + \left|t_{-1,+1}\right|^2 - 2\left|t_{0,0}\right|\left|t_{-1,+1}\right|,$$

$$e_{0,max} = \left|t_{0,0}\right|^2 + \left|t_{-1,+1}\right|^2 + 2\left|t_{0,0}\right|\left|t_{-1,+1}\right|.$$

Thus the visibility is given by

$$V_0 = \frac{2\left|t_{0,0}\right|\left|t_{-1,+1}\right|}{\left|t_{0,0}\right|^2 + \left|t_{-1,+1}\right|^2} \tag{XII.18}$$

This expression may be far interior to unity. For example, if the gratings produce transmitted orders with a ratio 10 in their efficiencies, then

$$\frac{\left|t_{-1}\right|^2}{\left|t_0\right|^2} = \frac{1}{10}.$$

Eq. (XII.18) leads to $V_0 = 0.18$, and the interference will be difficult to use. There is a solution to raise V_0, given by Eq. (XII.18), to unity; it consists of having $\left|t_{0,0}\right| = \left|t_{-1,+1}\right|$, which implies the equality of transmitted efficiencies stated in Eq. (XII.17). Again, this requires gratings with optimized deep grooves, a price to be paid by the customer. Anyway, when such gratings with equal transmitted efficiencies are available, their can ameliorate the device performance by using two detectors. The first one, receiving orders (-1,0) and (0,-1), detects an intensity varying as $1 + \sin\tilde{\varphi}$, a second detector, which receives orders (-1,+1) and (0,0), detects an intensity that varies as $1 - \sin\tilde{\varphi}$, because of the symmetry of the configuration. By use of both measurements it is then possible, with an adequate electronics, to eliminate random parasitic signals from a hostile environment and thus to improve the visibility of the fringes.

XII.4.3.2 Experimental results

A pair of inexpensive photoresist gratings with $d = 0.85\,\mu m$ and $a/d \approx 0.5$ was recorded holographically with grooves close to sinusoidal. When lighted with He-Ne red laser ($\lambda = 0.6328\,\mu m$) under Littrow incidence ($\theta_2 = 21.85^o$), the zeroth and -1st order efficiencies were, respectively, 47% and 13%. It shows that the groove depth was not at all optimized to obtain equal transmitted efficiencies.

The two gratings were then mounted upon separate optical micropositioners to permit independent motions for alignment. The second grating was mounted on a piezoelectric table, electrically controlled to within 40 nm/V, which permitted a precision of about 50 nm in relative positioning of the two gratings. The gratings were illuminated by an expanded beam with 18 mm diameter. A small 3 mm in diameter detector was then put in the central part of the interference pattern obtained after the beam passing the gratings. Fine positioning of the gratings is required to obtain the pattern. The detector signal was recorded while the upper grating was moved along an axis Ox perpendicular to its rulings. Fig. XII.28 shows as a function of the displacement the measured intensity of the interference field produced by the (-1,0) and (0,-1) orders. Despite that each grating has very different zeroth and -1st order efficiencies, the visibility is high, close to 67%.

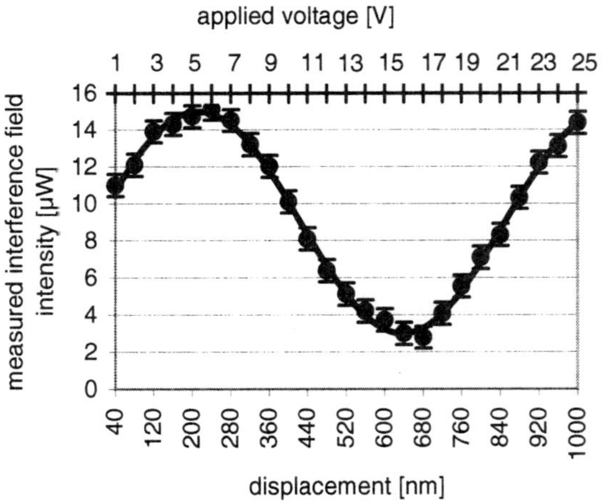

Fig. XII.28. Measured intensity at the center of the interference pattern as a function of displacement x and of corresponding voltage V of the piezoelectric table (after [XII.17]).

The minimum in Fig. XII.28 is different from zero, due to the relatively large size of the detector when compared with the width of the fringes of the interferogram [XII.17]. The signal has same period $(0.85\,\mu m)$ as the gratings. Since modern signal processing allows detecting variations of one hundredth of fringe, there is a possibility of producing a high accuracy (within few nanometers) translation encoder with cheap plastic gratings. As shown in [XII.17] the use of Littrow lighting condition ensures large tolerances with respect to the parallelism of the two gratings. More generally, this mounting shows large tolerances concerning the incidence and the distance between the two moving gratings. It can be adapted to detect rotations by using gratings illustrated in Fig. XII.29.

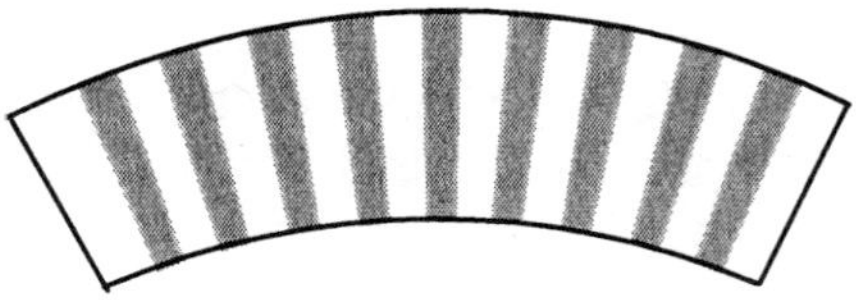

Fig. XII.29. Schematic of gratings used to detect rotation motion.

References

XII.1. E. G. Loewen, *Diffraction Grating Handbook*, Second Ed. (Bausch and Lomb, Rochester, N. Y., 1972)

XII.2. M. Nevière, "Electromagnetic study of transmission gratings," Appl. Opt. 30, 4540-4547, 1991

XII.3. H. Dekker, S. D'Odorico, and R. Arsenault, "First results with a transmission echelle grating on the ESO Faint Object Spectrograph: observations of the SN 1986a in NGC3367 and of the nucleus of the galaxy," Astron. Astrophys. 189, 353-360, 1988

XII.4. D. Maystre, M. Nevière, and R. Petit, "Experimental verifications and applications of the theory," in *Electromagnetic Theory of Gratings*, R. Petit, ed. (Springer-Verlag, Berlin, 1980)

XII.5. K. Yokomori, "Dielectric surface relief gratings with high diffraction efficiency," Appl. Opt. 23, 2303-2310, 1984

XII.6. A. Maréchal and G. W. Stroke, "Sur l'origine des effets de polarization et de diffraction dans les réseaux optiques," C. R. Acad. Sci. Paris, 249, 2042-2045, 1959

XII.7. M. Nevière, D. Maystre, and J. P. Laude, "Perfect blazing for transmission gratings," J. Opt. Soc. Am. A 7, 1736-1739, 1990

XII.8. A. Baranne, "Sur l'emploi des réseaux par transmission en optique astronomique," C. R. Acad. Sci. Ser. B 291, 205-207, 1980

XII.9. F. Montiel and M. Nevière, "Electromagnetic study of a photolithography setup for periodic masks and application to nonperiodic masks," J. Opt. Soc. Am. A 13, 1429-1438, 1996

XII.10. A. Sentenac and D. Maystre, "Symmetry of the field transmitted by metallic grids," J. Mod. Optics 45, 785-797, 1998

XII.11. J. L. Roumiguieres and M. Nevière, "Procédé pour reporter sur un support l'ombre fidèle d'un masque percé de fentes distribuées périodiquement, et application de ce procédé notamment en photolithographie," French patent 792254, 1979; European patent 80 401 211-0, 1980; Japanese patent 1625035, 1980; U.S.A. patent 4 389 094, 1983; Canadian patent 1140373, 1983

XII.12. M. Okai, S. Tsuji, N. Chinone, and T. Haruda, "Novel method to fabricate corrugation for a $\lambda/4$-shifted distributed feed-back laser using a grating photomask," Appl. Phys. Lett. 55, 415-417, 1989

XII.13. D. M. Tennant, T. L. Koch, P. P. Mulgrew, R. P. Gnall, F. Ostermeyer, and J. M. Verdiell, "Characterization of near-field holography grating masks for optoelectronics fabricated by electron beam lithography," J. Vac. Sci. Technol. B10, 2530-2535, 1992

XII.14. T. E. Yeo, N. J. Philips, S. J. Clements, and S. Ojha, "Photopolymer replication – a new technique for $0/25\mu m$ phase shifted DFB-LD grating manufacture?" Proc. IEE 379, 186-191, 1983

XII.15. G. Pakulski, R. Moore, C. Maritan, F. Shepher, F. Fallahi, L. Templeton, and G. Champion, "Fusel silica masks for distributed feedback lasers," Appl. Phys. Lett. 62, 222-224, 1993

XII.16. O. Parriaux and M. Nevière, "Dispositif de transfert photo-lithographique de réseaux non-périodiques," Swiss patent 022 63/94-0, 1994

XII.17. M. Nevière, E. Popov, B. Bojhkov, L. Tsonev, and S. Tonchev, "High-accuracy translation-rotation encoder with two gratings in a Littrow mount," Appl. Opt. 38, 67-76, 1999

XII.18. E. Popov and M. Nevière, "Dispositif optique de mesure interférométrique d'un déplacement relatif entre deux éléments dont l'un au moins est mobile relativement à l'autre," French patent 98/05729 (6 May 1998)

XII.19. J. Heidenhain, M. Allgaüer, and E. Spanner, "Measurement set-up," German patent EP 92 11 0237.2 (17 June 1992)

XII.20. O. Parriaux, M. Nevière, and E. Popov, "Dispositif optique de mesure d'un déplacement relatif entre deux éléments," European patent EP 07 41 282 A2 (30 April 1996)

XII.21. C. R. Steinmetz, "Sub-micron position measurement and control on precision machine tools with laser interferometry," Precision Eng. 22, 513-519, 1990

XII.22. R. J. Hocken and H. P. Layer, "Lasers for dimensional measurement," Ann. CIRP 28, 303-309, 1979

XII.23. W. T. Estler, "High-accuracy displacement interferometry in air," Appl. Opt. 24, 808-815, 1985

Chapter XIII

Grating Couplers and Resonant Excitation of Guided Modes

Grating couplers are dielectric gratings put over planar optical waveguides or slabs. They are used to couple an incident laser beam into a guided mode, or vice-versa. The gratings may be photoresist holographic gratings deposited and recorded on the slab, gratings etched in the guiding layer, or resin gratings replicated using standard techniques. Sometimes they are designed with curved or chirped grooves to focus light outside the device or to couple in a focused laser beam. Grating couplers can be used to couple different modes of the waveguide, or to change the direction of propagation of the same mode in the slab, such are the distributed Bragg mirrors used to reflect the mode.

Whatever the application, the curvature radius of the grooves can be considered much larger than the groove period, so that the light diffraction by the coupler can be considered locally as if the grooves were straight and equidistant. In addition, the incident beam can be represented as a sum of plane waves and the response of the device to a convergent beam becomes a superposition of responses to each beam component, as shown further in this chapter. Using a 2-D model and representing the incident field by a plane wave, the device can be considered as a stack of homogeneous and modulated layers, as illustrated in Fig. XIII.1.

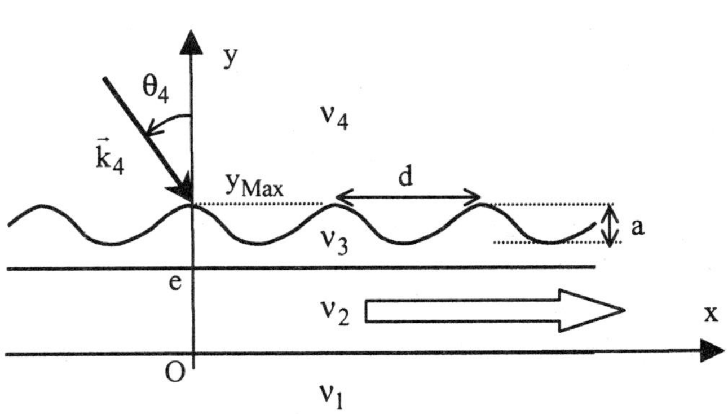

Fig. XIII.1. Schematic representation of a grating coupler.

For some particular values of the angle of incidence, linked with the grating period d and the optogeometrical parameters of the slab, a great percentage of the incident energy is transferred to a guided wave inside the slab. The process involves a resonance phenomenon, which is studied in detail in what follows.

XIII.1 The resonance phenomenon

Let us first consider a mere slab, consisting of two parallel planes separated by a distance e, which divides space into three homogeneous regions with permittivities ε_j (j = 1-3). The common theory of optical waveguides states that solutions of Maxwell equations and associated boundary conditions [VI.1], [VI.2] exist and can be written in the form:

$$u(x,y) = v(y)e^{i\alpha_0^p x}, \qquad \text{(XIII.1)}$$

where α_0^p are real numbers, called propagation constants, x is the direction of propagation of the guided wave, and axis $\overrightarrow{Oy}$ is normal to the slab. Such solutions tend to zero when $y \to \pm\infty$. In the same way as for an electron in a potential well, there exist (if any) only some discrete values of α_0^p and usually the solutions corresponding to these discrete propagation constants are called *modes*. They are homogeneous solutions of the boundary-value problem [XIII.1], which consists of Maxwell equations and the associated boundary conditions, i.e., solutions which may exist in the absence of any incident field. For an assymmetric waveguide ($\varepsilon_1 \neq \varepsilon_3$), they exist if the thickness e is greater than a cut-off thickness e_c.

The propagation constants must satisfy the double condition:

$$\mathrm{Sup}(k_1, k_3) \le \alpha_0^p \le k_2, \qquad \text{(XIII.2)}$$

which implies that the refractive index of the core is greater than those of both substrate and superstrate. The main consequence of Eq. (XIII.2) is that when an incident plane wave with wavevector $\vec{k}_3$ falls on the slab under incidence θ_3, its x-component equal to $k_3 \sin\theta$ is never greater than k_3, and thus can never become close to α_0^p. A resonance process, which would require k_{3x} to become as close as possible to α_0^p, never shows up. In order to excite resonantly a guided mode, an incident wave with an x-component of its wavevector greater than $Sup(k_1, k_3)$ must be used. Such a wave is produced by a prism coupler or a grating coupler. In order to understand in detail the resonance process, it is worth recalling a basic concept which applies to any domain in physics.

XIII.1.1 Basic concept of resonance

At an elementary level, let us first consider an electric network consisting of an inductance L and a capacitor C mounted in series and excited by a sinusoidal voltage $v(t)$ with circular frequency ω. The complex amplitude I of the current in the network will be related to the complex amplitude $\mathcal{V}$ of $v(t)$ by

$$I = \mathcal{V}/Z \qquad \text{(XIII.3)},$$

where Z is the impedance of the network given by:

$$Z = i(\omega L - \frac{1}{\omega C}). \qquad \text{(XIII.4)}$$

The impedance Z vanishes for a value ω_0 of ω given by: $\omega_0 = \frac{1}{\sqrt{LC}}$ which is real. Then, Eq. (XIII.3) shows that $|I| \to \infty$ when $\omega = \omega_0$.

Such a process never happens in real world, however. The network always presents a resistance R, even very low, so that its impedance indeed writes:

$$Z = R + i(\omega L - \frac{1}{\omega C})$$

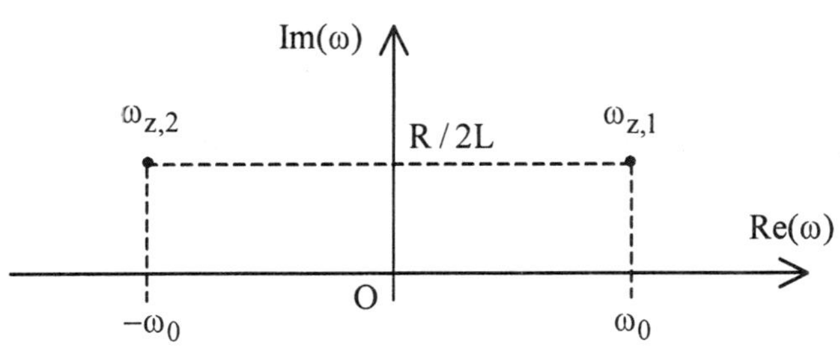

Fig. XIII.2. Complex zeros of $Z(\omega)$.

The study of this function of ω shows that it presents two complex zeros $\omega_{z,j} \simeq \pm\omega_0 + i\,R/(2L)$, $j = 1,2$, illustrated in the ω-complex plane in Fig. XIII.2. For an experimentalist, ω will only be allowed to be varied on the positive part of the real axis. When ω grows from 0 to infinity, it will never equal any of the two zeros. However, it will pass close to $\omega_{z,1}$, especially if the losses (R) are small. When $\omega = \mathrm{Re}(\omega_{z,1}) \simeq \omega_0$, $|Z|$ will present a minimum; the smaller the losses, the smaller the minimum. Eq. (XIII.2) shows that $|I| = |\mathcal{V}|/|Z|$ will then present a sharp maximum, called a resonance phenomenon.

Indeed, in the vicinity of $\omega_{z,1}$, Z can be written as

$$Z = (\omega - \omega_{z,1})\,c(\omega)\,,$$

where $c(\omega)$ is a slowly varying function without zero, which can be approximated by $c(\omega_{z,1})$. Thus

$$|I|^2 \simeq \frac{|\mathcal{V}|^2}{|c(\omega_{z,1})|^2} \frac{1}{(\omega - \omega_0)^2 + \omega_{z,1}''^2}\,,$$

where $\omega_{z,1}''^2 = \mathrm{Im}(\omega_{z,1})$. This equation shows that the curve plotting $|I|^2$ as a function of ω is a Lorenztian line centered at ω_0, having a width-at-half-maximum equal to $2\omega_{z,1}''$ and a height proportionnal to $(1/\omega_{z,1}'')^2$.

XIII.1.2 Leaky modes

When going back to the grating coupler problem, time t is replaced by x, and ω by α_0. The first point to clarify is the complex nature of the propagation constant.

As soon as a modulated interface is added on the top (or bottom, or inside) of a plane slab as illustrated in Fig. XIII.1, a homogeneous solution of the form given by Eq. (XIII.1) no longer satisfies the propagation equation. For TE polarization for example, the propagation equation (I.19)

$$\Delta u(x,y) + k^2 u(x,y) = 0$$

leads to

$$\frac{d^2 v(y)}{dy^2} + \left(k^2(x,y) - \left(\alpha_0^p\right)^2\right) v(y) = 0$$

or:

$$\left(\alpha_0^p\right)^2 - \frac{1}{v(y)}\frac{d^2v(y)}{dy^2} = k^2(x,y), \quad \forall x\,. \tag{XIII.5}$$

The left-hand side is a function of y only. It can be equal to the right-hand side only if $k^2(x,y)$ does not depend on x. This cannot happen with a grating. Eq. (XIII.5) requires that the function v depends on x, too, and has to be periodic with period d like $k^2(x,y)$. Representing $v(x,y)$ by its Fourier series, we state:

$$v(x,y) = \sum_{n=-\infty}^{+\infty} u_n(y)e^{inKx}\,.$$

Thus:

$$u(x,y) = \sum_{n=-\infty}^{+\infty} u_n(y)e^{i\alpha_n^p x} \tag{XIII.6}$$

where

$$\alpha_n^p = \alpha_0^p + nK\,. \tag{XIII.7}$$

At a first look, one could think that we obtain the same expression as in Eq. (I.29') of Section I.2.4. However, there is a big difference. In Section I.2.4, the coefficients α_n were real, since α_0 was the x-component of the incident wavevector $\vec{k}_2$. Now, no incident wavevector exists and α_n^p is the propagation constant of a mode inside the corrugated waveguide. Eq. (XIII.7) first shows that such a constant is not defined in an unique manner, but within integer times the grating constant $K = 2\pi/d$, as it also occurs in solid-state physics. Moreover, when Eq. (XIII.6) is put into the Helmholtz equation in the substrate or superstrate ($j = 1$ or 4) we obtain

$$u_n(y) = A_n^{(j)}\exp(-i\beta_n^{(j)}y) + B_n^{(j)}\exp(i\beta_n^{(j)}y)$$

with

$$\left(\beta_n^{(j)}\right)^2 = k_j^2 - \left(\alpha_n^p\right)^2. \tag{XIII.7}$$

As a result the field in the superstrate and substrate reads:

$$u(x,y) = \sum_{n=-\infty}^{+\infty}\left[A_n^{(j)}\exp(-i\beta_n^{(j)}y) + B_n^{(j)}\exp(i\beta_n^{(j)}y)\right]e^{i\alpha_n^p x}\,. \tag{XIII.8}$$

In these series, some terms correspond to outgoing waves that radiate energy out of the slab. Since this leakage of energy of the mode is not compensated by any incident energy, the mode is attenuated during its propagation. As a result, the propagation constants α_n^p (and thus α_0^p) of the modes of a corrugated waveguide are *complex numbers*. Eq. (XIII.8) gives the expression of what is usually called a "leaky mode," each term of the series being a *space harmonic*. It is worth noting that the determination of $\beta_n^{(j)}$ from Eq. (XIII.7) requires defining a suitable cut, because for complex values of α_n^p the square root is defined within a choice of its sign. Physical requirements that the field is bounded when $y \to \pm\infty$ imposes that $\operatorname{Im}\beta_n^{(j)} \geq 0$. Each space harmonic is the sum of two plane waves, which, in general, are not homogeneous, when α_0^p is complex. The second term in Eq. (XIII.8) can be written as $B_n^{(j)} e^{i\vec{k}_n^{(j)}\cdot\vec{r}}$, with $\vec{k}_n^{(j)} = \vec{k}_n^{(j)'} + i\vec{k}_n^{(j)''}$ where $k_n^{(j)'}$ and $k_n^{(j)''}$ are real numbers. Defining $\alpha_n^{p'}$ and $\alpha_n^{p''}$ to be the real and imaginary part of α_n^p , and similar notations for $\beta_n^{(j)}$, we have:

$$\vec{k}_n^{(j)} = \begin{pmatrix} \alpha_n^{p'} + i\alpha_n^{p''} \\ \beta_n^{(j)'} + i\beta_n^{(j)''} \\ 0 \end{pmatrix}, \qquad \text{(XIII.9)}$$

and the wavevector must satisfy:

$$\left|\vec{k}_n^{(j)}\right|^2 = \left|\vec{k}_n^{(j)'} + i\vec{k}_n^{(j)''}\right|^2 = k_j^2 \qquad \text{(XIII.10)}$$

In the superstrate, k_j is real, and Eq. (XIII.10) implies:

$$\vec{k}_n^{(j)'} \cdot \vec{k}_n^{(j)''} = 0,$$

which means that $\vec{k}_n^{(j)''}$ is orthogonal to $\vec{k}_n^{(j)'}$. Each term of a space harmonic thus appears as an inhomogeneous plane wave, the equiphase planes of which are orthogonal to the equiamplitude ones. They carry energy in direction $\vec{k}_n^{(j)'} = (\alpha_n', \beta_n^{(j)'}, 0)$ while their main attenuation direction is $\vec{k}_n^{(j)''} = (\alpha_n'', \beta_n^{(j)''}, 0)$, with

$$\alpha_n^{p'}\alpha_n^{p''}+\beta_n^{(j)'}\beta_n^{(j)''}=0\,.$$

When the groove depth a decreases to zero, all space harmonics vanish except the one (α_0^p) corresponding to a mode of the planar slab. Moreover α_0^p tends to a real number, if there are no absorption losses. For lossy materials, α_0^p has non-vanishing imaginary part even without corrugation.

XIII.1.3 Propagation constants of leaky modes and poles of the scattering matrix

As seen in Chapter III, the amplitudes of electromagnetic field components in the different media are linked through the S matrix of the stack of gratings and/or interfaces, defined by Eq. (III.28) in which $q = M = 4$. This matrix can be obtained using one of the grating theories, suitable for the specific grating geometry. As a result, the reflection matrix r, defined in Eq. (III.6) is known, and, as shown by Eq. (III.48), it turns out to be equal to $S_{1,2}^{(M)}$:

$$V_B^{(4)}=S_{1,2}^{(4)}V_A^{(4)}\,. \tag{XIII.11}$$

Here $V_A^{(4)}$ is a vector containing the incident field (see Section III.1), while $V_B^{(4)}$ contains the diffracted one and $S_{1,2}^{(4)}$ depends on $\alpha_0 = k_4\sin\theta_4$. Inverting Eq. (XIII.11) gives

$$\mathbf{M}V_B^{(4)}=V_A^{(4)} \tag{XIII.12}$$

where

$$\mathbf{M}=\left(S_{1,2}^{(4)}\right)^{-1} \tag{XIII.13}$$

When we search for leaky modes, i.e., homogeneous solutions, $V_A^{(4)}=0$. We thus have to look for solutions $V_B^{(4)}$ satisfying:

$$\mathbf{M}V_B^{(4)}=0 \tag{XIII.14}$$

This homogeneous linear equation has non-trivial solutions only if

$$\det\mathbf{M}=0, \tag{XIII.15}$$

where $\det \mathbf{M}$ is the determinant of $\mathbf{M}$. Of course matrix $\mathbf{M}$, as well as matrix S, depends on α_0, which is expected to be complex, as discussed in the previous section. Resolving Eq. (XIII.15) thus requires finding the complex zeros of a complex function, which is obtained numerically. The search can only be done on a computer. The complex solutions α_0^p (and $\alpha_n^p = \alpha_0^p + nK$) will be the propagation constants along x of the leaky modes of the corrugated waveguide. Their determination requires having a subroutine that is able to find the complex zeros of a complex function in the vicinity of a starting point, and which generalizes to the complex plane the dichotomia method used on the real axis. Beginning with $a \simeq 0$, the method starts in the vicinity of the propagation constant of a (lossless) mode of the planar slab. The groove depth is gradually increased to reach to desired value. It is found that during the process, α_0^p moves away from the real axis in the upper half of the complex plane. The physical meaning is that with increasing the groove depth, the diffraction becomes stronger and thus the radiation losses become larger, which is expressed as an increase of the imaginary part of α_0^p. Once a value α_0^p has been obtained for a given groove depth, an infinity of values α_n^p are deduced by translations with $\pm K$ along the real axis. If the period d is not too small, *several* values of α_n^p will have a real part inside the $[-k_4, +k_4]$ interval. In addition, as the mode can propagate in the negative x direction as well, there is another set of modes with propagation constants (i.e., set of poles of the scattering matrix) equal to $-\alpha_n^p$.

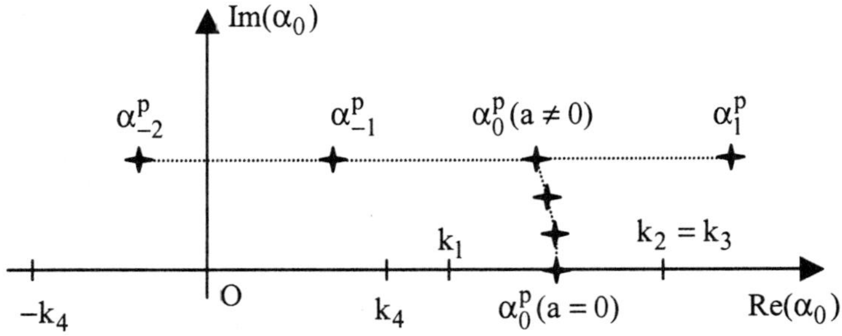

Fig. XIII.3. Trajectory of the propagation constants in the complex plane when a is increased from zero.

Equation (XIII.13) written in the reverse sense:

$$S_{1,2}^{(4)} = \boldsymbol{M}^{-1}$$

implies that

$$S_{1,2,n,m}^{(4)} = \frac{\left\|\boldsymbol{M}_{m,n}\right\|}{\det \boldsymbol{M}}$$

where $\left\|\boldsymbol{M}_{m,n}\right\|$ denotes the (m,n) minor of $\boldsymbol{M}$. In the vicinity of the position of each pole (e.g., the pole having a given subscript q), the determinant of $\boldsymbol{M}$ can be represented by the linear term in its Taylor expansion:

$$\det \boldsymbol{M}(\alpha_0) \propto \alpha_0 - \alpha_q^p .$$

As a result:

$$S_{1,2,n,m}^{(4)} \propto \frac{1}{\alpha_0 - \alpha_q^p} \qquad \forall n, m.$$

We then see that the propagation constants of the leaky modes are *poles* of all the elements of the scattering block matrix $S_{1,2}^{(4)}$, as well as of its determinant. Hence the superscript p in α_q^p.

XIII.1.4 The resonance phenomenon

In the vicinity of the pole α_q^p, each diffracted amplitude $B_n^{(4)}$ given by Eq. (XIII.11) thus can be approximated by:

$$B_n^{(4)}(\alpha_0) \approx \frac{C_{n,q}(\alpha_0)}{\alpha_0 - \alpha_q^p} A_0^{(4)} \qquad \text{(XIII.16)}$$

where $C_{nq}(\alpha_0)$ are slowly varying functions without poles, if we assume that α_q^p is a single pole. Taking unit amplitude for the incident field $A_0^{(4)}$, Eq. (XIII.16) leads to :

$$\left|B_n^{(4)}(\alpha_0)\right|^2 \simeq \frac{\left|C_{nq}\right|^2}{\left(\alpha_0 - \alpha_q^{p'}\right)^2 + \left(\alpha_q^{p''}\right)^2}. \quad \text{(XIII.17)}$$

The Lorentzian shape of this function when α_0 is varied on the real axis is illustrated in Fig. XIII.4.

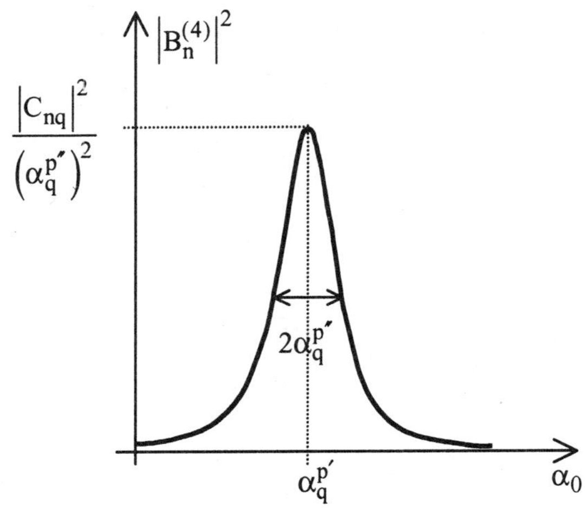

Fig. XIII.4. Resonance line of a Rayleigh coefficient in the vicinity of a pole.

XIII.1.5 The phase-matching condition

Eq. (XIII.17) and Fig. XIII.4 show that when α_0 is varied on the real axis, the square moduli of the Rayleigh coefficients are expected to present sharp maximum when $\alpha_0 = \alpha_q^{p'}$, the height of which is proportional to $(1/\alpha_q^{p''})^2$. The lower the losses, the sharper the resonance lines. As previously illustrated in Fig. XIII.3, if d is not too small, at least one pole α_q^p will have a real part inside the $[-k_4, +k_4]$ interval, namely the one obtained with $q = -1$. It is then possible to find at least one incidence angle θ_4 such that:

$$k_4 \sin\theta_4 = \mathrm{Re}(\alpha_{-1}^p),$$

so that we find a situation completely similar to the simple one described in Section XIII.1.1, with $\omega \rightleftarrows \alpha_0$, $\omega_{z,1} \rightleftarrows \alpha^{p}_{-1}$, $|Z| \rightleftarrows |\det \mathbf{M}|$, $I \rightleftarrows B_n^{(4)}$, $\mathcal{V} \rightleftarrows A_0^{(4)}$. The resonance phenomenon thus occurs when

$$k_4 \sin\theta_4 + nK = \mathrm{Re}(\alpha_0^p)\,, \tag{XIII.18}$$

where α_0^p denotes the value of the propagation constant of the leaky mode which remains when $a \to 0$. Eq. (XIII.18) is the so-called *phase-matching condition* required to excite a leaky mode inside a corrugated waveguide.

XIII.1.6 Complex zeros of propagating order Rayleigh coefficients

Let us assume that all media are lossless, so that the energy-balance criterion applies. The Rayleigh coefficients corresponding to propagating waves thus have to be bounded, which seems to be in conflict with Eq. (XIII.17). Moreover, without corrugation, the mode cannot be excited through a wave incident from the cladding or the substrate, as discussed in the beginning of Section XIII.1. This requires that the pole in Eq. (XIII.16) has to be compensated by a zero, i.e., the Rayleigh coefficients also present zeros $\alpha^z_{n,q}$ so that in Eq. (XIII.16), $C_{nq}(\alpha_0) = r_n(\alpha_0)(\alpha_0 - \alpha^z_{n,q})$, where $r_n(\alpha_0)$ are slowly varying functions without zero. For any grating, when $a \to 0$, $\alpha^z_{0,q} \to \alpha^p_q$, since $B_0 \to r_0(\alpha_0)$. When $a \neq 0$, for a perfectly conducting grating (single interface) diffracting only the 0th order, the unitarity of the diffraction operator implies that $\alpha^z_{0,q} = \bar{\alpha}^p_q$. For lossy or dielectric media, or for grating supporting greater number of diffraction orders, there is no simple link between the values of $\alpha^z_{n,q}$ and α^p_q and the Rayleigh coefficients of propagating orders can be written as:

$$B_n^{(4)}(\alpha_0) = r_n(\alpha_0)\frac{\alpha_0 - \alpha^z_{n,q}}{\alpha_0 - \alpha^p_q}\,. \tag{XIII.19}$$

The differential theories allow calculating both α^p_q and $\alpha^z_{n,q}$, and studying their trajectories in the complex plane when any grating parameter is varied. This method, nowadays called *polology* [XIII.2], is able to account for many strange phenomena in both linear and nonlinear optics, among which are grating anomalies, total absorption phenomenon, enhanced second harmonic generation, etc.

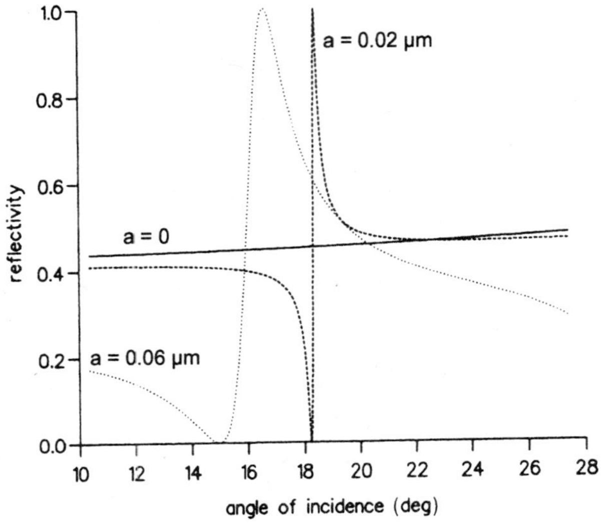

Fig. XIII.5. Zeroth order efficiency as a function of incidence for three different groove depths. The grating surface is sinusoidal with $d = 0.37\ \mu m$, $\nu_1 = 1$; $\nu_2 = \nu_3 = 2.3$, $\nu_4 = 1$, $\lambda = 0.6328$ μm. The thickness is $0.19\ \mu m$ measured from the bottom of the grooves.

An example of the angular dependence of $|B_0|^2$ of a corrugated dielectric waveguide is shown in Fig. XIII.5 for three different values of the groove depth. It fully confirms the existence of distinct pole and zero which separate when a is increased. We note that for a particular incidence, which must be strictly controled, the zeroth order efficiency reaches 100%, while it goes to zero for a neighboring incidence.

XIII.2 Coupling coefficient

A plane wave does not correctly represent laser beams. A laser beam has a limited width L and presents some angular divergence. These features have to be taken into account in view of determining the coupling coefficient, i.e., the ratio between the guided power to the incident one. We will also see that the divergence of the beam affects the use of grating couplers as optical filters [XIII.3, XIII.4]. In representing a limited beam, we will stick to a 2-D model, i.e., we will consider a field invariant in the z direction, but limited in the xOy plane.

XIII.2.1 Description of the incident beam

A limited beam with a small angular divergence can be represented by a superposition of plane waves with a density function $p(\alpha_0)$ and width at half height $\Delta\alpha_0 << k_4$. Here again $\alpha_0 = k_4 \sin\theta_4$. If the incidence angle of the beam axis is $\theta_{c,4}$, we state $\alpha_c = k_4 \sin\theta_{c,4}$. Remembering that for each α_0, $\beta_0^{(4)}$ is defined by $\beta_0^{(4)}(\alpha_0) = \sqrt{k_4^2 - \alpha_0^2}$, the incident field reads:

$$u^{(i)}(x,y) = \int_{-k_4}^{+k_4} p(\alpha_0 - \alpha_c)\exp(i\alpha_0 x)\exp\left(-i\beta_0^{(4)}(\alpha_0)y\right)d\alpha_0 \quad \text{(XIII.20)}$$

The limits of the integral can be extended to infinity if we remember that $\Delta\alpha_0 << k_4$. Thus we write:

$$u^{(i)}(x,y) = \int_{-\infty}^{+\infty} p(\alpha_0 - \alpha_c)\exp(i\alpha_0 x)\exp\left(-i\beta_0^{(4)}(\alpha_0)y\right)d\alpha_0 \,. \quad \text{(XIII.21)}$$

In the vicinity of α_c, we may approximate function $\beta_0^{(4)}(\alpha_0)$ by the first two terms of its Taylor series:

$$\beta_0^{(4)}(\alpha_0) \simeq \beta_0^{(4)}(\alpha_c) - \frac{\alpha_c}{\beta_0^{(4)}(\alpha_c)}(\alpha_0 - \alpha_c)\,, \quad \text{(XIII.22)}$$

so that Eq. (XIII.21) becomes:

$$u^{(i)}(x,y) \simeq e^{i\alpha_c x - i\beta_0^{(4)}(\alpha_c)y} \int_{-\infty}^{+\infty} p(\alpha_0 - \alpha_c)e^{i(\alpha_0 - \alpha_c)\left[x + \frac{\alpha_c}{\beta_0^{(4)}(\alpha_c)}y\right]} d\alpha_0 \,. \quad \text{(XIII.23)}$$

Eq. (XIII.23) suggests to introduce the Fourier transform $q(x)$ of the density $p(\alpha_0)$, which describes the spacial structure of the incident beam:

$$q(x) = \int_{-\infty}^{+\infty} p(\xi)\exp(i\xi x)d\xi . \quad \text{(XIII.24)}$$

The incident beam then can be simply expressed as:

$$u^{(i)}(x,y) \simeq q(x + \frac{\alpha_c}{\beta_0^{(4)}(\alpha_c)} y)\exp\left[i\left(\alpha_c x - \beta_0^{(4)}(\alpha_c)y\right)\right]. \quad \text{(XIII.25)}$$

Eq. (XIII.25) means that the limited beam is the product of a plane wave with wavevector along the beam axis by its spacial structure $q(x_1/\cos\theta_{c,4})$ along an axis Ox_1 normal to the beam axis.

XIII.2.2 Response of the grating coupler to an incident plane wave

When an incident plane wave with unit amplitude given by $\exp(i\alpha_0 x)\exp\left[-i\beta_0^{(4)}(\alpha_0)y\right]$ strikes the grating coupler, it generates a total field given by Eq. (I. 29'). Thus the diffracted field $u^{(d)}$ reads:

$$u^{(d)}(x,y,\alpha_0) = \sum_{n=-\infty}^{+\infty} u_n^{(d)}(y,\alpha_0)e^{i\alpha_n x} \quad \text{(I.29')}$$

Above the modulated region,

$$u_n^{(d)}(y,\alpha_0) = B_n^{(4)}e^{i\beta_n^{(4)}y} . \quad \text{(XIII.26)}$$

Eq. (XIII.26) shows at evidence that the poles α_n^p of $B_n^{(4)}$ are also poles of $u_n^{(d)}(y,\alpha_0)$ for $y \geq y_{Max}$. Since the field in the modulated region and below it lineary depends on the field at $y \geq y_{Max}$, α_n^p will be poles of $u_n^{(d)}(y,\alpha_0)$ $\forall y$. Let us assume that the central incidence $\theta_{c,4}$ is chosen to satisfy the phase-matching condition (XIII.18). It means that there exists an integer n_0 such that:

$$\alpha_c + n_0 K = \mathrm{Re}\left(\alpha_0^p\right) \tag{XIII.27}$$

and we recall that $\alpha_0 \simeq \alpha_c$, since $\Delta\alpha_0 << k_0$. We then speak of coupling by means of the n_0^{th} order, and with wavelengths and groove spacings generally used, $n_0 = \pm 1$. Defining $\alpha_{-n_0}^p$ by:

$$\alpha_{-n_0}^p = \alpha_0^p - n_0 K\,, \tag{XIII.28}$$

$u_n^{(d)}(y,\alpha_0)$ can be written in the vicinity of $\alpha_{-n_0}^p$ as:

$$u_n^{(d)}(y,\alpha_0) = \frac{v_n(y,\alpha_0)}{\alpha_0 - \alpha_{-n_0}^p}\,, \tag{XIII.29}$$

where $v_n(y,\alpha_0)$ are slowly varying functions without poles. Thus $u_n^{(d)}(y,\alpha_0)$ can be approximated by:

$$u_n^{(d)}(y,\alpha_0) \simeq \frac{v_n(y,\alpha_c)}{\alpha_0 - \alpha_{-n_0}^p}\,. \tag{XIII.30}$$

Finally:

$$u^{(d)}(x,y,\alpha_0) \simeq \sum_{n=-\infty}^{+\infty} \frac{v_n(y,\alpha_c)}{\alpha_0 - \alpha_{-n_0}^p} \exp\left[i(\alpha_0 + nK)x\right]. \tag{XIII.31}$$

Eq. (XIII.31) gives the response of the device to incident plane waves with unit amplitude when their propagation constant α_0 are close to the value α_c satisfying the phase-matching condition.

XIII.2.3 Response of the grating coupler to a limited beam

As soon as the response to a plane wave is known, the linearity of the problem allows determining the response to a superposition of plane waves by:

$$u^{(d)}(x,y) = \int_{-\infty}^{+\infty} p(\alpha_0 - \alpha_c) \sum_{n=-\infty}^{+\infty} \frac{v_n(y,\alpha_c)}{\alpha_0 - \alpha_{-n_0}^p} \exp\left[i(\alpha_0 + nK)x\right] d\alpha_0\,. \tag{XIII.32}$$

If we state that $\alpha_{-n_0}^p = \alpha^{p'} + i\alpha^{p''}$, the phase-matching condition (XIII.27) leads to $\alpha_c = \alpha^{p'}$, and Eq. (XIII.32) reads:

$$u^d(x,y) = \int_{-\infty}^{+\infty} p(\alpha_0 - \alpha_c) \sum_{n=-\infty}^{+\infty} \frac{v_n(y,\alpha_c)}{\alpha_0 - \alpha_c - i\alpha^{p''}} \exp\left[i(\alpha_0 + nK)x\right] d\alpha_0 . \quad \text{(XIII.33)}$$

This expression provides the diffracted field as soon as the functions $v_n(y,\alpha_c)$ have been computed for the central value α_c of α_0. It is possible to simplify this expression by introducing a convenient function $q^{(d)}(x)$, similar to the function $q(x)$ of the spatial distribution of the incident beam. Let us define:

$$q^{(d)}(x) = -i \int_{-\infty}^{+\infty} p(\xi) \frac{\exp(i\xi x)}{\xi - i\alpha^{p''}} d\xi . \quad \text{(XIII.34)}$$

The diffracted field then reduces to:

$$u^{(d)}(x,y) = iq^{(d)}(x) \sum_{n=-\infty}^{+\infty} v_n(y,\alpha_c) \exp\left[i(\alpha_c + nK)x\right] . \quad \text{(XIII.35)}$$

Eq. (XIII.35) avoids the numerical integral which was required by Eq. (XIII.33). And indeed, the determination of function $q^{(d)}(x)$ can be done explicitly if we note that:

$$\frac{dq^{(d)}(x)}{dx} = \int_{-\infty}^{+\infty} \frac{\xi p(\xi) \exp(i\xi x)}{\xi - i\alpha^{p''}} d\xi$$

or

$$\frac{dq^{(d)}(x)}{dx} = \int_{-\infty}^{+\infty} \frac{(\xi - i\alpha^{p''}) p(\xi) e^{i\xi x}}{\xi - i\alpha^{p''}} d\xi + \int_{-\infty}^{+\infty} \frac{i\alpha^{p''} p(\xi) e^{i\xi x}}{\xi - i\alpha^{p''}} d\xi$$

or

$$\frac{dq^{(d)}(x)}{dx} = q(x) - \alpha^{p''} q^{(d)}(x) .$$

Thus $q^{(d)}(x)$ is solution of the first-order nonhomogeneous differential equation:

$$\frac{dq^{(d)}(x)}{dx} + \alpha^{p''} q^{(d)}(x) = q(x) . \quad \text{(XIII.36)}$$

From the classical theorem, which states that the Fourier transform of an integrable function vanishes at infinity, $q^{(d)}(x)$ verifies:

$$q^{(d)}(-\infty) = 0 \,. \qquad \text{(XIII.37)}$$

The function $q(x)$ describes the spatial structure of the incident beam; thus it is a real function. Eqs. (XIII.36) and (XIII.37) show that $q^{(d)}(x)$ is a real function, too, and they allow an easy determination of $q^{(d)}(x)$.

XIII.2.4 Definition and calculation of the coupling coefficient

The coupling coefficient is defined as the ratio of the guided power to the power of the incident beam. By guided power, we mean the power that is carried by the wave having a subscript n equal to n_0 defined by the phase-matching condition (XIII.27). Except under very special circumstances, terms with $n \neq n_0$ in the Fourier series of the field carry a negligible amount of energy. In other words, only one space harmonic is excited by the incident beam, usually. The calculation of the coupling coefficient will be developed in TE polarization.

The Poynting vector $\vec{P}^{(i)}$ of the incident field $\vec{E}^{(i)} \equiv u^{(i)}(x,y)\,\vec{z}$ is given by:

$$\vec{P}^{(i)} = \frac{\vec{E}^{(i)} \times \vec{H}^{(i)*}}{2} = \frac{u^{(i)} \operatorname{grad} u^{(i)*}}{-2i\omega\mu_0} \,.$$

If $\hat{\vec{k}}^{(i)}$ is the unit vector of the axis of the beam:

$$\vec{P}^{i} \cdot \hat{\vec{k}}^{(i)} = \frac{i u^{(i)} \operatorname{grad} u^{(i)*}}{2\omega\mu_0} \cdot \hat{\vec{k}}^{(i)} \qquad \text{(XIII.38)}$$

where $u^{(i)}$ is given by Eq. (XIII.25).

In a new coordinate system (x_1, y_1) rotated from the (x,y) system by an angle $\theta_{c,4}$:

$$x_1 = x\cos\theta_{c,4} + y\sin\theta_{c,4}$$
$$y_1 = -x\sin\theta_{c,4} + y\cos\theta_{c,4} \,,$$

$u^{(i)}$ is simply given by:

$$u^{(i)} = q\left(\frac{x_1}{\cos\theta_{c,4}}\right)\exp(-ik_4 y_1) \qquad \text{(XIII.39)}$$

Thus Eq. (XIII.38) gives:

$$\vec{P}^{(i)} \cdot \hat{\vec{k}}^{(i)} = \frac{i u^{(i)}}{2\omega\mu_0}\frac{\partial u^{(i)*}}{\partial y_1} = \frac{k_4}{2\omega\mu_0}\left|q\left(\frac{x_1}{\cos\theta_{c,4}}\right)\right|^2 .$$

The incident power $\Phi^{(i)}$ is given by:

$$\Phi^{(i)} = \int_{-\infty}^{+\infty} \vec{P}^{(i)} \cdot \hat{\vec{k}}^{(i)} dx_1 = \frac{k_4 \cos\theta_{c,4}}{2\omega\mu_0} \int_{-\infty}^{+\infty} |q(x)|^2 \, dx . \qquad \text{(XIII.40)}$$

In a similar way, the calculation of the guided power $\Phi^{(g)}$ implies the calculations of $\vec{P}^{(g)} \cdot \vec{x}$, where $\vec{P}^{(g)}$ is the Poynting vector of the n_0^{th} Fourier component of $u^{(d)}$. If we denote this component by $u^{(g)}$, Eq. (XIII.35) shows that

$$u^{(g)} = q^{(d)}(x) i v_{n_0}(y, \alpha_c) \exp\left[i(\alpha_c + n_0 K)x\right], \qquad \text{(XIII.41)}$$

while

$$\vec{P}^{(g)} \cdot \vec{x} = \frac{i u^{(g)} \operatorname{grad} u^{(g)*}}{2\omega\mu_0} \cdot \vec{x} = \frac{i u^{(g)}}{2\omega\mu_0} \frac{\partial u^{(g)*}}{\partial x} . \qquad \text{(XIII.42)}$$

From Eq. (XIII.41), we obtain:

$$\frac{\partial u^{(g)*}}{\partial x} = -i v_{n_0}^{*}(y, \alpha_c) \left[\frac{dq^{(d)}}{dx} - i(\alpha_c + n_0 K) q^{(d)}(x) \right] \exp\left[-i(\alpha_c + n_0 K)x\right]$$

and recalling Eq. (XIII.36), Eq. (XIII.42) leads to:

$$\vec{P}^{(g)} \cdot \vec{x} = \left| \frac{v_{n_0}(y, \alpha_c)}{2\omega\mu_0} \right|^2 \left\{ i \left[q^{(d)}(x) q(x) - \alpha^{p''} q^{(d)^2}(x) \right] + (\alpha_c + n_0 K) q^{(d)^2}(x) \right\} .$$

The guided power $\Phi^{(g)}$ is the real part of the flux of $\vec{P}^{(g)}$ through a plane perpendicular to the x-axis:

$$\Phi^{(g)} = \operatorname{Re} \int_{-\infty}^{+\infty} \vec{P}^{(g)} \cdot \vec{x} \, dy .$$

Thus:

$$\Phi^{(g)} = \frac{(\alpha_c + n_0 K)}{2\omega\mu_0} q^{(d)^2}(x) \int_{-\infty}^{+\infty} \left| v_{n_0}(y, \alpha_c) \right|^2 dy$$

The coupling coefficient $C(x)$ thus reads:

$$C(x)=\frac{\Phi^{(g)}}{\Phi^{(i)}}=\frac{(\alpha_c+n_0K)q^{(d)^2}(x)\int_{-\infty}^{+\infty}\left|v_{n_0}(y,\alpha_c)\right|^2dy}{k_4\cos\theta_{c,4}\int_{-\infty}^{+\infty}q(x)^2dx} \qquad \text{(XIII.43)}$$

As expected, C both depends on the structure of the incident beam and on the quality of the resonance line.

XIII.2.5 Application to a boxcar input beam

We illustrate the general theory on the simple case of a rectangular (boxcar) incident beam, the spatial structure of which is given by:

$$q(x)=1 \quad \text{if } |x|\le l/2$$
$$q(x)=0 \quad \text{otherwise},$$

where l is related to the beam width L by: $l=L/\cos\theta_{c,4}$. The solution of Eqs. (XIII.36) and (XIII.37) then gives:

$$\begin{aligned} q^{(d)}(x)&=0, \qquad \text{if } x<-\frac{l}{2}\\ &=\frac{1}{\alpha^{p''}}\left\{1-\exp\left[-\alpha^{p''}\left(x+\frac{l}{2}\right)\right]\right\}, \qquad \text{if } |x|<\frac{l}{2}\\ &=\frac{1}{\alpha^{p''}}\left[1-\exp(-\alpha^{p''}l)\right]\exp\frac{\alpha^{p''}l}{2}\exp(-\alpha^{p''}x), \qquad \text{if } x>\frac{l}{2}\end{aligned}$$

This function presents a maximum for $x=l/2$. Thus, from Eq. (XIII.43) the coefficient $C(x)$ presents a maximum value C_{Max} for $x=l/2$ given by:

$$C_{Max}=\frac{(\alpha_c+n_0K)\left[1-\exp(-\alpha^{p''}l\right]^2\int_{-\infty}^{+\infty}\left|v_{n_0}(y,\alpha_c)\right|^2dy}{\alpha^{p''2}k_4 l\cos\theta_{c,4}}.$$

It is interesting to split this expression into two factors

$$C_{Max}=F\cdot B_R, \qquad \text{(XIII.44)}$$

where

$$F = \frac{2\left[1-\exp(\alpha^{p''} l)\right]^2}{\alpha^{p''} l} \tag{XIII.44'}$$

and

$$B_R = \frac{(\alpha_c + n_0 K) \int_{-\infty}^{+\infty} \left| v_{n_0}(y, \alpha_c) \right|^2 dy}{2\alpha^{p''^2} k_4 \cos\theta_{c,4}} . \tag{XIII.44''}$$

The first term F is the function described by Ulrich [XIII.5]. It depends both on the grating coupler (through $\alpha^{p''} l$) and on the beam (through l). It presents a maximum value equal to 0.814 when

$$\alpha^{p''} l = 1.25 . \tag{XIII.45}$$

The second factor B_R is called the branching ratio. It is always less than unity [XIII.6] and only depends on the grating coupler. The branching ratio can only be determined numerically and often turns out to be far inferior to unity.

The optimization of the coupling coefficient C_{Max} must maximize the two terms F and B_R . Eq. (XIII.45) immediately gives the optimal beam width L_0:

$$L_0 = l \cos\theta_{c,4} = \frac{1.25}{\alpha^{p''}} \cos\theta_{c,4} . \tag{XIII.46}$$

With the common values of $\alpha^{p''}$, L_0 is found to be about 1 mm.

A particular situation named as *reverse coupler* is obtained when the incident wave comes from the substrate with an incidence angle high enough so that no transmitted order exists, and with a groove spacing such that only the zeroth reflected order propagates. An extension of the above developed formalism [XIII.7] shows that the coupling coefficient is then independent of the groove shape. With a convenient incident beam, the upper limit of 0.814 for the coupling coefficient C_{Max} can be reached.

Let us point out that the results obtained in this section are not too much affected when the structure of the input beam is changed. For a Gaussian beam for example, the upper limit of 0.814 would be changed into 0.80.

XIII.3 Grating coupler with double-surface corrugation

Another interesting way of getting a high coupling coefficient which avoids the constraints of the reverse coupling, is to use a grating coupler with double surface corrugation, as illustrated in Fig. XIII.6.

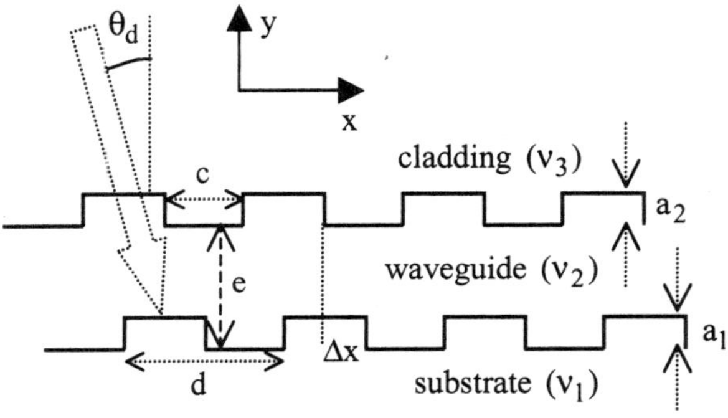

Fig. XIII.6. Schematic representation of a grating coupler having double-surface corrugation.

Experiments have shown [XIII.8] that a lateral shift Δx between the two gratings greatly influences the branching ratio, due to the fact that it modifies the interference phenomenon between the first orders diffracted by the two gratings. This effect was fully accounted for by a rigorous analysis [XIII.9].

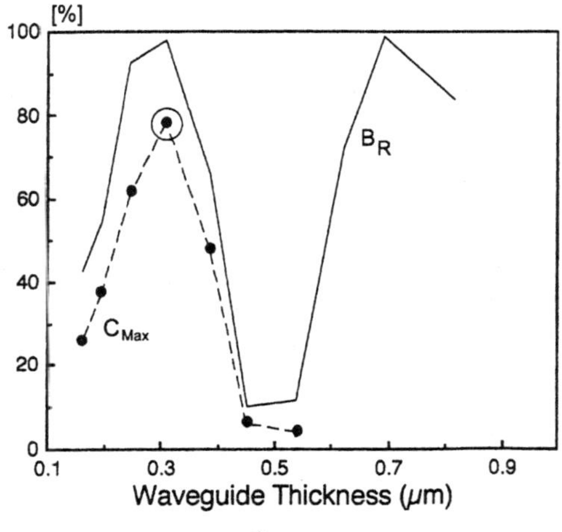

Fig. XIII.7. Measured B_R and C_{Max} of a waveguide with double corrugation as a function of thickness (after [XIII.8]).

The double-corrugation waveguide was produced by thermal evaporation in vacuum of an organic glass, under a deposition angle θ_d. The process produced a lateral shift $\Delta x = e\tan(\theta_d)$, proportional to the waveguide thickness e. Fig. XIII.7 shows the measured branching ratio and coupling coefficient as a function of thickness, with given the parameters of the structure, $\nu_1 = 1.472$, $\nu_2 = 1.60$, $\nu_3 = 1$, $\theta_d = 7.5^o$, $d = 0.347\,\mu m$, $a_1 = 0.074\,\mu m$, $a_2 = 0.074\,\mu m$, $\lambda = 0.6328\,\mu m$, and duty cycles c/d of the lamellar gratings close to 0.5. For two particular thicknesses, 98% branching ratios where measured, which resulted in a measured 78% coupling coefficient for the device lighted by a gaussian beam. This is very near the theoretical limit of 80%.

XIII.4 Corrugated waveguides as resonance optical filters

Section XIII.1.6 has shown that propagating reflected orders present both poles and zeros, which result in sharp anomalies in the zeroth order efficiency, called also *reflectivity* in what follows. And indeed, when plotting $|B_0^{(4)}|^2$ as a function of the wavelength λ, with θ_4 kept constant, we may expect anomalies as sharp as those in Fig. XIII.5, since $B_0^{(4)}$ is a function of

$$\alpha_0 = \frac{2\pi}{\lambda}\nu_4 \sin\theta_4 . \tag{XIII.47}$$

Such narrow peaks in reflectivity with 100% maximum have been a topic of interest in optical filtering. In so far as the propagation constants of the TE and TM guided waves are different, the position of the peaks strongly depends on polarization. By changing the angle of incidence in Eq. (XIII.47), the spectral position of the resonance can be changed, too. Thus, the transmission of the filter can be tuned.

However, experiments have shown that the maximum is quite far from the desired 100% [XIII.10, XIII.11]. On top of losses, which may play a significant role [XIII.10], the finite input-beam width, which results in the angular divergence, has to be taken into account [XIII.3]. The calculation of the reflectiviy corresponding to a limited incident beam is done along similar lines as the calculations of the coupling coefficient.

XIII.4.1 Reflectivity of a corrugated waveguide lighted by a limited input beam

Like in Section XIII.2, the linearity of the problem leads to a response that is a linear combination of responses to each of the incident-field components, with a density function $p(\alpha_0 - \alpha_c)$. If we denote by $u^{(r)}(x, y)$ the zeroth term in the Rayleigh expansion of the diffracted field, we thus have:

$$u^{(r)}(x, y) = \int_{-\infty}^{+\infty} B_0^{(4)}(\alpha_0) p(\alpha_0 - \alpha_c) \exp\left[i\left(\alpha_0 x + \beta_0^{(4)}(\alpha_0) y\right)\right] d\alpha_0 \,. \quad \text{(XIII.48)}$$

Let us introduce a translated function ${}^t B_0^{(4)}(\alpha_0)$ defined by:

$${}^t B_0^{(4)}(\alpha_0 - \alpha_c) = B_0^{(4)}(\alpha_0) \,,$$

and the Fourier transform s(x) of the function ${}^t B_0^{(4)}(\alpha_0) p(\alpha_0)$:

$$s(x) = \int_{-\infty}^{+\infty} {}^t B_0^{(4)}(\xi) p(\xi) \exp(i\xi x) dx \,. \quad \text{(XIII.49)}$$

A calculation similar to the one which led to Eq. (XIII.25) gives:

$$u^{(r)}(x, y) = s\left(x - \frac{\alpha_c y}{\beta_0^{(4)}(\alpha_c)} \right) \exp\left[i(\alpha_c x + \beta_0^{(4)}(\alpha_c) y) \right]. \quad \text{(XIII.50)}$$

By introducing the coordinate system (x_2, y_2) rotated from the (x, y) system by an angle $-\theta_{c,4}$, we thus have:

$$u^{(r)} = s\left(\frac{x_2}{\cos\theta_{c,4}} \right) \exp(ik_4 y_1) \,. \quad \text{(XIII.51)}$$

If $\vec{P}^{(r)}$ is the Poynting vector associated to $u^{(r)}$, the reflected power is:

$$\Phi^{(r)} = \frac{k_4 \cos\theta_{c,4}}{2\omega\mu_0} \int_{-\infty}^{+\infty} |s(x)|^2 dx \,. \quad \text{(XIII.52)}$$

The reflectivity of the device illuminated by an incident beam is simply given by:

$$e_0^{(r)} = \frac{\Phi^{(r)}}{\Phi^{(i)}} = \frac{\int_{-\infty}^{+\infty} |s(x)|^2 \, dx}{\int |q(x)|^2 \, dx} \tag{XIII.53}$$

From Parseval theorem, Eq. (XIII.53) can be also written using functions in α_0 space:

$$e_0^{(r)} = \frac{\int_{-\infty}^{+\infty} \left| {}^t B_0^4(\alpha_0) \right|^2 |p(\alpha_0)|^2 \, d\alpha_0}{\int_{-\infty}^{+\infty} |p(\alpha_0)|^2 \, d\alpha_0} . \tag{XIII.54}$$

This result agree with Eq. 10 of [XIII.3] if we take into account the way we approximated $\beta_0^{(4)}(\alpha_0)$ by Eq. (XIII.22).

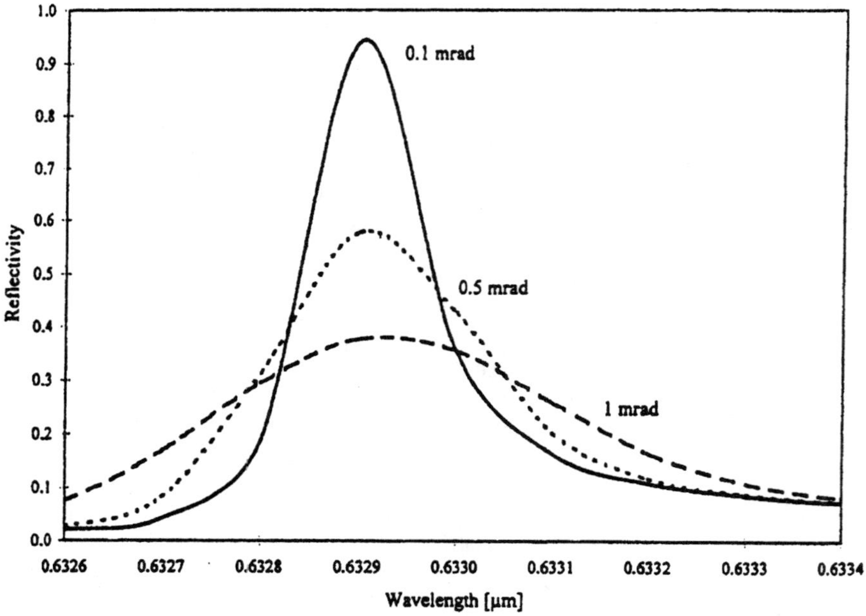

Fig. XIII.8 Spectral dependence of the reflectivity of a lossless waveguide for three different input-beam-divergence values. $\theta_4 = 34.785^{\circ}$; $d = 0.303\,\mu m$, $a = 0.12\,\mu m$, $e = 0.7\,\mu m$ (thickness at the mid-modulation height), $\nu_1 = 1.5115$, $\nu_2 = 1.542 = \nu_3$, $\nu_4 = 1$, TE polarization (after [XIII.3]).

The reflectivity of a corrugated waveguide that is expected to reach 100% when illuminated by an incident plane wave is presented in Fig. XIII.8 for three different beam divergences $\Delta\alpha_0$ corresponding to $\Delta\theta_4$ between 0.1 and 1 mrad. A spectacular reduction of the maximum value and a broadening of the curves are observed. Both effects depend, of course, on the properties of the anomaly which leads to perfect blazing. The smaller the groove depth, the smaller the anomaly width and the greater the influence of the incident-beam divergence. For many practicle devices, the divergence of laser beams does not allow approaching 100% efficiency. An original solution to approach this goal is to use a doubly-periodic grating.

XIII.4.2 Reflectivity of a doubly-periodic grating coupler lighted by a limited input beam

The device considered in this section should not be confused with the grating couper with double-surface corrugation considered in Section XIII.3, or with the crossed gratings of Chapter IX. A doubly periodic grating coupler can be obtained by superposing two gratings which both are periodic along the x direction, but have different periods d_1 and d_2. It can be also made by producing a single modulated interface doubly periodic in the x direction.

A study conducted in the Brillouin diagram [XIII.4] then shows that when illuminated under normal incidence, with an incident vector $\vec{k}_M$ close to a Bragg plane ($k_M = (2n+1)\pi/d_1$), two guided modes propagating in opposite directions can be simultaneously excited at resonance if $K_1 = \dfrac{2\pi}{d_1} \simeq \mathrm{Re}(\alpha_q^p)$.

This can be understood if in Fig. XIII.3 we add the positions of the poles corresponding to $-\alpha_q^p$. If one of the poles, say α_q^p, has real part close to zero, the same is valid for its counterpropagating homolog $-\alpha_q^p$. The interaction of the two modes leads to an energy transfer between them, which is expressed by an increase of the imaginary part of the propagation constant. This broadens the resonance line in the angular dependance of the reflectivity. The study of the dispersion diagrams shows that the interaction causes a part of the dispersion curve close to the normal incidence $(k_x = 0)$ to become almost parallel to the k_x-axis (Fig. XIII.9). In that region the spectral position of the resonance depends only slightly on k_x (i.e., on the angle of incidence). The stronger the interaction between the counter-propagating modes, the larger the plateau region of the dispersion curve. By chosing the second period d_2 such that $d_2 = d_1/2$, the interaction is enhanced. Both theoretical study and numerical results [XIII.4] have shown that the angular width of the resonance line can be increased by a factor 10. Fig.XIII.10 shows the results for a single modulated grating coupler and two doubly periodic ones. When increasing the refractive index contrast of the second grating (responsible for the interaction between the counter-propagating modes), the influence on the incident beam divergence decreases. The increase in angular tolerance, without shifting the spectral response, allows producing narrowband filters which are compatible with the divergence of common laser beams.

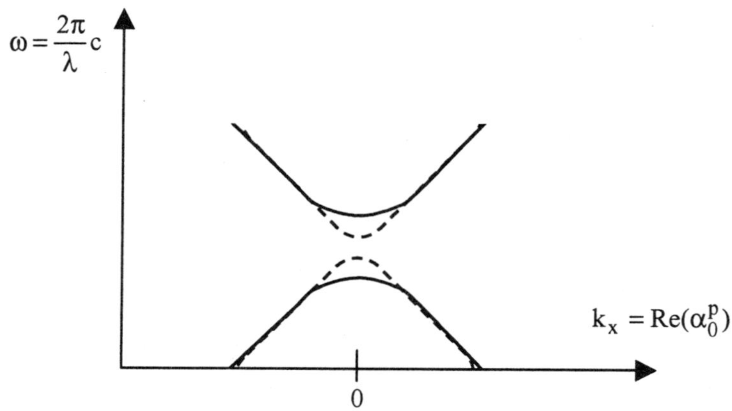

Fig. XIII.9. A portion of the dispersion diagram of a waveguide mode close to normal incidence for two different values of the mode interaction: weaker (dashed line) and stronger (solid line).

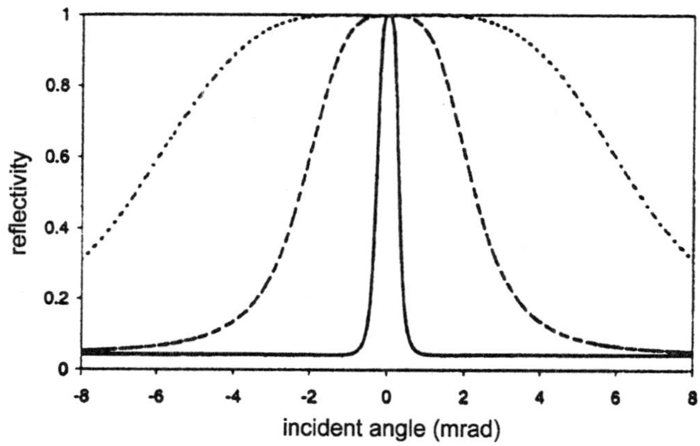

Fig. XIII.10. Angular behavior of the reflectivity of a single modulated grating coupler (solid line) and two doubly-periodic grating couplers. Dashed line, lower contrast of refractive index modulation of the second grating, dotted line, higher contrast (after [XIII.4]).

References

XIII.1. M. Nevière, "The homogeneous problem," in *Electromagnetic Theory of Gratings*, R. Petit ed. (Springer-Verlag, Berlin, 1980), pp. 123-157

XIII.2. M. Nevière, E. Popov, R. Reinisch, and G. Vitrant, *Electromagnetic Resonances in Nonlinear Optics* (Gordon and Breach Science Publishers, Amsterdam, 2000), Ch. VI, pp. 120-164

XIII.3. E. Popov and B. Bozhkov, "Corrugated waveguides as resonance optical filters – advantages and limitations," J. Opt. Soc. Am. A18, 1758-1764, 2001

XIII.4. F. Lemarchand, A. Sentenac, E. Cambril, and H. Giovannini, "Study of the resonant behaviour of waveguide grating: increasing the angular tolerance of guided-mode filters," J. Opt. A: Pure Appl. Opt. 1, 545-551, 1999

XIII.5. R. Ulrich, "Efficiency of optical-grating couplers," J. Opt. Soc. Am. 63, 1419-1431, 1973

XIII.6. M. Nevière, P. Vincent, R. Petit, and M. Cadilhac, "Determination of the coupling coefficient of a holographic thin film coupler," Opt. Commun. 9, 240-243, 1973

XIII.7. P. Vincent, "Contribution à l'étude théorique et numérique de la propagation des ondes électromagnétiques en présence d'un objet cylindrique : tiges et réseaux," Thesis, University of Aix-Marseille III, France, 1978

XIII.8. J. C. Brazas, L. Li, and A. L. McKeon, "High efficiency input coupling into optical waveguides using gratings with double-surface corrugation," Appl. Opt. 34, 604-609, 1995

XIII.9. L. Li, "Analysis of planar waveguide grating couplers with double surface corrugations of identical periods," Opt. Commun. 114, 406-412, 1995

XIII.10. S. M. Norton, G. M. Morris, and T. Erdogan, "Experimental investigation of resonant-grating filter line shapes in comparison with theoretical models," J. Opt. Soc. Am. A 15, 464-472, 1998

XIII.11. L. Mashev and E. Popov, "Zero order anomaly of a dielectric coated grating," Opt. Commun. 55, 377-380, 1985

Chapter XIV

Differential Theory of Non-Periodic Media

A key feature of all differential theories of diffraction is the representation of the total field by a Fourier series, as it was done in Eq. (I.29'). The process assumes a periodicity of the diffracting device, which obviously is not fulfilled if we analyze the diffraction by an isolated bump lying on a flat interface. However, there exists a possibility of using the differential theory for non-periodic systems.

XIV.1 Diffraction by an isolated bump

An example of nonperiodic object is an isolated bump lying on a single flat interface or on the top interface of a slab or a stack of slabs. Such a problem has been extensively studied 20 to 25 years ago [XIV.1-XIV.3]. Fig. XIV.1 represents a bump with height a, which is a localized cylindrical deformation of the Oxz plane, invariant in the z-direction and extending in x in the region between $-L$ and $+L$. This deformation can be considered as a single groove of a classical gratings, which is z-invariant, too. When a polarized field falls on the device, it generates diffracted fields in both superstrate and substrate that are no longer pseudoperiodic.

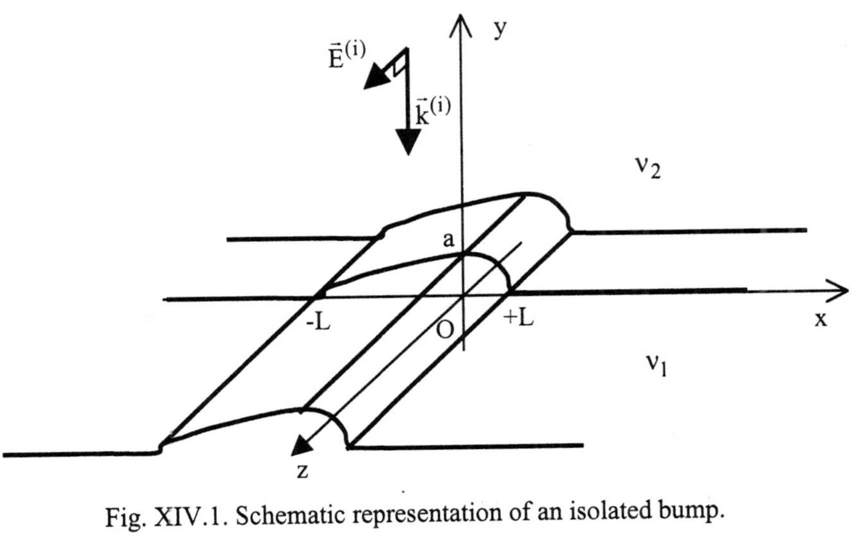

Fig. XIV.1. Schematic representation of an isolated bump.

XIV.1.1 Expression of the total field

Let us assume for simplicity a normal incidence and TE polarization. Instead of representing the total field $u(x,y)$ by the discrete Fourier series as given in Eq. (I.29), it can be represented by a Fourier integral:

$$u(x,y) = \int_{-\infty}^{+\infty} u_f(\alpha, y) e^{i\alpha x} d\alpha . \qquad \text{(XIV.1)}$$

In view of numerical computation, however, the integral in Eq. (XIV.1) will have to be restricted over an interval of limited length, say $2\alpha_{max}$, by assuming that $u_f(\alpha, y)$ is negligible outside this interval:

$$u(x,y) \approx \int_{-\alpha_{max}}^{+\alpha_{max}} u_f(\alpha, y) e^{i\alpha x} d\alpha . \qquad \text{(XIV.2)}$$

Then, the integral will have to be discretized, i.e., replaced by its Riemann series. If we cut the integration domain into 2N+1 intervals of equal length, we substitute:

$$d\alpha \to \Delta\alpha = 2\alpha_{max}/(2N+1) .$$

Let us define:

$$\alpha_n = \alpha_{-N} + (n+N)\Delta\alpha ,$$

where $\alpha_{-N} = -\alpha_{max}$.

As a result, Eq. (XIV.2) leads to:

$$\begin{aligned} u(x,y) &\approx \sum_{n=-N}^{+N} u_f(\alpha_n, y) \Delta\alpha e^{i\alpha_n x} \\ &\approx \sum_{n=-N}^{+N} u_f(\alpha_n, y) \Delta\alpha e^{in\Delta\alpha x} \end{aligned} \qquad \text{(XIV.3)}$$

Defining $K = \Delta\alpha$ and $u_n(y) = u_f(\alpha_n, y)\Delta\alpha$, we see that the discretized Fourier transform reduces to the following truncated Fourier series:

$$u(x, y) \simeq \sum_{n=-N}^{+N} u_n(y) e^{inKx} . \tag{XIV.4}$$

This is the Fourier series of a function periodic with respect to x with period

$$d = \frac{2\pi}{K} = \frac{\pi(2N+1)}{\alpha_{max}} .$$

From a classical theorem of Fourier transform, the product of the widths of a function and its Fourier transform is greater than or equal to 2, so that:

$$L\alpha_{max} \geq 1/2 .$$

As a result:

$$2N+1 \geq \frac{d}{2\pi L} . \tag{XIV.5}$$

Thus the numerical calculation of a Fourier transform, which requires to discretize and to bound the integration interval, automatically transforms the integral representation of Eq. (XIV.1) into the Fourier series of Eq. (XIV.4). This introduces in the problem a period d verifying Eq. (XIV.5).

XIV.1.2 Practical considerations

The preceding section opens the way for applying the differential theory to nonperiodic objets. For example, the field map in the vicinity of a rectangular bump can be obtained by replacing the bump by a lamellar grating with period d much greater than the bump width, so that the coupling between two adjoining bumps become negligible. The periodic device is then studied with the differential theory, which requires using a high number of Fourier components of the field. This idea was already developed in [XIV.4] for rectangular dielectric bumps and grooves lighted in TE polarization. The devices were transformed into lamellar gratings studied with the modal theory [I.16]. It was observed that when the period of the grating is about 13 times the wavelength and bump width, the field map near one bump of the lamellar grating practically coincides with the one of an isolated bump. Nowadays, thanks to the S-matrix propagation algorithm, to the FFF method and to the memory storage capacity and speed of present computers, the same analysis can be conducted for TM polarization, including metallic bumps or grooves having arbitrary shape.

XIV.2 Near-field maps in photon scanning tunneling microscopy

XIV.2.1 Principle of photon scanning tunneling microscopy

An interesting situation appears when considering the interaction between periodic and nonperiodic diffracting objects. This is the case with the photon scanning tunneling microscopy [XIV.5-XIV.12]. The technique is often used to investigate objects which are periodic at a submicronic scale in one or two directions. The object is lighted through a cylindrical prism at an incidence angle beyond the angle of total reflection. The evanescent field is probed by a thin tip of an optical fiber, which is mounted on a piezoelectrical scanner. The light collected by the fiber is transmitted to a photomultiplier tube and is transformed into an electrical current. The feedback circuit adjusts the tip-sample separation so as to maintain constant current. A computer is then used to process the data and form a crude gray scale rendering of the sample.

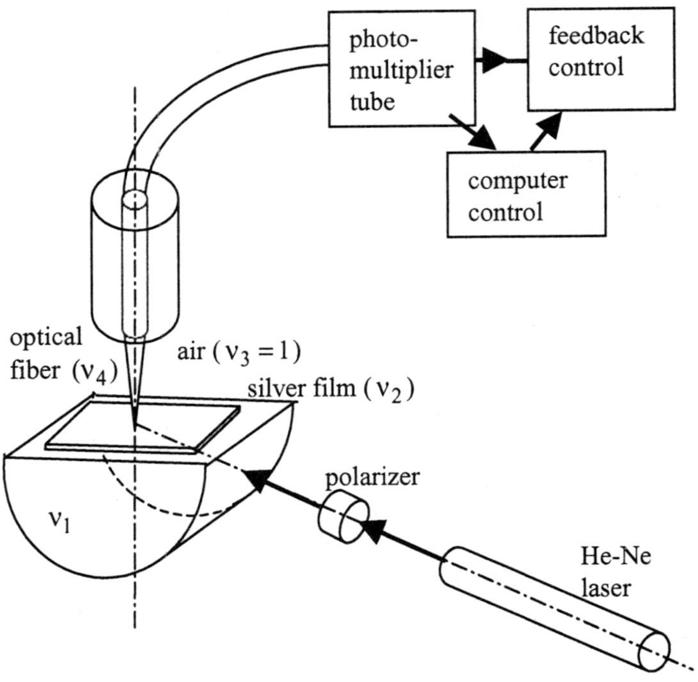

Fig. XIV.2. Block diagram schematic of the photon scanning tunneling microscope (after [XIV.15]).

The images contain a mixture of topographical information, scattering effects, and variations in optical properties across the sample, and therefore proper interpretation of images depends on the knowledge of the field-surface interaction that gives rise to these features [XIV.13]. Cites et al. [XIV.14] calculated image features obtained by a photon scanning tunneling microscope and arising from the topography of the sample, by using a model of sinusoidal grating structures and the simple small-perturbation theory. A more refined analysis [XIV.15] requires taking into account the perturbation of the near field introduced by the probe and studying the influence of the tip-sample distance on the image. The diffracting device then consists of a periodic surface plus a nonperiodic tip in its neighborhood.

A study of the tip of a fiber shows that it can be represented by a half sphere with 50 nm radius put on top of a truncated conical fiber tip, with refractive index $\nu = 1.458$, as shown in Fig. XIV.3. Since the field probed by the fiber is evanescent, only the end of the fiber will participate in the interaction, and that part can be limited to 150 nm length, including the spherical tip.

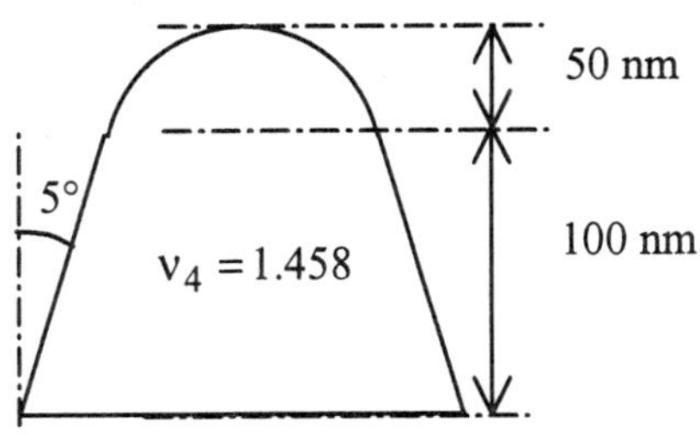

Fig. XIV.3. The tip of a fiber.

XIV.2.2 Near-field maps in the presence of tips

In order to include the tip of the fiber in the theory, we first introduce an infinite collection of tips lying on a glass interface with the same refractive index ν as the fiber. We are then confronted with the problem of diffraction by a stack of two gratings with the same periodicity, as shown in the insert of Fig. XIV.4. The upper grating is the sample to be scanned and the lower grating contains the fiber tips.

Fig. XIV.4 shows the field map obtained at a distance 1 nm above the top of the tips when the lamellar glass grating ($d = 0.416\ \mu m$, $a = 0.006\ \mu m$) is illuminated under 60° incidence, in TE polarization, with $0.6328\ \mu m$ wavelength. The distance between the bottom of the grooves and the top of the tips is 50 nm. A minimum in the peaks of the field map appears when the tips are placed in the middle of the grooves. The space between grating and tips is air.

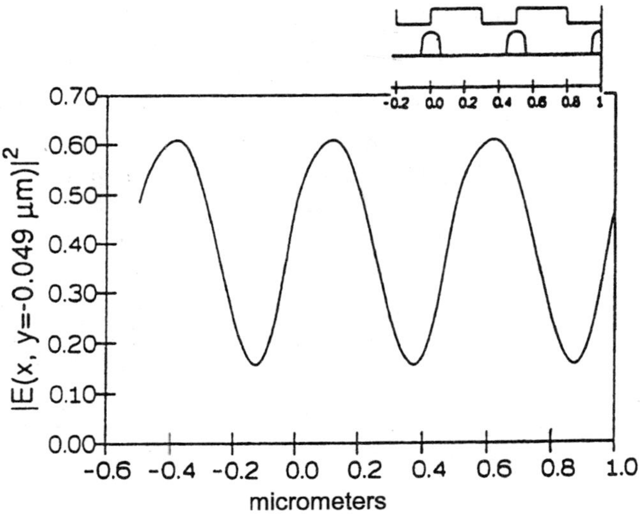

(a)

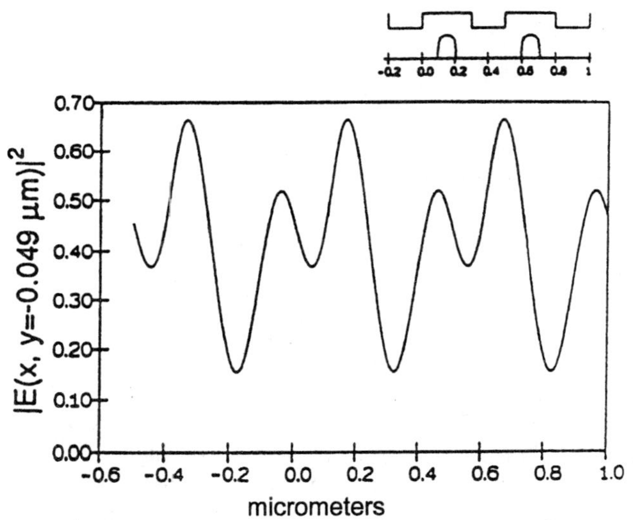

(b)

Fig. XIV.4. x-variation of the intensity of the light below a lamellar grating in front of a series of tips for two different positions of the tips, as shown in the inserts. The gratings have the same period; the index of refraction of both grating and tips is $\nu = 1.458$ (after [XIV.15]).

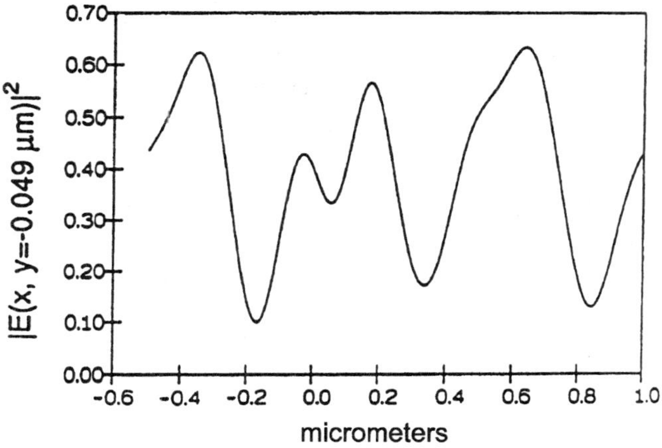

Fig. XIV.5. x-variation of the intensity when one tip out of two has been maintained on an extended scale (after [XIV.15]).

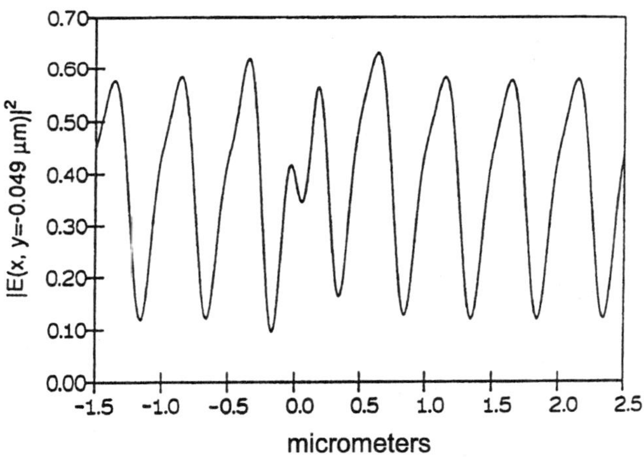

Fig. XIV.6. x-variation of the intensity when one tip out of eight has been maintained on an extended scale (after [XIV.15]).

The influence of a single tip is studied as a second step. To that end, we only maintain one tip out of two in the periodic collection, then one out of three, one out of five, etc., and compare the results. The other parameters are those in Fig. XIV.4, except for the incidence which is 45°. The field intensity map when one tip out of two has been maintained are represented in Fig. XIV.5. The groove without tips in front have a field map with a single maximum. A local minimum only appears in the field map when a tip is present in front of the groove. This tendency is further confirmed in the other cases when one tip out of three, one tip out of five, and one tip out of eight have been maintained. In the latter case presented on an extended scale in Fig. XIV.6, it can be clearly seen that the interaction of the tip and the sample is localized mostly in the vicinity of the tip. By comparing results obtained when keeping different number of tips, it was found that the field map in the vicinity of each periodically found tip are those created by an isolated tip.

A detailed comparison between the numerical results obtained on nonperiodic objects with the differential theory after introducing an artificial large period and the rigorous approach of Hugonin and Petit [XIV.1-XIV.3] has been done in [XIV.16]. A discrepancy of less than 0.5% was found.

XIV.3 Bragg-Fresnel linear zone plates

The rapid development of diffractive optics and optical information processing has brought new requirements to diffractive elements. Not only are they required to diffract light, but they can also be required to redirect it, to separate, reshape, or focus beams. The capacity of simultaneously diffracting and focusing is important in the visible and the near-infrared domains for realization of concave diffraction gratings as single optical components in the flat-field spectrographs, but also to the design of flat optical antennas concentrating a laser beam onto a detector, as well as in the short wavelength end of the spectrum for the development of X-ray imaging microscopy. The focusing property can arise by using a concave substrate, but can also arise on a flat surface by patterning the diffractive element as a Fresnel zone plate. In the near-infrared and the visible domain the zone plate will be ruled in a homogeneous material. In the X-ray domain the low reflectivity of all materials requires combining the Bragg reflection from multilayer structures with the diffraction from the Fresnel zone plates, forming a new domain called Bragg-Fresnel multilayer optics [XIV.17a and b].

XIV.3.1 Description of a Bragg-Fresnel linear zone plate

Fig. XIV.7 shows the central part of a Bragg-Fresnel linear zone plate used in reflection, and introduces the notations used in the section. The zone plate is etched inside a stack of metallic-dielectric bilayers (e.g., carbon and tungsten layers), invariant along the z-axis, hence the designation of the zone plate as linear. The total number of media is M. The profile of the zone plate is a set of plateaus whose widths and locations along the x-axis are determined by the values of x_n, where n is a positive integer less than a maximum value P_{max}. The abscissas x_n follow the rule

$$x_n = \sqrt{n}\,\Xi, \tag{XIV.6}$$

where Ξ is related to the focal distance F, the wavelength λ (in vacuum), and the incidence θ_M by the equation $\Xi = \sqrt{F\lambda}/\cos\theta_M$. In order to obtain high reflectivity, we choose θ_M equal to the Bragg angle given, in first approximation, by

$$\cos\theta_M = \lambda/(2D), \tag{XIV.7}$$

where D is the period of the multilayer along the vertical direction.

A perfect Fresnel zone is assumed to have infinite extension along the x-direction. Technological considerations limit the grooves that can be etched to widths not thinner than $0.1\,\mu m$, which leads to a number of grooves much greater than 1000. Let us designate by Ξ_L the half-width of the finite linear Fresnel zone. It is directly related to the number L of lines (or valleys) etched symmetrically on both sides of the central part (lying in region $[-x_1, +x_1]$) through Eq. (XIV.6), which gives $\Xi_L = \Xi\sqrt{2L+1}$.

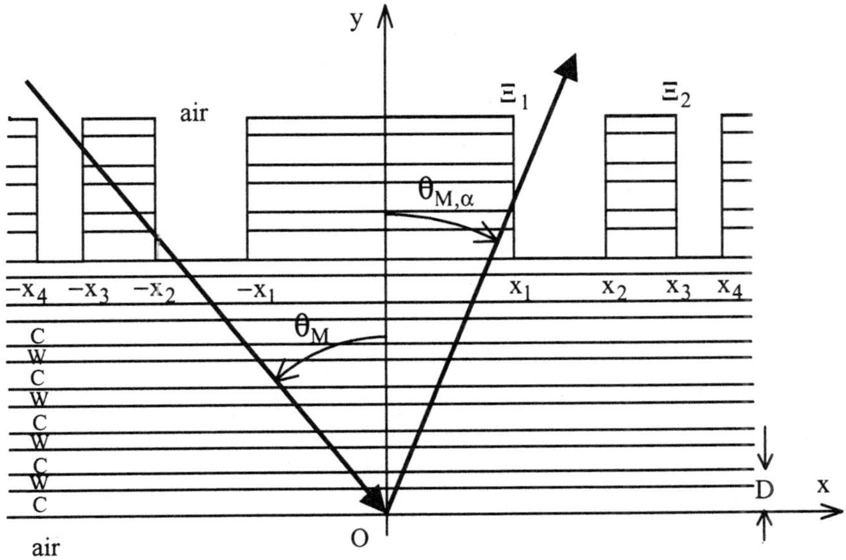

Fig. XIV.7 Schematic representation of a Bragg-Fresnel linear zone plate and notation.

It is worth noting that in addition to the principal focus, there are other foci in which much lower amounts of energy are concentrated. They lie along the direction of specular reflection at focal distances $F_m = F/m$, where m is an odd integer.

XIV.3.2 Electromagnetic study

Stratified Fresnel linear zone plates have already attracted the interest of the theoreticians [XIV.18-XIV.20]. Different approaches are used, based on coupled-wave theory [XIV.18], or the Born-approximation and Fourier-optics [XIV.19 and XIV.20] methods, which bring their own limitations. Indeed, since the linear Fresnel zone is a limited, nonperiodic element, a rigorous analysis should follow the path opened by Hugonin and Petit [XIV.1-XIV.3] which represents the field by its Fourier transform. This approach would lead to the resolution of an integral equation whose numerical treatment is impossible in the X-ray domain because of the low λ/Ξ_L ratio, typically of the order of 10^{-4}. The result established in Section XIV.1.1 open way to another approach, based on the differential theory.

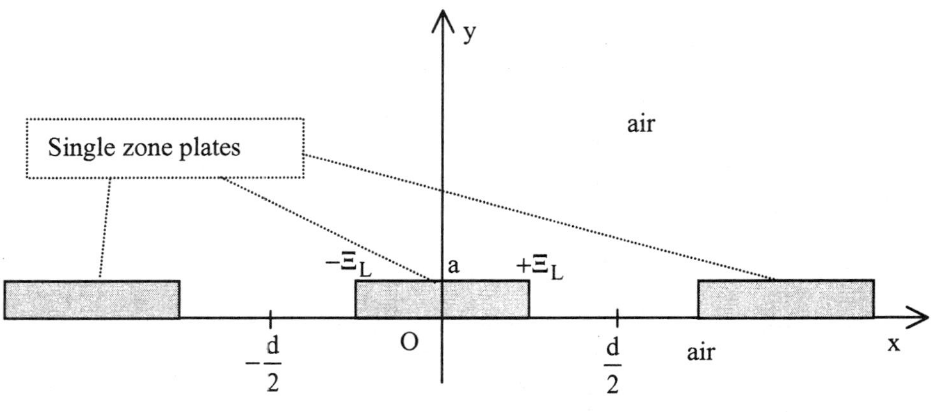

Fig. XIV.8. Schematic representation of a Bragg-Fresnel zone plate grating.

The linear Bragg-Fresnel zone plate, described in Fig. XIV.7, for which we want to determine the intensity of the field reflected at the focus, is introduced as a single bump (but having a complex structure) of a lamellar grating as shown in Fig. XIV.8. The structure period d of the grating will have to be large enough to avoid coupling from neighboring zone plates, a requirement that will have to be tested numerically. To that end, we introduce a convergence criterion parameter δ, defined as $\delta = d/2\Xi_L$. Because of memory storage capabilities and time computation limitations, the total number P_{max} of zones in each zone plate will be restricted to be less than 12. In the numerical example, the total number of C-W bilayers will be 60, half of them etched to form the Fresnel zone plate. The vertical period D will be equal to 3 nm (2 nm for C layers, 1 nm for W layers) and the focal length F will be 15 cm. The wavelength λ will be 0.1 nm, at which carbon and tungsten have refractive indices $0.99999690 + i0.1935194\times10^{-8}$ and $1.000088 + i0.8429687\times10^{-6}$, respectively.

As soon as the initial Bragg-Fresnel zone plate problem is transformed into the problem of diffraction by the grating illustrated in Fig. XIV.8, there is no difference in the theory with a classical grating, except that a high number 2N+1 of Fourier components is required to accurately describe $\varepsilon(x, y)$ in the modulated area. The resolution of the boundary-value problem gives the field $u^{(r)}(x, y)$ diffracted in reflection at the top of the grooves $(y = a)$ over the entire period d. This field will be a good representation of the field at the top of an isolated zone plate if (1) it converges to a limit value when N is increased and (2) it vanishes at $x = \pm d/2$. Numerical experiments show that the latter condition is obtained as soon as δ is increased to ~ 1.5, a value that turns out to be large enough to avoid interferences from two adjacent zone plates.

XIV.3.3 Determination of the far field

When the diffracted field at the top of the grooves has been rigorously determined, the approximation of Fourier optics can be used to derive the far field near the focus. This approximation requires two conditions:

1. The wavelength must be much smaller than the size of the diffracting object.
2. The diffracted field must be computed far from the diffracting objects.

The first condition is obviously satisfied in the X-ray domain, for which $\lambda/\Xi_L \sim 10^{-4}$. Concerning the second condition, it is obviously satisfied, as far as we want to determine the field near the focus, i.e., at distances greater than billions of wavelengths.

Recalling that the incident field with a unit amplitude $u^{(i)}(x, y)$ has an $e^{i\alpha_0 x}$ dependence, which is reproduced in the reflected and transmitted field, we omit it in what follows by introducing the field $u^{(g)}(x)$ at the top of the grating defined by:

$$u^{(g)}(x) = u^{(r)}(x, h)\exp(-i\alpha_0 x). \qquad \text{(XIV.8)}$$

Let us introduce its Fourier transform $U(\alpha)$ defined by

$$U(\alpha) = \int_{-\infty}^{+\infty} u^{(g)}(x) e^{-i\alpha x} dx \quad \text{(XIV.9)}$$

which leads to:

$$u^{(g)}(x) = \frac{1}{2\pi} \int_{-\infty}^{+\infty} U(\alpha) e^{i\alpha x} d\alpha . \quad \text{(XIV.10)}$$

Since $u^{(g)}(x)$ vanishes at $x = \pm d/2$, Eq. (XIV.9) can be reduced to:

$$U(\alpha) = \int_{-d/2}^{+d/2} u^{(g)}(x) e^{-i\alpha x} dx . \quad \text{(XIV.11)}$$

This allows the numerical determination of the function $U(\alpha)$ from the numerical values of $u^{(r)}(x,h)$ obtained through the electromagnetic analysis discussed in the previous section. The far field is then expressed everywhere in space through the Fourier transform of the function $U(\alpha)\exp[i\beta(\alpha)(y-h)]$. This leads to:

$$E(x,y) = \frac{e^{i\alpha_0 x}}{2\pi} \int_{-\infty}^{+\infty} U(\alpha) \exp\{i[\alpha x + \beta(\alpha)(y-h)]\} d\alpha \quad , \quad \text{(XIV.12)}$$

where $\beta(\alpha) = \sqrt{k_M^2 - (\alpha_0 + \alpha)^2}$ with $\text{Im}\,\beta(\alpha) > 0$, and k_M is the modulus of the incident wavevector.

XIV.3.4 Example: A Bragg-Fresnel zone plate at grazing incidence

The method presented in the previous sections gives fast converging results for linear Bragg-Fresnel zone plates with a low number of grooves [XIV.21]. As expected, the average intensity I, defined by:

$$I = \frac{1}{d} \int_{-d/2}^{+d/2} \left| u^{(g)}(x) \right|^2 dx ,$$

presents a sharp maximum when the incidence satisfies the Bragg condition, Eq. (XIV.7). More interesting is the study of $|U(\alpha)|$ as a function of the angle of diffraction $\theta_{M,\alpha}$. Similarly to what was done for gratings, we define $\theta_{M,\alpha}$ by:

$$k_M \sin\theta_{M,\alpha} = \alpha_0 + \alpha\,, \tag{XIV.13}$$

which leads to

$$\sin\theta_{M,\alpha} = \sin\theta_0 + \alpha/k_M\,.$$

Fig. XIV.9 shows the variation of $|U(\theta_{M,\alpha})|$ around the Bragg incidence $\theta_M = 89.045^o$. As expected, a sharp peak is obtained at the Bragg angle. $|U|$ approaches zero as soon as $\theta_{M,\alpha}$ is shifted by 0.01^o from that value. This fact allows reducing the integration limits in Eq. (XIV.12) to a narrow finite domain, namely to $[\alpha^-,\alpha^+]$, deduced from Eq. (XIV.13) by replacing $\theta_{M,\alpha}$ by $\theta_M \pm 0.01^o$. This brings a strong reduction in the computation time.

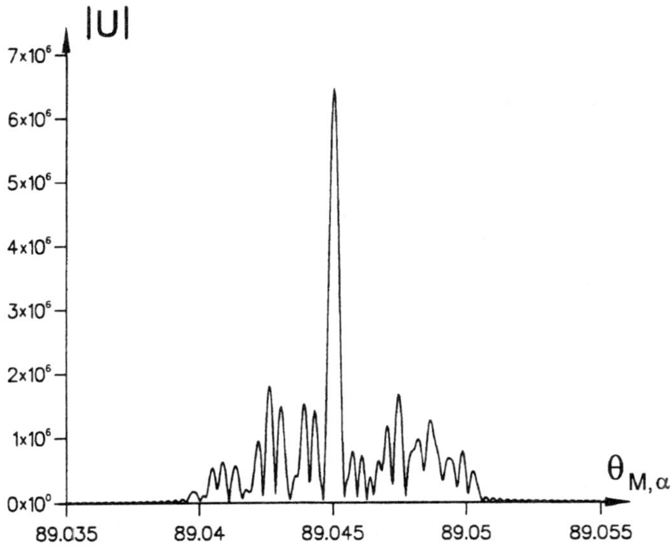

Fig. XIV.9. Behavior of $|U|$ as function of the angle of diffraction (after [XIV.21]).

In view of studying the focusing potentiabilities of the device, it is worth plotting the field intensity as a function of the radial coordinate r in the specular direction $\theta_{M,\alpha} = \theta_M$. Figure XIV.10 shows the results for four different negative zone plates with increasing numbers of elements $(L \in [2,8])$; a negative zone plate is derived from the one illustrated in Fig. XIV.7 by turning each bump into a groove and vice versa. The figure shows the reduction of the undesired maximums outside the focus, as well as the increase in intensity at the focus, when the zone plate is enlarged.

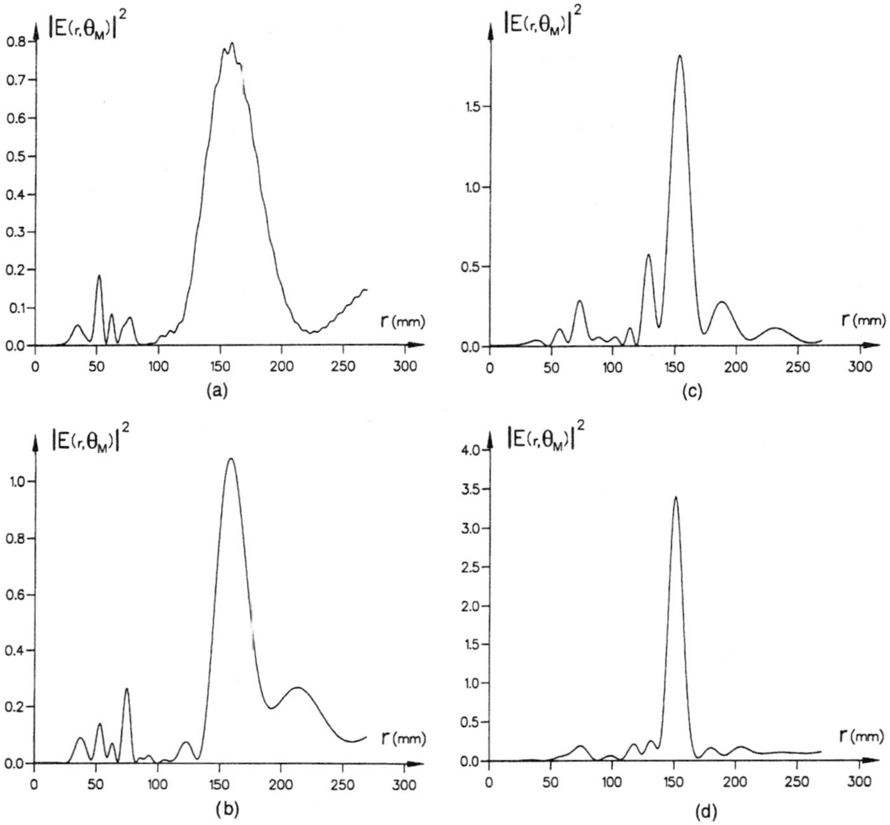

Fig. XIV.10. Focusing capabilities of negative Fresnel zone plates with a 150 mm expected focal length: (a) $L = 2$; (b) $L = 4$; (c) $L = 6$; (d) $L = 8$ (after [XIV.21]).

XIV.4 Random rough surfaces and random media

The electromagnetic study of random rough surfaces or random media requires an analysis of a specific set of a large number of arbitrary-shaped and randomly-distributed objects, which constitute the structure. Each set (speciment of the random medium) is a perfectly known nonperiodic object that can be studied through various approximate or rigorous methods [XIV.22]. Among them, the differential theory has already been applied to 2-D diffraction problems [XIV.23], i.e., rough surfaces described by functions of x only, and to 3-D diffraction problems [XIV.22], for which the surface is given by $y = g(x, z)$. For 2-D problems, the situation is the same as in Section XIV.3, i.e., for linear Bragg-Fresnel zone plates, except that the wavelength is not small if we work in visible region. The representation of the field by a single Fourier integral that has to be discretized is equivalent to introducing a periodicity along the x direction, having a period large enough to avoid interaction between neighboring elements. The numerical implementation of the theory has provided stable results for dielectric objects lighted in TE and TM polarization. Spectacular agreement with the integral theory was observed [XIV.23]. However, when the dielectric contrast is high (for metallic surfaces, for instance), it was observed that the method presented in [XIV.22] is plagued by the slow convergence of the Fourier integrals of the fields. The FFF method gets rid of the difficulty and allows studying random objects in both TE and TM polarization.

XIV.5 Diffraction by a 2-D isolated object

Let us consider an arbitrary shaped cylindrical isolated object, and describe its surface in a cylindrical coordinate system (r, θ, z) by the equation $r = g(\theta)$ or, more generally, by $f(r, \theta) = 0$ (Fig. XIV.11). The surface divides space into two regions, with permitivities ε_j, $j = 1, 2$. An incident plane wave with wavevector $\vec{k}_2$ falls on the device. Inside the surface, ε_1 may be real or complex and is assumed to be scalar, although the generalization to a tensor (i.e., anisotropic media) is straightforward. Outside the object, ε_2 is real. The orientation of $\vec{k}_2$ is arbitrary with respect to the Cartesian coordinate system. The incident field reads:

$$\vec{E}^{(i)} = \vec{A} \exp\left[i(\alpha_0 x - \beta_0 y + \gamma_0 z)\right],$$

where α_0, $-\beta_0$ and γ_0 are the Cartesian components of $\vec{k}_2$. We recall (see Appendix 1) that, since the diffraction device is z-independent, the total field will have the same $e^{i\gamma_0 z}$ dependence as the incident one. As a result all the z derivatives in Maxwell equations reduce to multiplications by $i\gamma_0$.

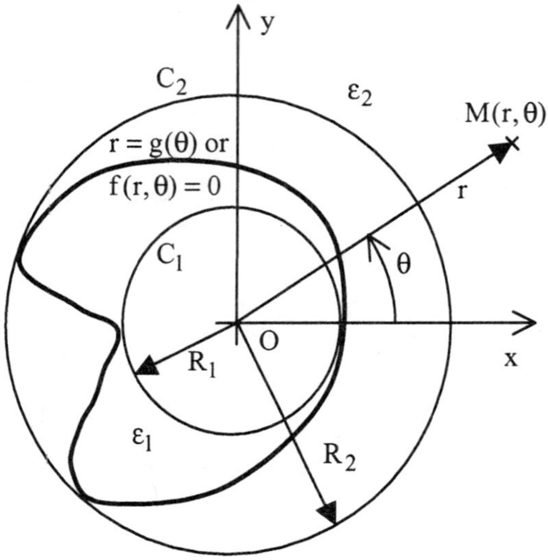

Fig. XIV.11. Arbitrary shaped cylindrical object.

XIV.5.1 Propagation equations

The isolated cylindrical object is typically a nonperiodic device, when described in Cartesian coordinates. However, it is much more convenient to describe it in cylindrical coordinates. In a such coordinate system, starting from a point $M(r,\theta,z)$, when we increase θ by 2π while other coordinates are unchanged, we go back to the same point M. As a result, all field components, as well as the function $g(\theta)$ or $f(r,\theta)$ will be periodic with respect to θ with a period 2π. For example, the component $E_z(r,\theta)$ of the total field will be written as:

$$E_z(r,\theta) = \sum_{n=-N}^{+N} E_{z,n}(r)e^{in\theta}, \qquad \text{(XIV.14)}$$

and we will define a column vector $[E_z]$ having as components the $2N+1$ elements $E_{z,n}(r)$.

Similar notations apply to the other field components and to the surface equation. Maxwell equations written in the cylindrical coordinate system read, taking into account the $e^{i\gamma_0 z}$ dependence of the field:

$$\frac{1}{r}\frac{\partial E_z}{\partial \theta} - i\gamma_0 E_\theta = i\omega\mu_0 H_r \qquad \text{(XIV.15)}$$

$$i\gamma_0 E_r - \frac{\partial E_z}{\partial r} = i\omega\mu_0 H_\theta \qquad \text{(XIV.16)}$$

$$\frac{E_\theta}{r} + \frac{\partial E_\theta}{\partial r} - \frac{1}{r}\frac{\partial E_r}{\partial \theta} = i\omega\mu_0 H_z \qquad \text{(XIV.17)}$$

$$\frac{1}{r}\frac{\partial H_z}{\partial \theta} - i\gamma_0 H_\theta = -i\omega D_z \qquad \text{(XIV.18)}$$

$$i\gamma_0 H_r - \frac{\partial H_z}{\partial r} = -i\omega D_\theta \qquad \text{(XIV.19)}$$

$$\frac{H_\theta}{r} + \frac{\partial H_\theta}{\partial r} - \frac{1}{r}\frac{\partial H_r}{\partial \theta} = -i\omega D_z . \qquad \text{(XIV.20)}$$

Let us introduce a diagonal matrix α_θ with elements $(\alpha_\theta)_{n,m} = n\delta_{n,m}$. Expanding all field components in their Fourier series and noting that $\frac{\partial}{\partial \theta}$ reduces to a multiplication by the factor *in*, we obtain:

$$i\alpha_\theta \frac{[E_z]}{r} - i\gamma_0 [E_\theta] = i\omega\mu_0 [H_r] \qquad \text{(XIV.21)}$$

$$i\gamma_0 [E_r] - \frac{d[E_z]}{dr} = i\omega\mu_0 [H_\theta] \qquad \text{(XIV.22)}$$

$$\frac{E_\theta}{r} + \frac{d}{dr}[E_\theta] - i\alpha_\theta \frac{[E_r]}{r} = i\omega\mu_0 [H_z] \qquad \text{(XIV.23)}$$

$$i\alpha_\theta \frac{[H_z]}{r} - i\gamma_0 [H_\theta] = -i\omega[D_r] \tag{XIV.24}$$

$$i\gamma_0 [H_r] - \frac{d[H_z]}{dr} = -i\omega[D_\theta] \tag{XIV.25}$$

$$\frac{[H_\theta]}{r} + \frac{d[H_\theta]}{dr} - i\alpha_\theta \frac{[H_r]}{r} = -i\omega[D_z]. \tag{XIV.26}$$

In Eqs. (XIV.24)-(XIV.26), $[D_r]$, $[D_\theta]$ and $[D_z]$ must be factorized in order to be expressed as functions of $[E_r]$, $[E_\theta]$ and $[E_z]$. This will be done by using the FFF method. The vector $[\vec{D}]$ will be made with three blocks $[D_r]$, $[D_\theta]$ and $[D_z]$, each block having 2N+1 Fourier components. In view of defining the normal and tangential components of the field, we first calculate in the cylindrical coordinate system the components of the unit vector $\vec{N}$ normal to the surface:

$$\vec{N} = \frac{\overrightarrow{\text{grad}}\, f}{\left|\overrightarrow{\text{grad}}\, f\right|} \Rightarrow N_r = \frac{\dfrac{\partial f}{\partial r}}{\left|\overrightarrow{\text{grad}}\, f\right|}, \quad N_\theta = \frac{1}{r}\frac{\dfrac{\partial f}{\partial r}}{\left|\overrightarrow{\text{grad}}\, f\right|}, \quad N_z = 0,$$

where

$$\left|\overrightarrow{\text{grad}}\, f\right| = \sqrt{\left(\frac{\partial f}{\partial r}\right)^2 + \left(\frac{1}{r}\frac{\partial f}{\partial \theta}\right)^2}.$$

When the surface is defined by the equation $r = g(\theta)$, we can write

$$N_r = \frac{1}{\sqrt{1+\left(\dfrac{1}{r}\dfrac{dg}{d\theta}\right)^2}}, \quad N_\theta = -\frac{1}{r}\frac{\dfrac{dg}{d\theta}}{\sqrt{1+\left(\dfrac{1}{r}\dfrac{dg}{d\theta}\right)^2}}.$$

Defining $\vec{E}_N = \vec{N}(\vec{N}\cdot\vec{E})$ and $\vec{E}_T = \vec{E} - \vec{E}_N$, we express $[\vec{D}]$ as:

$$[\vec{D}] = [\varepsilon\vec{E}] = [\varepsilon\vec{E}_T + \varepsilon\vec{E}_N],$$

and the factorization rules, described in detail in Chapters IV and V give:

$$\left[\vec{D}\right] = [\![\varepsilon]\!]\left[\vec{E}_T\right] + \left[\!\left[\frac{1}{\varepsilon}\right]\!\right]^{-1}\left[\vec{E}_N\right].$$

The calculations continue in the same manner as in Section IV.3. We obtain:

$$\left[\vec{D}\right] = [\![\varepsilon]\!]\left[\vec{E}\right] - \left([\![\varepsilon]\!] - \left[\!\left[\frac{1}{\varepsilon}\right]\!\right]\right)\left(\vec{N}\vec{N}\right)\left[\vec{E}\right], \tag{IV.32}$$

where $\left(\vec{N}\vec{N}\right)_{ij} = N_i N_j$ with $i, j = r, \theta, z$.

If we define matrix Q_ε by the relation

$$\left[\vec{D}\right] = Q_\varepsilon\left[\vec{E}\right], \tag{IV.32'}$$

we thus obtain:

$$Q_\varepsilon = \begin{pmatrix} [\![\varepsilon]\!]\left[\!\left[N_\theta^2\right]\!\right] + \left[\!\left[\frac{1}{\varepsilon}\right]\!\right]^{-1}\left[\!\left[N_r^2\right]\!\right] & -\left([\![\varepsilon]\!] - \left[\!\left[\frac{1}{\varepsilon}\right]\!\right]^{-1}\right)[\![N_r N_\theta]\!] & 0 \\ -\left([\![\varepsilon]\!] - \left[\!\left[\frac{1}{\varepsilon}\right]\!\right]^{-1}\right)[\![N_r N_\theta]\!] & [\![\varepsilon]\!]\left[\!\left[N_r^2\right]\!\right] + \left[\!\left[\frac{1}{\varepsilon}\right]\!\right]^{-1}\left[\!\left[N_\theta^2\right]\!\right] & 0 \\ 0 & 0 & [\![\varepsilon]\!] \end{pmatrix}. \tag{XIV.27}$$

From Eqs. (IV.32') and (XIV.27), we see that

$$[D_r] = Q_{\varepsilon,rr}[E_r] + Q_{\varepsilon,r\theta}[E_\theta],$$

so that

$$[E_r] = Q_{\varepsilon,rr}^{-1}\left\{[D_r] - Q_{\varepsilon,r\theta}[E_\theta]\right\}.$$

Then Eq. (XIV.24) leads to

$$[E_r] = Q_{\varepsilon,rr}^{-1}\left\{\frac{\gamma_0[H_\theta] - \frac{\alpha_\theta}{r}[H_z]}{\omega} - Q_{\varepsilon,r\theta}[E_\theta]\right\}. \qquad \text{(XIV.28)}$$

This expression is put into Eqs. (XIV.22) and (XIV.23) for eliminating $[E_r]$. In a similar way, Eq. (XIV.21) gives:

$$[H_r] = \frac{\alpha_\theta}{r\omega\mu_0}[E_z] - \frac{\gamma_0}{\omega\mu_0}[E_\theta], \qquad \text{(XIV.29)}$$

and this expression is put into Eqs. (XIV.25) and (XIV.26). We finally get a set of four blocks of first-order differential equations of the form:

$$\frac{d}{dr}\begin{pmatrix}[E_\theta]\\ [E_z]\\ [H_\theta]\\ [H_z]\end{pmatrix} = M\begin{pmatrix}[E_\theta]\\ [E_z]\\ [H_\theta]\\ [H_z]\end{pmatrix}, \qquad \text{(XIV.30)}$$

where

$$M = \qquad \text{(XIV.31)}$$

$$i\begin{pmatrix} -\frac{\alpha_\theta}{r}Q_{\varepsilon,rr}^{-1}Q_{\varepsilon,r\theta} + \frac{i}{r} & 0 & \frac{\alpha_\theta}{\omega r}Q_{\varepsilon,rr}^{-1}\gamma_0 & \omega\mu_0 - \frac{\alpha_\theta}{\omega r^2}Q_{\varepsilon,rr}^{-1}\alpha_\theta \\ -\gamma_0 Q_{\varepsilon,rr}^{-1}Q_{\varepsilon,r\theta} & 0 & \frac{\gamma_0}{\omega}Q_{\varepsilon,rr}^{-1}\gamma_0 - \omega\mu_0 & -\gamma_0 Q_{\varepsilon,rr}^{-1}\frac{\alpha_\theta}{\omega r} \\ \frac{-\alpha_\theta\gamma_0}{\omega\mu_0 r} & \frac{\alpha_\theta^2}{\omega\mu_0 r^2} - \omega Q_{\varepsilon,zz} & \frac{i}{r} & 0 \\ \omega(Q_{\varepsilon,\theta\theta} - Q_{\varepsilon,\theta r}Q_{\varepsilon,rr}^{-1}Q_{\varepsilon,r\theta}) - \frac{\gamma_0^2}{\omega\mu_0} & \frac{\gamma_0\alpha_\theta}{\omega\mu_0 r} & Q_{\varepsilon,\theta r}Q_{\varepsilon,rr}^{-1}\gamma_0 & -Q_{\varepsilon,\theta r}Q_{\varepsilon,rr}^{-1}\frac{\alpha_\theta}{r} \end{pmatrix}$$

Eqs. (XIV.30) and (XIV.31) are the basic propagation equations for out-of-cross-section plane incidence. When $\vec{k}_2$ lies in the cross-section plane of the grating, matrix M reduces to:

$$M = \quad \text{(XIV.32)}$$

$$i\begin{pmatrix} -\frac{\alpha_\theta}{r}Q_{\varepsilon,rr}^{-1}Q_{\varepsilon,r\theta}+\frac{i}{r} & 0 & 0 & \omega\mu_0-\frac{\alpha_\theta}{\omega r^2}Q_{\varepsilon,rr}^{-1}\alpha_\theta \\ 0 & 0 & -\omega\mu_0 & 0 \\ 0 & \frac{\alpha_\theta^2}{\omega\mu_0 r^2}-\omega Q_{\varepsilon,zz} & \frac{i}{r} & 0 \\ \omega(Q_{\varepsilon,\theta\theta}-Q_{\varepsilon,\theta r}Q_{\varepsilon,rr}^{-1}Q_{\varepsilon,r\theta}) & 0 & 0 & -Q_{\varepsilon,\theta r}Q_{\varepsilon,rr}^{-1}\frac{\alpha_\theta}{r} \end{pmatrix}.$$

As a result, Eq. (IV.30) splits into two independent sets:

$$\frac{d}{dr}[E_\theta]=i\left(-\frac{\alpha_\theta}{r}Q_{\varepsilon,rr}^{-1}Q_{\varepsilon,r\theta}+\frac{i}{r}\right)[E_\theta]+\left(i\omega\mu_0-\frac{i\alpha_\theta}{\omega r^2}Q_{\varepsilon,rr}^{-1}\alpha_\theta\right)[H_z] \quad \text{(XIV.33)}$$

$$\frac{d}{dr}[H_z]=-i\omega\left(Q_{\varepsilon,\theta r}Q_{\varepsilon,rr}^{-1}Q_{\varepsilon,r\theta}-Q_{\varepsilon,\theta\theta}\right)[E_\theta]-iQ_{\varepsilon,\theta r}Q_{\varepsilon,rr}^{-1}\frac{\alpha_\theta}{r}[H_z] \quad \text{(XIV.34)}$$

and

$$\frac{d}{dr}[E_z]=-i\omega\mu_0[H_\theta] \quad \text{(XIV.35)}$$

$$\frac{d}{dr}[H_\theta]=i\left(\frac{\alpha_\theta^2}{\omega\mu_0 r^2}-\omega[\![\varepsilon]\!]\right)[E_z]-\frac{1}{r}[H_\theta]. \quad \text{(XIV.36)}$$

Eqs. (XIV.35) and (XIV.36) are the propagation equations written in a truncated Fourier space for a TE polarized field ($\vec{E}=E_z\vec{z}$). They are equivalent to the second-order equation:

$$\frac{d^2[E_z]}{dr^2}+\frac{1}{r}\frac{d[E_z]}{dr}-\frac{\alpha_\theta^2}{r^2}[E_z]+[\![k^2]\!][E_z]=0, \quad \text{(XIV.37)}$$

which is the basic equation already derived thirty years ago for cylindrical objects [XIV.24]. On the other hand, Eqs. (XIV.33) and (XIV.34) are new. They are the propagation equations written in a truncated Fourier space of a TM polarized field ($\vec{H}=H_z\vec{z}$), obtained with the FFF method.

XIV.5.2 Expression of the field inside homogeneous regions

As in Cartesian coordinates, one basic idea of the differential theory is to split space into three regions instead of two, by distinguishing two homogeneous regions and one which is periodically modulated with respect to θ. To that aim, we introduce two circles C_1 and C_2 (see Fig. XIV.11) with radius R_1 and R_2 given by:

$$R_1 = \underset{\theta}{\mathrm{Min}}\, g(\theta)$$

$$R_2 = \underset{\theta}{\mathrm{Max}}\, g(\theta)\,.$$

If $r < R_1$, $\varepsilon(r,\theta) = \varepsilon_1$, and if $r > R_2$, $\varepsilon(r,\theta) = \varepsilon_2$. Thus inside the circle C_1 and outside the circle C_2, $\left[\!\left[k^2\right]\!\right]$ reduces to $k_j^2 \mathbb{1}$, with $j = 1, 2$, and the Fourier components of each field component satisfy the Helmholtz equation, which leads to

$$\frac{d^2 u_n}{dr^2} + \frac{1}{r}\frac{du_n}{dr} + \left(k_j^2 - \gamma_0^2 - \frac{n^2}{r^2}\right) u_n = 0\,. \qquad \text{(XIV.38)}$$

As a result, u_n can be expressed by suitable combinations of Bessel (J_n) and Hankel (H_n^+) functions, chosen in order to remain bounded when $r \to 0$ and to satisfy the outgoing-wave condition.

For sake of simplicity, let us chose the axis $\overrightarrow{Ox}$ parallel to the projection $\vec{k}_2^*$ of $\vec{k}_2$ in the cross-section plane. Dropping the $e^{i\gamma_0 z}$ dependence of the incident field which will be common to all total field components, the incident electric field vector takes the form:

$$\vec{E}^{(i)}(r,\theta) = \vec{A}\exp(ik_2^* x) = \vec{A}\exp(ik_2^* r\cos\theta).$$

From the classical properties of Bessel functions [XIV.25], this expression can be turned into:

$$\vec{E}^{(i)}(r,\theta) = \vec{A}\sum_{n=-\infty}^{+\infty} i^n J_n(k_2^* r)\exp(in\theta) \qquad \text{(XIV.39)}$$

where J_n is the nth Bessel function.

In Eq. (XIV.39), $\vec{A}$ specifies the polarization of the incident plane wave and lies in the yOz plane. Its θ and z components are known. So are the θ and z components of the associated incident magnetic field $\vec{H}^{(i)}$. Outside the circle C_2, $E_{z,n}$ will be written:

$$E_{z,n} = A_z i^n J_n(k_2^* r) + B_{e,n}^{(2)} H_n^+(k_2^* r), \qquad \text{(XIV.40)}$$

and similar form is obtained for $H_{z,n}$, with Rayleigh coefficients $B_{h,n}^{(2)}$.

Inside the circle C_1, in order to avoid singularities coming from Neumann functions, we represented E_z by Eq. (XIV.14) with $E_{z,n}$ having the form:

$$E_{z,n} = B_{e,n}^{(1)} J_n(k_1^* r), \qquad \text{(XIV.41)}$$

and similar notation for $H_{z,n}$ with Rayleigh coefficients $B_{h,n}^{(1)}$ and incident amplitude equal to $(\vec{k}_2 \times \vec{A})_z / \omega\mu_0$.

Of course, in the part of the plane lying between the circles C_1 and C_2, called modulated region, the Fourier components of the field satisfy Eq. (XIV.30).

XIV.5.3 Resolution of the boundary-value problem

The shooting method used to resolve the problem is quite similar to the one used for conical diffraction of classical gratings (Section IV.3). Each field component being represented by 2N+1 Fourier coefficients, the dimension of the set in Eq. (XIV.30) is $4(2N+1)$. However, in the homogeneous regions we know the explicit expression of the field components which allows obtaining their derivatives, so that E_θ and H_θ can be derived from E_z and H_z. The general formulae (A.III.12,13) established in Appendix III lead to:

$$E_\theta = \frac{1}{k^2 - \gamma_0^2}\left[i\gamma_0 \frac{1}{r}\frac{\partial E_z}{\partial \theta} - i\omega\mu \frac{\partial H_z}{\partial r}\right] \qquad \text{(XIV.42)}$$

$$H_\theta = \frac{1}{k^2 - \gamma_0^2}\left[i\gamma_0 \frac{1}{r}\frac{\partial H_z}{\partial \theta} + i\omega\varepsilon \frac{\partial E_z}{\partial r}\right]. \qquad \text{(XIV.43)}$$

Inside the circle C_1, k^2 is a constant equal to k_1^2. Thus there is no need of factorizing the products that appear in Eqs. (XIV.42) and (XIV.43), when calculating their Fourier components, and no convergence problems appear. We obtain:

$$E_{\theta,n} = \frac{1}{(k_1^*)^2}\left[-n\gamma_0 \frac{E_{z,n}}{r} - i\omega\mu B_{h,n}^{(1)} \frac{dJ_n(k_1^* r)}{dr}\right] \tag{XIV.44}$$

$$H_{\theta,n} = \frac{1}{(k_1^*)^2}\left[-n\gamma_0 \frac{H_{z,n}}{r} + i\omega\varepsilon_1 B_{e,n}^{(1)} \frac{dJ_n(k_1^* r)}{dr}\right], \tag{XIV.45}$$

where $\vec{k}_1^*$ is the projection of $\vec{k}_1$ in the cross-section plane, and

$$\frac{dJ_n(k_1^* r)}{dr} = \frac{nJ_n(k_1^* r)}{r} - k_1^* J_{n+1}(k_1^* r) \tag{XIV.46}$$

In order to solve the boundary-value problem, we start a first series of (2N+1) numerical integrations with starting vectors numbered by their subscript p ($p \in [-N, +N]$) by choosing:

$$\begin{aligned} E_{z,n,p} &= \delta_{n,p} J_p(k_1^* R_1) \\ H_{z,n,p} &= 0 \end{aligned}. \tag{XIV.47}$$

Putting these values in Eqs. (XIV.42) and (XIV.43) gives:

$$\begin{aligned} E_{\theta,n,p} &= \delta_{n,p} \frac{-p\gamma_0}{(k_1^*)^2 R_1} J_p(k_1^* R_1) \\ H_{\theta,n,p} &= \delta_{n,p} \frac{i\omega\varepsilon_1}{(k_1^*)^2} \left.\frac{dJ_p(k_1^* r)}{dr}\right|_{r=R_1} \end{aligned}. \tag{XIV.48}$$

Using the starting values provided by Eqs. (XIV.47) and (XIV.48), the differential set (IV.30) is integrated with a standard algorithm from $r = R_1$ up to $r = R_2$. We then start a second series of (2N+1) numerical integrations ($p \in [N+1, 3N+1]$) with:

$$\begin{aligned} E_{z,n,p} &= 0 \\ H_{z,n,p} &= \delta_{n,p-(2N+1)} J_p(k_1^* R_1), \end{aligned}$$

from which, the use of Eqs. (XIV.42) and (XIV.43) gives:

$$\begin{aligned} E_{\theta,n,p} &= \delta_{n,p} \frac{-i\omega\mu}{(k_1^*)^2} \frac{d}{dr} J_p(k_1^* r)\Big|_{r=R_1} \\ H_{\theta,n,p} &= \delta_{n,p} \frac{-p\gamma_0}{(k_1^*)^2 R_1} J_p(k_1^* R_1) \end{aligned} \tag{XIV.49}$$

We finally obtain $2(2N+1)$ numerical values of $E_{\theta,n,p}(R_2)$, $E_{z,n,p}(R_2)$, $H_{\theta,n,p}(R_2)$ and $H_{z,n,p}(R_2)$. The exact Fourier components are obtained by a linear superposition of the particular values, as stated by Eqs. (XIV.14), (XIV.41) and (XIV.47) for the E_z component of the field. This linear superposition will be matched with the Rayleigh expansion for $r = R_2$, given by Eq. (XIV.40). For example, for $E_{z,n}$ and $H_{\theta,n}$ we will write:

$$\sum_{p=-N}^{+N} B_{e,p}^{(1)} E_{z,n,p}(R_2) = A_z i^n J_n(k_2^* R_2) + B_{e,n}^{(2)} H_n^+(k_2^* R_2)$$

and

$$\begin{aligned} &\sum_{p=-N}^{+N} B_{e,p}^{(1)} H_{\theta,n,p}(R_2) + \sum_{p=N+1}^{3N+1} B_{h,p}^{(1)} H_{\theta,n,p}(R_2) = \\ &\qquad \frac{i\omega\varepsilon_2}{(k_2^*)^2} \left[A_z i^n \frac{dJ_n(k_2^* r)}{dr}\Big|_{r=R_2} + B_{e,n}^{(2)} \frac{dH_n^+(k_2^* r)}{dr}\Big|_{r=R_2} \right] \\ &\qquad + \frac{1}{(k_2^*)^2} \left(\frac{-n\gamma_0}{R_2} \right) \left[\frac{(\vec{k}_2 \times \vec{A})_z}{\omega\mu_0} i^n J_n(k_2^* R_2) + B_{h,n}^{(2)} H_n^+(k_2^* R_2) \right], \end{aligned}$$

which gives $2(2N+1)$ equations. The matching of $H_{z,n}$ and $E_{\theta,n}$ in a similar way again provides $2(2N+1)$ equations. We finally obtain a set of $4(2N+1)$ linear equations with $4(2N+1)$ unknowns $B_{e,p}^{(1)}$, $B_{e,n}^{(2)}$, $B_{h,p}^{(1)}$ and $B_{h,n}^{(2)}$. Its resolution by a matrix inversion provides all Rayleigh coefficients of E_z and H_z. The Rayleigh coefficients of E_θ and H_θ are derived through Eqs. (XIV.42) and (XIV.43).

Of course, the concept of S-matrix can be applied to this diffraction problem, and the S-matrix propagation algorithm can be used to avoid numerical instabilities coming from growing exponential functions during the integration process. It also simplifies the matching with Rayleigh expansions at the end of the integration process. The study of complex poles and zeros of the S operator has given rise to what is called the *singularity expansion method* [XIV.26], which allows a good description of the time-response of the scatterer from a few number of complex parameters.

As soon as the Rayleigh coefficients of the field are known, all quantities of interest for the experimentalist, such as the differential cross-section σ_d, can be computed. We recall that σ_d is defined as:

$$\sigma_d(\theta) = \lim_{r \to \infty} 2\pi r \left| E^d(r,\theta) \right|^2 \Big/ \left| E^{(i)}(r) \right|^2 .$$

In TE polarization for $r \to \infty$, $E_z(r,\theta)$ takes an asymptotic form. It contains the incident field and the diffracted field which reduces to a cylindrical wave with amplitude $b(k_2,\theta)$, so that Eq. (XIV.40) reduces to:

$$E_z(r,\theta) \simeq e^{ik_2 r\cos\theta} + b(k_2,\theta)\frac{e^{ik_2 r}}{\sqrt{r}}.$$

The differential cross-section is then given by:

$$\sigma_d(k_2,\theta) = 2\pi \left| b(k_2,\theta) \right|^2 .$$

The study of the poles of the S-matrix allows predicting resonance peaks in function $\sigma_d(k_2)$ for a given angle θ [XIV.27], as well as resonance peaks in the total cross-section σ_T defined as the ratio of the modulus of the Poynting vectors of the diffracted and incident field.

Similar extension of the FFF approach can be developed for an isolated arbitrary-shaped three-dimensional object, described in spherical coordinates using the double periodicity with respect to the two spherical angular coordinates in the same manner as for crossed gratings. This will require the extension of the factorization rules of Fourier series to a series of spherical harmonics.

References

XIV.1. J. P. Hugonin and R. Petit, "Numerical study of the problem of diffraction at a non-periodic obstacle," Opt. Commun. 20, 360-364, 1977

XIV.2. J. P. Hugonin and R. Petit, "Theoretical and numerical study of locally deformed stratified medium," J. Opt. Soc. Am. 71, 664-674, 1981

XIV.3. J. P. Hugonin, "Sur l'étude théorique et numérique de la propagation des ondes électromagnétiques en milieu inhomogène non périodique," Ph.D. thesis, University of Paris –Sud, 1983

XIV.4. J. T. Sheridan and C. J. R. Sheppard, "Coherent imaging of periodic thick fine isolated structures," J. Opt. Soc. Am. A 10, 614-632, 1993

XIV.5a. D. W. Pohl, *Advances in Optical and Electron Microscopy* (Academic, London, 1991)

XIV.5b. K. E. Betzig and J. K. Trautman, "Near field optics: microscopy and surface modification beyond the diffraction limit," Science 257, 189, 1992

XIV.6. R. C. Reddick, R. J. Warmack, and T. L. Ferrell, "New form of scanning optical microscopy," Phys. Rev. B 39, 767-770, 1989

XIV.7. F. de Fornel, J. P. Goudonnet, L. Salomon, and E. Lesniewska, "An evanescent field optical microscope," in *Optical Storage and Scanning Technology*, T. Wilson, ed., Proc. Soc. Photo-Opt. Instrum. Eng. 1139, 77-84, 1989

XIV.8. D. Courjon, K. Sarayeddine, and M. Spajer, "Scanning tunneling optical microscopy," Opt. Commun. 71, 23-28, 1989

XIV.9. N. F. Van Hulst, N. P. De Boer, and B. Bolger, "An evanescent field optical microscope," J. Microsc. 163, 117-129, 1991

XIV.10. J. P. Goudonnet, F. de Fornel, L. Salomon, E. Bourillot, and P. Adam, "Recent progress in photon scanning tunneling microscopy," in *Scanning Probe Microscopies*, S. Manne, ed., Proc. Soc. Photo-Opt. Instrum. Eng. 1639, 12-23, 1992

XIV.11. G. Binning and H. Rohrer, "Scanning tunneling microscopy," Surf. Sci. 126, 236-244, 1983

XIV.12. L. Salomon, F. de Fornel, and J. P. Goudonnet, "Sample-tip coupling efficiencies of the photon scanning tunneling microscope," J. Opt. Soc. Am. A 8, 2009-2015, 1991

XIV.13. X. Bouju, C. Girard, and D. Courjon, "Spectroscopie locale d'une surface par détection de champ proche: étude théorique comparative des métaux nobles," J. Phys. 1 (Paris) 2, 1431-1444, 1992

XIV.14. J. Cites, M. F. M. Sanghadese, C. C. Sung, R. C. Reddick, R. J. Warmack, and T. L. Ferrell, "Analysis of photon scanning tunneling microscope images," J. Appl. Phys. 71, 7-17, 1992

XIV.15. J. P. Goudonnet, E. Bourillot, P. M. Adam, F. de Fornel, L. Salomon, P. Vincent, M. Nevière, and T. L. Ferrell, "Imaging of test quartz gratings with a photon scanning tunneling microscope. Experiment and theory," J. Opt. Soc. Am. A12, 1749-1764, 1995

XIV.16. J. C. Weber, "Diffraction en champ proche optique. Analyse des images de microscopie à effet tunnel optique," Thesis, University of Bourgogne, Dijon, France, 1996

XIV.17a. V. V. Aristov, A. I. Erko, and V. V. Martinov, "Principles of Bragg-Fresnel multilayer optics," Rev. Phys. Appl. 23, 1623-1630, 1988

XIV.17b. S. Babin and A. Erko, "Fabrication of diffractive x-ray optical elements," Nucl. Instrum. Methods Phys. Res. A 282, 529-535, 1989

XIV.18. J. Maser, "Evaluation of the efficiency of zones plates with high aspect ratios by application of coupled wave theory,"in *X-ray Microscopy III*, A. G. Michette, G. R. Morrison, and G. J. Buckley, eds. (Springer-Verlag, Berlin, 1992), pp.104-106

XIV.19. A. Sammar and J. M. André, "Diffraction of multilayer gratings and zone plates in the x-ray region using the Born approximation," J. Opt. Soc. Am. A 10, 600-613, 1993

XIV.20a. A. Sammar and J. M. André, "Dynamical theory of stratified Fresnel linear zone plates," J. Opt. Soc. Am. A 10, 2324-2337, 1993

XIV.20b A. Mirone, M. Idir, P. Dhez, G. Soullie, and A. Erko, "Dynamical theory for Bragg-Fresnel multilayer lenses for X-UV and X-ray range," Opt. Commun. 111, 191-198, 1994

XIV.21. F. Montiel and M. Nevière, "Electromagnetic theory of Bragg-Fresnel linear zone plates," J. Opt. Soc. Am. A, 2672-2678, 1995

XIV.22. M. Saillard and A. Sentenac, "Rigorous solutions for electromagnetic scattering from rough surfaces," Waves in Random Media 11, R103-R107, 2001

XIV.23. H. Giovannini, M. Saillard, and A. Sentenac, "Numerical study of scattering from rough inhomogeneous films," J. Opt. Soc. Am. A 15, 1182-1191, 1998

XIV.24. P. Vincent and R. Petit, "Sur la diffraction d'une onde plane par un cylindre diélectrique," Opt. Commun. 5, 261-266, 1972

XIV.25. M. Abramovitz and I. E. Stegun, *Handbook of Mathematical Functions* (National Bureau of Standards, Washington D.C., USA, 1964)

XIV.26. C. E. Baum, *The Singularity Expansion Method*, Topics in Applied Physics Transient Electromagnetic Fields, L. B. Felsen ed., (Springer-Verlag, Berlin, New York, 1976), pp. 129-179

XIV.27. P. Vincent and M. Nevière, "Matrice de diffraction et fréquences naturelles de résonance d'un cylindre," Annales des télécommunications 31, 1-6, 1976

Chapter XV

Fourier Factorization of Maxwell Equations in Nonlinear Optics

The use of periodic structures in nonlinear optics has many interesting aspects. A detailed presentation goes beyond the goal of this book. Interested readers can find details in the literature [XV.1-XV.10]. Our aim here is to discuss the problem of factorization of truncated Fourier expansions of fields put in nonlinear Maxwell equations, in view of the numerical modeling of the diffraction by periodic structures in nonlinear optics.

XV.1 Nonlinear Maxwell equations

The general form of Maxwell equations is

$$\begin{aligned} \operatorname{curl}\vec{E}(\vec{r},t) &= -\frac{\partial\vec{B}(\vec{r},t)}{\partial t} \\ \operatorname{curl}\vec{H}(\vec{r},t) &= \frac{\partial\vec{D}(\vec{r},t)}{\partial t} \end{aligned} \tag{XV.1}$$

Let us consider media without magnetic nonlinearity and time-independent properties. The consecutive relations then are [XV.10]:

$$\begin{aligned} &\vec{B}(\vec{r},t) = \overline{\overline{\mu}}(\vec{r})\vec{H}(\vec{r},t) \\ &\vec{D}(\vec{r},t) = \varepsilon_0\vec{E}(\vec{r},t) + \vec{P}(\vec{r},t) = \varepsilon_0\vec{E}(\vec{r},t) + \vec{P}^{L}(\vec{r},t) + \vec{P}^{NL}(\vec{r},t) \end{aligned} \tag{XV.2}$$

where the polarization vector $\vec{P}$ is split into a linear $\vec{P}^{L}$ and nonlinear $\vec{P}^{NL}$ part. Further simplifications can be made, including locality, instantaneous response, lack of hysteresis phenomena, steady-state conditions and exponential time-dependence $\exp(-i\omega t)$ with a discrete spectral representation to finally express the Maxwell equations as a set of coupled time-independent equations at each discrete circular frequency ω_m:

$$\begin{aligned} \operatorname{curl}\vec{E}_{\omega_m}(\vec{r}) &= i\omega_m\overline{\overline{\mu}}_{\omega_m}\vec{H}_{\omega_m}(\vec{r}) \\ \operatorname{curl}\vec{H}_{\omega_m}(\vec{r}) &= -i\omega_m\overline{\overline{\varepsilon}}_{\omega_m}\vec{E}_{\omega_m}(\vec{r}) - i\omega_m\vec{P}^{NL}_{\omega_m}(\vec{r}) \end{aligned} \tag{XV.3}$$

where for each vector $\vec{E}$, $\vec{H}$ and $\vec{P}$ the following relation is valid:

$$\vec{E}_{\omega_m}(\vec{r}) = \frac{1}{2\pi}\int_{-\infty}^{\infty} \vec{E}(\vec{r},t)\exp(i\omega_m t)dt \tag{XV.4}$$

and

$$\bar{\bar{\varepsilon}}_{\omega_m} = \varepsilon_0\left(\mathbb{1} + \chi^L_{\omega_m}\right) \tag{XV.5}$$

is the electric permittivity at circular frequency ω_m. The linear susceptibility $\chi^L_{\omega_m}$ is a second-rank tensor which links the electric field and the linear polarization vector:

$$\vec{P}^L_{\omega_m}(\vec{r}) = \varepsilon_0 \chi^L_{\omega_m} \vec{E}_{\omega_m}(\vec{r}) . \tag{XV.6}$$

The frequency coupling in Eqs. (XV.3) is produced by the vector of nonlinear polarization $\vec{P}^{NL}_{\omega_m}(\vec{r})$, which depends on the electric field vector inside optically nonlinear media through tensors of nonlinear susceptibility $\chi^{NL}_{\omega_m}$. Different optical phenomena are described by different dependences of $\vec{P}^{NL}_{\omega_m}(\vec{r})$ on $\vec{E}$. *Linear* Maxwell equations are obtained by assuming that the nonlinear part $\vec{P}^{NL}_{\omega_m}(\vec{r})$ vanishes, which decouples the equations (XV.3) written at different circular frequences ω_m. *Second harmonic generation (SHG)* is described by two circular frequences $\omega_1 = \omega$ and $\omega_2 = 2\omega$, and two nonlinear polarization vectors:

$$\begin{aligned} \vec{P}^{SHG}_{\omega} &= \varepsilon_0 \chi^{SHG}_{\omega} : \bar{\vec{E}}_{\omega} \vec{E}_{2\omega} \\ \vec{P}^{SHG}_{2\omega} &= \varepsilon_0 \chi^{SHG}_{2\omega} : \vec{E}_{\omega} \vec{E}_{\omega} \end{aligned} \tag{XV.7}$$

where the overbar means complex conjugation, χ^{SHG}_{ω} and $\chi^{SHG}_{2\omega}$ are third-rank tensors of nonlinear susceptibility, and the colon stands for a contraction with respect to two of the vector indices. *Optical Kerr effect* is characterized by a single circular frequency ω and a single nonlinear polarization vector

$$\vec{P}^{Kerr}_{\omega} = \varepsilon_0 \chi^{Kerr}_{\omega} : \vec{E}_{\omega} \bar{\vec{E}}_{\omega} \vec{E}_{\omega} \tag{XV.8}$$

with a four-rank tensor of nonlinear susceptibility $\chi_{\omega}^{\text{Kerr}}$.

Unless a surface nonlinearity is present, the boundary conditions for the electromagnetic field components require the continuity of the tangential components of $\vec{E}$ and $\vec{H}$, and of the normal components of $\vec{B}$ and $\vec{D}$ across each interface at which some elements of $\overline{\overline{\varepsilon}}$, $\overline{\overline{\mu}}$, or $\chi_{\omega_m}^{\text{NL}}$ are discontinuous. Moreover, it can be shown using Maxwell equations that the continuity requirements have to be independently fulfilled at each circular frequency ω:

$$E_{T,\omega}, D_{N,\omega}, H_{T,\omega}, B_{N,\omega} \text{ - continuous } \forall\omega . \tag{XV.9}$$

XV.2 Methods of solution in nonlinear optics of periodic media

Due to the boundary conditions (XV.9), modeling of periodic media in linear and nonlinear optics requires integrating boundary-value differential equations. In linear optics this could be done using a shooting method, as described in detail in Section II.1. The unknown field components at a given boundary are expressed as a *linear* combination of the unknown components on another interface, as presented, for example, in Eq. (III.14). A simple matrix inversion (or a more complicated but straightforward procedure, when numerical problems arise during the integration) is then necessary to determine the unknown field components. Unfortunately, Eqs. (XV.3) contain a nonlinear term with respect to $\vec{E}$, given, for example, by Eqs. (XV.7) or (XV.8). Due to the nonlinearity, the shooting method is no longer valid, as the linear superposition of solutions is not necessarily a solution of the nonlinear equations. Moreover, the nonlinearity puts under question the possibility of expanding the solution into Fourier series, because they represent a kind of linear combination of solutions.

XV.2.1 Modal methods

Most of the methods of solving the nonlinear problem are based on the approximation that the nonlinearity changes the field distribution only slightly, i.e., that the solution of the nonlinear problem can be represented in the form, usuall called *modal* representation:

$$\vec{E}_{\omega_m}(\vec{r}) \approx a_{\omega_m}(\vec{r}) \exp(i\vec{k}_{\omega_m}.\vec{r}) \vec{E}_{\omega_m}^{\{0\}}(\vec{r}) \tag{XV.10}$$

where $\vec{E}^{\{0\}}_{\omega_m}(\vec{r})$ is a solution of the linear problem (Eq. (XV.3) with $\vec{P}^{NL}_{\omega_m}(\vec{r})=0$) and $a_{\omega_m}(\vec{r})$ is a slowly varying unknown function, representing the field (*mode*) amplitude. The propagation vector $\vec{k}_{\omega_m}$ is another unknown to be determined, when the nonlinear part is taken into account. The expression (XV.10) is substituted into Eq. (XV.3) with the nonlinear term(s) given by Eqs. (XV.7) or (XV.8). The second-order derivatives of $a_{\omega_m}(\vec{r})$ can be omitted due to its slow variation. As a result, one arrives at a set of differential equations for $a_{\omega_m}(\vec{r})$, which can further be simplified by taking into account the geometry of the problem. For example, for a planar or corrugated waveguide with optically nonlinear properties, the amplitudes depend on a single direction of propagation. Close to electromagnetic resonances (surface or guided wave excitation) it is sufficient to consider only the amplitudes of the guided waves [XV.10], propagating in a single direction, say x. This leads to a single (for the optical Kerr effect) or double ordinary first-order differential equation(s) for the *mode* amplitude(s) $a_{\omega_m}(x)$.

XV.2.2 Rigorous methods

There are not so many methods for solving nonlinear partial differential equations. The only possibility is to make iterations with respect to some relatively small parameter. In the case of optical nonlinearity, this parameter is naturally the nonlinearity and, in particular, the change in the electromagnetic field distribution due to the nonlinear terms. From that point of view, the modal methods represent the second iteration step, solved in the modal approximation, Eq. (XV.10), with the first step representing the linear problem. However, if the nonlinearity is strong enough to change the field map, or when staying away from an electromagnetic resonance region, the modal approximation is not enough and one has to make more iterations to obtain satisfactory results. To that sake, it is necessary to linearize the nonlinear set of equations, so that at each iteration step one can use the shooting method to compute the solution of Maxwell equations and the boundary conditions.

XV.3 Linearization of the nonlinear Maxwell equations

There are different ways of linearizing Eqs. (XV.3) with the nonlinear parts given by Eqs. (XV.7) or (XV.8). To start the iteration process, it is necessary to assume that during the first iteration step the nonlinear term is null. It is sufficient, in what follows, to state that the 0-th approximation of the field is null, i.e., $\vec{E}^{\{0\}}_{\omega_m}(\vec{r})=0$. Let us assume that during the (j-1)th iteration, one can obtain the electric field components $\vec{E}^{\{j-1\}}_{\omega_m}(\vec{r})$. During the next (j-th) iteration, the linear approximation of the vector of nonlinear polarization can be represented as a *linear* function of the *unknow* field $\vec{E}^{\{j\}}_{\omega_m}(\vec{r})$ with coefficient, depending on the *known* functions $\vec{E}^{\{j-1\}}_{\omega_m}(\vec{r})$. Optical Kerr effect allows for the following approximation of Eq. (XV.8):

$$\vec{P}^{\mathrm{Kerr}\{j\}}_{\omega} = \varepsilon_0 \chi^{\mathrm{Kerr}}_{\omega} \vdots \vec{E}^{\{j-1\}}_{\omega} \bar{\vec{E}}^{\{j-1\}}_{\omega} \vec{E}^{\{j\}}_{\omega}. \qquad \text{(XV.11)}$$

Similar expressions can be established for Eqs. (XV.7) in the case of second-harmonic generation:

$$\begin{aligned} \vec{P}^{\mathrm{SHG}\{j\}}_{\omega} &= \varepsilon_0 \chi^{\mathrm{SHG}}_{\omega} : \bar{\vec{E}}^{\{j-1\}}_{\omega} \vec{E}^{\{j\}}_{2\omega} \\ \vec{P}^{\mathrm{SHG}\{j\}}_{2\omega} &= \varepsilon_0 \chi^{\mathrm{SHG}}_{2\omega} : \vec{E}^{\{j-1\}}_{\omega} \vec{E}^{\{j\}}_{\omega} \end{aligned} \qquad \text{(XV.12)}$$

Here the choice is made that during each iteration, the nonlinear polarization vector $\vec{P}^{\mathrm{SHG}\{j\}}_{\omega}$ at ω contains the unknown field $\vec{E}^{\{j\}}_{2\omega}$ at 2ω and vice versa. This ensures the coupling between the fields at ω and 2ω at each iteration step. Then the linearized Maxwell equations at the j-th iteration step are written in the form:

$$\begin{aligned} \mathrm{curl}\,\vec{E}^{\{j\}}_{\omega}(\vec{r}) &= i\omega\mu(\vec{r})\vec{H}^{\{j\}}_{\omega}(\vec{r}) \\ \mathrm{curl}\,\vec{H}^{\{j\}}_{\omega}(\vec{r}) &= -i\omega\bar{\bar{\varepsilon}}_{\omega}(\vec{r})\vec{E}^{\{j\}}_{\omega}(\vec{r}) - i\omega\bar{\bar{\varepsilon}}^{\mathrm{Kerr}\{j-1\}}_{\omega}(\vec{r})\vec{E}^{\{j\}}_{\omega}(\vec{r}) \end{aligned} \qquad \text{(XV.13)}$$

for the optical Kerr effect, and

$$\begin{aligned} \mathrm{curl}\,\vec{E}^{\{j\}}_{\omega}(\vec{r}) &= i\omega\mu_{\omega}\vec{H}^{\{j\}}_{\omega}(\vec{r}) \\ \mathrm{curl}\,\vec{H}^{\{j\}}_{\omega}(\vec{r}) &= -i\omega\bar{\bar{\varepsilon}}_{\omega}(\vec{r})\vec{E}^{\{j\}}_{\omega}(\vec{r}) - i\omega\bar{\bar{\varepsilon}}^{\mathrm{SHG}\{j-1\}}_{\omega}(\vec{r})\vec{E}^{\{j\}}_{2\omega}(\vec{r}) \\ \mathrm{curl}\,\vec{E}^{\{j\}}_{2\omega}(\vec{r}) &= i2\omega\mu_{2\omega}\vec{H}^{\{j\}}_{2\omega}(\vec{r}) \\ \mathrm{curl}\,\vec{H}^{\{j\}}_{2\omega}(\vec{r}) &= -i2\omega\bar{\bar{\varepsilon}}_{2\omega}(\vec{r})\vec{E}^{\{j\}}_{2\omega}(\vec{r}) - i2\omega\bar{\bar{\varepsilon}}^{\mathrm{SHG}\{j-1\}}_{2\omega}(\vec{r})\vec{E}^{\{j\}}_{\omega}(\vec{r}) \end{aligned} \qquad \text{(XV.14)}$$

for the SHG. The effective nonlinear permittivities $\overline{\overline{\tilde{\varepsilon}}}^{\{j-1\}}(\vec{r})$ are, in general, second rank tensors, which are calculated from the electric field vector of the previous iteration step and the nonlinear susceptibility tensors:

$$\overline{\overline{\tilde{\varepsilon}}}_{\omega}^{\mathrm{Kerr}\{j-1\}}(\vec{r}) = \varepsilon_0 \chi_{\omega}^{\mathrm{Kerr}} : \vec{E}_{\omega}^{\{j-1\}} \vec{\tilde{E}}_{\omega}^{\{j-1\}} \tag{XV.15}$$

and

$$\begin{aligned} \overline{\overline{\tilde{\varepsilon}}}_{\omega}^{\mathrm{SHG}\{j-1\}} &= \varepsilon_0 \chi_{\omega}^{\mathrm{SHG}} : \vec{\tilde{E}}_{\omega}^{\{j-1\}} \\ \overline{\overline{\tilde{\varepsilon}}}_{2\omega}^{\mathrm{SHG}\{j-1\}} &= \varepsilon_0 \chi_{2\omega}^{\mathrm{SHG}} : \vec{E}_{\omega}^{\{j-1\}} \end{aligned} \tag{XV.16}$$

For magnetic anisotropic media, μ in Eqs. (XV.13) and (XV.14) must be replaced by a second-rank tensor $\overline{\overline{\mu}}$. During the first integration, as already said, $\vec{E}_{\omega_m}^{\{0\}}(\vec{r}) = 0$.

Eqs. (XV.13) have the same form as linear Maxwell equations, except for the substitution of ε with $\overline{\overline{\varepsilon}}_{\omega} + \overline{\overline{\tilde{\varepsilon}}}_{\omega}^{\mathrm{Kerr}\{j-1\}}$. Thus, the solution at each iteration step is searched in the same way as in the linear problem, but ε takes a more complicated nonconstant form.

SHG leads to Eqs. (XV.14), where the two frequencies ω and 2ω are coupled, but in a *linear* form with respect to the unknown field components. Section XV.5.1 presents a generalized approach to deal with the problem.

XV.4 Nonlinear Maxwell equations in nontruncated Fourier space

In what follows, we assume that the solution of the nonlinear problem can be represented in a Fourier basis. This is the case for 1-D or 2-D periodic media illuminated with a plane wave, because the plane wave can be considered pseudoperiodic with the same period as the geometry of the system. Let us consider a crossed grating with periods d_x and d_y along x and z, respectively, as described in Chapter IX, illuminated by a field incident from the superstrate and given in the form:

$$\vec{E}_{\omega}^{(i)} = \vec{A}\exp[i(\alpha_{\omega,0}x - \beta_{\omega,0}y + \gamma_{\omega,0}z)], \tag{XV.17}$$

which propagates in air or vacuum. We are looking for a solution of Eqs. (XV.3) and the boundary conditions (XV.9) having the form:

$$\vec{E}(x,y,z)=\sum_{p}\vec{E}_{\omega_p}(x,y,z)=\sum_{p}\sum_{m,n}\vec{E}_{\omega_p,m,n}(y)\exp[i(\alpha_{\omega_p,m}x+\gamma_{\omega_p,n}z)]\,, \quad \text{(XV.18)}$$

where

$$\begin{aligned}\alpha_{\omega_p,m}&=\alpha_{\omega_p,0}+m\frac{2\pi}{d_x}\\ \gamma_{\omega_p,n}&=\gamma_{\omega_p,0}+n\frac{2\pi}{d_z}\end{aligned}\,. \quad \text{(XV.19)}$$

The subscript p indicates the circular frequency and the subscripts m and n stands for the double Fourier series. The x and z components of the propagation constants at different circular frequencies ω_p in Eq. (XV.19) are linked through the nonlinear polarization vector(s). In the case of optical Kerr effect, there is a single-frequency field, so that the subscript ω_p can be omitted. The i-th component of $\vec{P}^{Kerr}$ can be written in the form:

$$P_i^{Kerr}(x,y,z)=\varepsilon_0\sum_{j,k,\ell}\chi_{ijk\ell}^{Kerr}E_j\overline{E}_kE_\ell\,, \quad i,j,k,l=x,\ y,\ z. \quad \text{(XV.20)}$$

When a single-frequency component of the sum in Eq. (XV.18) is substituted into Eq. (XV.20), one can easily observe that the x- and z-dependence of $\vec{P}^{Kerr}$ is given in the form

$$P_i^{Kerr}(x,y,z)=\sum_{m,n}P_{i,m,n}^{Kerr}(y)\exp[i(\alpha_m x+\gamma_n z)]\,,\ i=x,\ y,\ z, \quad \text{(XV.21)}$$

with α_m and γ_n being of the same set as given by Eq. (XV.19), because of the relations:

$$\begin{aligned}i\alpha_m+\overline{(i\alpha_n)}+i\alpha_q&=i\alpha_{m-n+q}\\ i\gamma_m+\overline{(i\gamma_n)}+i\gamma_q&=i\gamma_{m-n+q}\end{aligned}\,. \quad \text{(XV.22)}$$

The Fourier components of $P_i^{Kerr}(x,y,z)$ in Eq. (XV.21) depend on y and have the form

$$\begin{aligned}P_{i,m,n}^{Kerr}(y)=\varepsilon_0\sum_{j,k,l}\ &\sum_{\left\{\begin{smallmatrix}m',m'',m'''\\ n',n'',n'''\end{smallmatrix}\right\}=-\infty}^{\infty}(\chi_{i,j,k,l,m-m',n-n'}^{Kerr}\\ &E_{j,m'-m'',n'-n''}\overline{E}_{k,m''-m''',n''-n'''}E_{l,m''',n'''})\end{aligned}\,. \quad \text{(XV.23)}$$

Here i, j, k, and l = x, y, z, and m and n with primes, etc., are the subscripts labelling the Fourier components. Eq. (XV.23) represents the Fourier series of the i-th vector component of the product $\vec{P}_{\omega}^{Kerr} = \varepsilon_0 \chi_{\omega}^{Kerr} : \vec{E}_{\omega} \vec{\bar{E}}_{\omega} \vec{E}_{\omega}$ as a convolution of four Fourier series.

Second-harmonic generation couples two frequencies. The nonlinear polarization vector at 2ω, Eq. (XV.12), is proportional to the product of two electric field vectors at ω, so that its x-dependence is obtained by adding two exponentials:

$$\alpha_{\omega,m} + \alpha_{\omega,n} = 2\alpha_{\omega,0} + (m+n)\frac{2\pi}{d_x}. \tag{XV.24}$$

We can express it in the following form

$$\alpha_{2\omega,m} = 2\alpha_{\omega,0} + m\frac{2\pi}{d_x}, \tag{XV.25}$$

i.e., as the exponential factor of the electric-field x-dependence at 2ω. The same considerations are valid for the z-dependence, wherefrom we obtain the following relations for SHG:

$$\begin{aligned} \alpha_{2\omega,0} &= 2\alpha_{\omega,0} \\ \gamma_{2\omega,0} &= 2\gamma_{\omega,0} \end{aligned}. \tag{XV.26}$$

By using Eqs. (XV.26), presentation similar to Eqs. (XV.21) and (XV.23) can be established for SHG:

$$\left.\begin{aligned} P_{\omega,i}^{SHG}(x,y,z) &= \sum_{m,n} P_{\omega,i,m,n}^{SHG}(y)\exp[i(\alpha_{\omega,m}x + \gamma_{\omega,n}z)] \\ P_{2\omega,i}^{SHG}(x,y,z) &= \sum_{m,n} P_{2\omega,i,m,n}^{SHG}(y)\exp[i(\alpha_{2\omega,m}x + \gamma_{2\omega,n}z)] \end{aligned}\right., \; i = x,y,z, \tag{XV.27}$$

with

$$\begin{aligned} P_{\omega,i,m,n}^{SHG}(y) &= \varepsilon_0 \sum_{j,k} \sum_{\left\{\begin{smallmatrix} m',m'', \\ n',n'' \end{smallmatrix}\right\}=-\infty}^{\infty} \chi_{\omega,i,j,k,m-m',n-n'}^{SHG} \bar{E}_{\omega,j,m'-m'',n'-n''} E_{2\omega,k,m'',n''} \\ P_{2\omega,i,m,n}^{SHG}(y) &= \varepsilon_0 \sum_{j,k} \sum_{\left\{\begin{smallmatrix} m',m'', \\ n',n'' \end{smallmatrix}\right\}=-\infty}^{\infty} \chi_{2\omega,i,j,k,m-m',n-n'}^{SHG} E_{\omega,j,m'-m'',n'-n''} E_{\omega,k,m'',n''}. \end{aligned} \tag{XV.28}$$

As can be observed, when the Fourier representation of the electric field, Eq. (XV.18), is introduced into the nonlinear polarization vector at ω, Eqs. (XV.7), the exponential factors of $\vec{E}_\omega$ and $\vec{E}_{2\omega}$ mutually compensate to give the correct x and z dependences of $\vec{P}_\omega^{SHG}$, as given in Eqs. (XV.27). And indeed, the (x, z) dependence of $\vec{P}_\omega^{SHG}$ is given by

$$\begin{aligned}\vec{P}_\omega^{SHG} &= \varepsilon_0 \chi_\omega^{SHG} : \vec{\bar{E}}_\omega \vec{E}_{2\omega} \\ &\sim \exp[im\frac{2\pi}{d_x} + in\frac{2\pi}{d_y} - i(\alpha_{\omega,m'}x + \gamma_{\omega,n'}z) + i(\alpha_{2\omega,m''}x + \gamma_{2\omega,n''}z)] \\ &= \exp[i(\alpha_{\omega,m-m'+m''}x + \gamma_{\omega,n-n'+n''}z)]\end{aligned} \tag{XV.29}$$

which is the same as the (x, z) dependence of $\vec{E}_\omega$.

XV.5 Fourier factorization of the linearized Maxwell equations in truncated Fourier space

Numerical integration of the Fourier transformed linearized Maxwell equations in nonlinear optics has been done successfully using different techniques [XV.11-XV.13]. They all present the same numerical difficulties, as in linear optics, enhanced by the requirement to correctly represent the electric field inside the optically-nonlinear region. Contamination due to the growing exponents, discussed in detail in Chapters II and III, needs utilization of special algorithms, like the S-matrix propagation algorithm. The computer implementation requires truncation of the Fourier series to a finite number of terms. When the numerical method uses a suitable coordinate transformation, for example, the transformation described in Section VII.4, which maps the modulated surface(s) onto planes, Fourier transformation of discontinuous field components and of their products is avoided. However, this is not the case with the differential method. As explained in Chapters IV and V, the differential method requires the Fourier factorization of the continuous and discontinuous components of $\vec{D}$ (see Section V.1.1), which must be correctly made according to the factorization rules described in Section IV.2.1. For magnetic media, the same is valid for $\vec{B}$.

Nonlinearity introduces additional difficulties, due to the vectors of nonlinear polarization. Contrary to $\vec{D}$ and $\vec{B}$, which have continuous normal components all over the space, there is no such physical requirement for $\vec{P}^{NL}$. Moreover, as the nonlinear susceptibility χ^{NL} vanishes outside the nonlinear materials, the components of $\vec{P}^{NL}$, when non-null, are always discontinuous across the boundaries of the nonlinear medium. This means that if the electric field components, responsible for the nonlinearity, Eqs. (XV.7) or (XV.8), are discontinuous, neither of rules **1** or **2** (Section IV.2.1) of Fourier factorization in the truncated Fourier space can be applied. The only situation causing no problem is the classical one-dimensional grating with a corrugation in a single (say, x) direction, made of linearly and nonlinearly isotropic materials when used in TE polarization with in-plane incidence. In that case, the only non-zero components of $\vec{E}$ are E_z, tangential to the corrugated surface and thus continuous. The Fourier factorization of $\vec{P}^{NL}$ in Eqs. (XV.7) or (XV.8) then involves a product of continuous functions E_z and a discontinuous function χ^{NL}. Thus the Laurent's rule applies, and no numerical difficulty due to the truncation is observed [XV.11, XV.12]. However, even that simplest structure, when used in TM polarization, imposes spectacular slowdown of convergence rates of Fourier series of the field in the classical differential method. Detailed description of how to generalize the classical differential method and the C-method to nonlinear optics can be found in Chapter IV of [XV.10].

Contrary to the linear case, in the nonlinear case it is impossible to work with products of continuous and discontinuous functions of the two types that permit rules of correct factorization. To briefly remind, type **1** concerns products of continuous functions or of a continuous function and a discontinuous one. Type **2** involves discontinuous functions having a continuous product. The nonlinear polarization, in general, falls into the type **3**, for which neither the functions nor their products are continuous. The aim of the next section is to apply the correct factorization rules to the nonlinear case, when this is possible, and to show how to avoid factorization of products of type **3**.

XV.5.1 Factorization of linearized Maxwell equations

In order to be able to follow the same lines as in Chapter V, let us rewrite the linearized Eqs. (XV.13) and (XV.14) using more general notations. Let us define the unknown field vectors during the j-th iteration step in the form:

$$\vec{E}_{NL}^{\{j\}} = \begin{cases} \begin{pmatrix} \vec{E}_{\omega}^{\{j\}} \\ \vec{E}_{2\omega}^{\{j\}} \end{pmatrix} & \text{for SHG} \\ \\ \vec{E}_{\omega}^{\{j\}} & \text{for Kerr effect} \end{cases} , \qquad \text{(XV.30)}$$

$$\vec{D}_{NL}^{\{j\}} = \begin{cases} \begin{pmatrix} \vec{D}_{\omega}^{\{j\}} \\ \vec{D}_{2\omega}^{\{j\}} \end{pmatrix} & \text{for SHG} \\ \\ \vec{D}_{\omega}^{\{j\}} & \text{for Kerr effect} \end{cases} , \tag{XV.31}$$

$$\vec{H}_{NL}^{\{j\}} = \begin{cases} \begin{pmatrix} \vec{H}_{\omega}^{\{j\}} \\ \vec{H}_{2\omega}^{\{j\}} \end{pmatrix} & \text{for SHG} \\ \\ \vec{H}_{\omega}^{\{j\}} & \text{for Kerr effect} \end{cases} , \tag{XV.32}$$

$$\vec{B}_{NL}^{\{j\}} = \begin{cases} \begin{pmatrix} \vec{B}_{\omega}^{\{j\}} \\ \vec{B}_{2\omega}^{\{j\}} \end{pmatrix} & \text{for SHG} \\ \\ \vec{B}_{\omega}^{\{j\}} & \text{for Kerr effect} \end{cases} . \tag{XV.33}$$

The components, which are continuous across each interface are given by the boundary conditions (XV.9). As in Chapter V, let us introduce the vectors containing the continuous components:

$$\vec{F}_{\varepsilon,NL}^{\{j\}} = \begin{pmatrix} E_{NL,T_1}^{\{j\}} \\ D_{NL,N}^{\{j\}} \\ E_{NL,T_2}^{\{j\}} \end{pmatrix} . \tag{XV.34}$$

For media with periodical magnetic properties, a similar vector must be introduced for the continuous magnetic components.

Generalized notations are necessary for the dielectric permittivity tensor. Let us define for the linear part:

$$\overline{\overline{\varepsilon}}_L = \begin{cases} \begin{pmatrix} \overline{\overline{\varepsilon}}_\omega & 0 \\ 0 & \overline{\overline{\varepsilon}}_{2\omega} \end{pmatrix} & \text{for SHG} \\ \overline{\overline{\varepsilon}}_\omega & \text{for Kerr effect} \end{cases}, \tag{XV.35}$$

and for the nonlinear part, which has to be calculated during the previous iteration step:

$$\overline{\overline{\varepsilon}}_{NL}^{\{j-1\}} = \begin{cases} \begin{pmatrix} 0 & \tilde{\overline{\overline{\varepsilon}}}_\omega^{SHG\{j-1\}} \\ \tilde{\overline{\overline{\varepsilon}}}_{2\omega}^{SHG\{j-1\}} & 0 \end{pmatrix} & \text{for SHG} \\ \tilde{\overline{\overline{\varepsilon}}}_\omega^{Kerr\{j-1\}} & \text{for Kerr effect} \end{cases} \tag{XV.36}$$

The propagation Eqs. (XV.14) together with Eqs. (XV.15) and (XV.16) suggest to introduce a generalized permittivity:

$$\tilde{\overline{\overline{\varepsilon}}}^{\{j-1\}} = \overline{\overline{\varepsilon}}_L + \overline{\overline{\varepsilon}}_{NL}^{\{j-1\}} \tag{XV.37}$$

We also define

$$\omega_{NL} = \begin{cases} \begin{pmatrix} \omega & 0 \\ 0 & 2\omega \end{pmatrix} & \text{for SHG} \\ \omega & \text{for Kerr effect} \end{cases} \tag{XV.38}$$

and diagonal matrix α_{NL} with diagonal elements $\alpha_{\omega_p,m}$:

$$\alpha_{NL} = \begin{cases} \begin{pmatrix} \alpha_{\omega,m} & 0 \\ 0 & \alpha_{2\omega,m} \end{pmatrix} & \text{for SHG} \\ \left(\alpha_{\omega,m}\right) & \text{for Kerr effect} \end{cases}, \tag{XV.39}$$

and the same for γ_{NL} with diagonal components $\gamma_{\omega_p,m}$:

$$\gamma_{NL} = \begin{cases} \begin{pmatrix} \gamma_{\omega,m} & 0 \\ 0 & \gamma_{2\omega,m} \end{pmatrix} & \text{for SHG} \\ \\ \left(\gamma_{\omega,m}\right) & \text{for Kerr effect} \end{cases} \tag{XV.40}$$

Using the notations (XV.30)-(XV.37), the equations (XV.13) and (XV.14) can be represented for non-magnetic media in the form

$$\begin{aligned} \mathrm{curl}\vec{E}_{NL}^{\{j\}} &= i\omega_{NL}\mu_0 H_{NL}^{\{j\}} \\ \mathrm{curl}\vec{H}_{NL}^{\{j\}} &= -i\omega_{NL}\tilde{\bar{\bar{\varepsilon}}}^{\{j-1\}}\vec{E}_{NL}^{\{j\}} \end{aligned} . \tag{XV.41}$$

The form of Eqs. (XV.41) is the same as the common form of the linear Maxwell equations and can be Fourier factorized in the same manner as was done in Chapter V. This requires introducing two matrices $A_{\bar{\bar{\varepsilon}}}^{\{j-1\}}$ and $C_{\bar{\bar{\varepsilon}}}^{\{j-1\}}$, defined, respectively, by Eqs. (V.9) and (V.11) in which $\bar{\bar{\varepsilon}}$ is replaced by $\tilde{\bar{\bar{\varepsilon}}}^{\{j-1\}}$. The vector $F_{\varepsilon}^{\{j\}}$, as defined in Eq. (XV.34), is determined through the relation, $F_{\varepsilon,NL}^{\{j\}} = A_{\bar{\bar{\varepsilon}}}^{\{j-1\}} E_{NL}^{\{j\}}$, while the inverse transformation is given by $E_{NL}^{\{j\}} = C_{\bar{\bar{\varepsilon}}}^{\{j-1\}} F_{\varepsilon,NL}^{\{j\}}$

Finally, the Fourier transformation and the corresponding factorization of the product $\tilde{\bar{\bar{\varepsilon}}}^{\{j-1\}}\vec{E}_{NL}^{\{j\}}$ in Eq. (XV.41) is done exactly in the same way as in Eq. (V.12) by introducing a matrix

$$Q_{\bar{\bar{\varepsilon}}}^{\{j-1\}} = \left[\!\left[\tilde{\bar{\bar{\varepsilon}}}^{\{j-1\}} \bullet C_{\bar{\bar{\varepsilon}}}^{\{j-1\}}\right]\!\right] \left[\!\left[C_{\bar{\bar{\varepsilon}}}^{\{j-1\}}\right]\!\right]^{-1} . \tag{XV.42}$$

The difference with the linear case is that the nonlinearity requires iterations and that during each iteration it is necessary to evaluate the quantity $\tilde{\bar{\bar{\varepsilon}}}^{\{j-1\}}$, defined in Eq. (XV.37), the matrix $C_{\bar{\bar{\varepsilon}}}^{\{j-1\}}$, their product and their Fourier transform, in order to obtain the matrix $Q_{\bar{\bar{\varepsilon}}}^{\{j-1\}}$. How to correctly make that is explained in detail in Section XV.5.3. Before discussing this, it is necessary to determine the set of differential equations.

XV.5.2 The set of differential equations in truncated Fourier space

The Fourier transformation of Eqs. (XV.41) leads to a set of ordinary differential equations, quite similar to the set (V.14'-19') when $[\vec{D}]$ is replaced by $\left[\!\left[\tilde{\bar{\varepsilon}}^{\{j-1\}}.C_{\bar{\bar{\varepsilon}}}^{\{j-1\}}\right]\!\right]\left[\!\left[C_{\bar{\bar{\varepsilon}}}^{\{j-1\}}\right]\!\right]^{-1}\left[\vec{E}_{NL}^{\{j\}}\right] \equiv Q_{\bar{\bar{\varepsilon}}}^{\{j-1\}}\left[\vec{E}_{NL}^{\{j\}}\right]$. Let us recall that the square braces denote a column-vector containing the truncated Fourier components of the quantity in the braces. Section V.1.3 explains in detail that, independent of the geometry, in order to solve the set of ordinary differential equations in the Fourier space, it is necessary to eliminate one electric and one magnetic field component. Let us assume, as it has been done there, that we can eliminate the y components $\left[E_{y,NL}^{\{j\}}\right]$ and $\left[H_{y,NL}^{\{j\}}\right]$ by using Eqs. (V.20) and (V.21), where Q_ε is replaced with $Q_{\bar{\bar{\varepsilon}}}^{\{j-1\}}$:

$$\begin{aligned}\left[E_{y,NL}^{\{j\}}\right] = -\left(Q_{\bar{\bar{\varepsilon}},yy}^{\{j-1\}}\right)^{-1}\Big\{&\omega_{NL}^{-1}\left(\gamma_{NL}\left[H_{x,NL}^{\{j\}}\right] - \alpha_{NL}\left[H_{z,NL}^{\{j\}}\right]\right) \\ &+ Q_{\bar{\bar{\varepsilon}},yx}^{\{j-1\}}\left[E_{x,NL}^{\{j\}}\right] + Q_{\bar{\bar{\varepsilon}},yz}^{\{j-1\}}\left[E_{z,NL}^{\{j\}}\right]\Big\}.\end{aligned} \quad \text{(XV.43)}$$

A similar work for $\left[H_{y,NL}^{\{j\}}\right]$ leads to:

$$\begin{aligned}\left[H_{y,NL}^{\{j\}}\right] = Q_{\mu,yy}^{-1}\Big\{&\omega_{NL}^{-1}\left(\gamma_{NL}\left[E_{x,NL}^{\{j\}}\right] - \alpha_{NL}\left[E_{z,NL}^{\{j\}}\right]\right) \\ &- Q_{\mu,yx}\left[H_{x,NL}^{\{j\}}\right] - Q_{\mu,yz}\left[H_{z,NL}^{\{j\}}\right]\Big\}.\end{aligned} \quad \text{(XV.44)}$$

The resulting set of equations to integrate can be expressed in the same form as Eq. (V.22):

$$\frac{d}{dy}\begin{pmatrix}\left[E^{\{j\}}_{x,NL}\right]\\ \left[E^{\{j\}}_{z,NL}\right]\\ \left[H^{\{j\}}_{x,NL}\right]\\ \left[H^{\{j\}}_{z,NL}\right]\end{pmatrix} = M^{\{j-1\}}_{NL}\begin{pmatrix}\left[E^{\{j\}}_{x,NL}\right]\\ \left[E^{\{j\}}_{z,NL}\right]\\ \left[H^{\{j\}}_{x,NL}\right]\\ \left[H^{\{j\}}_{z,NL}\right]\end{pmatrix}, \tag{XV.45}$$

where $M^{\{j-1\}}_{NL}$ is a known matrix, calculated at the end of the (j-1)-th iteration step and having the same components as given in Eq. (V.22') in which we replace

$$\begin{aligned}
&Q_\varepsilon \text{ with } Q^{\{j-1\}}_{\tilde{\tilde{\varepsilon}}},\\
&Q_\mu \text{ with } \mu_0 \mathbb{1},\\
&\omega \text{ with } \omega_{NL},\\
&\alpha \text{ with } \alpha_{NL}, \text{ and}\\
&\gamma \text{ with } \gamma_{NL}.
\end{aligned} \tag{XV.46}$$

Moreover, the form of $M^{\{j-1\}}_{NL}$ can be further simplified, depending on the geometry and optical properties of the materials involved. A classical grating with optically isotropical (both linear and nonlinear) properties allows for polarization separation and the corresponding matrix $M^{\{j-1\}}_{NL}$ can be obtained from Eq. (V.28) using the substitution (XV.46). For crossed gratings, the same substitution has to be applied to Eq. (IX.12), etc. However, all the particular cases require calculations of matrix $Q^{\{j-1\}}_{\tilde{\tilde{\varepsilon}}}$, defined in Eq. (XV.42).

XV.5.3 Fourier transform of a discontinuous product of discontinuous functions

There is no way of correctly making the factorization of products of type 3, so that such factorization must be avoided. The calculation of Toeplitz matrices in Eq. (XV.42) requires Fourier transformation of matrices $\tilde{\tilde{\varepsilon}}^{\{j-1\}}\cdot C^{\{j-1\}}_{\tilde{\tilde{\varepsilon}}}$ and $C^{\{j-1\}}_{\tilde{\tilde{\varepsilon}}}$, which contain non-factorizable terms. It is possible to correctly Fourier transform them by using, for example, the Fast Fourier

Transformation algorithm, if their coordinate dependencies are known. To this aim, it is necessary to obtain $\tilde{\bar{\bar{\varepsilon}}}^{\{j-1\}}$ as a function of space coordinates from Eqs. (XV.15) and (XV.16) using the coordinate dependencies of electric field components $\vec{E}_{NL}^{\{j-1\}}$ at the end of each iteration step. In general, this is possible by using the inverse Fourier transform, once the Fourier components are known. However, the reconstruction of the spatial dependence from the truncated Fourier series can introduce Gibb's phenomena, if the functions to reconstruct are discontinuous. This is the case with some of the field components. To avoid Gibb's phenomena, it is necessary to work with continuous functions only.

As discussed in the previous section, the differential set to integrate, Eqs. (XV.45), concerns only four of the six field components. Let us assume that at the end of the (j-1)-th iteration step we have obtained the Fourier components $\left[E_{x,NL}^{\{j-1\}}\right]$, $\left[E_{z,NL}^{\{j-1\}}\right]$, $\left[H_{x,NL}^{\{j-1\}}\right]$, and $\left[H_{z,NL}^{\{j-1\}}\right]$. The other two series $\left[E_{y,NL}^{\{j-1\}}\right]$ and $\left[H_{y,NL}^{\{j-1\}}\right]$ are obtained from Eqs. (XV.43) and (XV.44) using the values of $Q_{\bar{\bar{\varepsilon}}}^{\{j-2\}}$. During the first integration step $E_{x,y,z,NL}^{\{0\}} = H_{x,y,z,NL}^{\{0\}} \equiv 0$, so that $\tilde{\bar{\bar{\varepsilon}}}^{\{0\}} = \bar{\bar{\varepsilon}}_L$ and the matrix $Q_{\bar{\bar{\varepsilon}}}^{\{0\}} = \left[\!\left[\bar{\bar{\varepsilon}}_L \bullet C_\varepsilon\right]\!\right]\left[\!\left[C_\varepsilon\right]\!\right]^{-1}$ is obtained using the matrix C_ε given in Eq. (V.11).

Once the components $\left[E_{x,y,z,NL}^{\{j-1\}}\right]$ and $\left[H_{x,y,z,NL}^{\{j-1\}}\right]$ are given, one can calculate the Fourier components of the field vector $\left[\vec{F}_{\varepsilon,NL}^{\{j-1\}}\right]$ using Eq. (XV.34). This can be done, because $\vec{F}_{\varepsilon,NL}^{\{j-1\}}$ contains only continuous vector components and is defined as a *continuous* product of *discontinuous* functions. And indeed, in the iteration process, it is calculated by:

$$\vec{F}_{\varepsilon,NL}^{\{j-1\}} = A_{\bar{\bar{\varepsilon}}}^{\{j-2\}}\vec{E}_{NL}^{\{j-1\}},$$

which requires the use of the inverse rule of factorization:

$$\left[\vec{F}_{\varepsilon,NL}^{\{j-1\}}\right] = \left[\!\left[\left(A_{\bar{\bar{\varepsilon}}}^{\{j-2\}}\right)^{-1}\right]\!\right]^{-1}\left[E_{NL}^{\{j-1\}}\right] \equiv \left[\!\left[C_{\bar{\bar{\varepsilon}}}^{\{j-2\}}\right]\!\right]^{-1}\left[\vec{E}_{NL}^{\{j-1\}}\right], \qquad \text{(XV.47)}$$

so that one can use the matrix $\left[\!\left[C_{\bar{\bar{\varepsilon}}}^{\{j-2\}} \right]\!\right]^{-1}$, already known from the previous iteration. The summation of the truncated Fourier series to obtain $\vec{F}_{\varepsilon,NL}^{\{j-1\}}$:

$$\vec{F}_{\varepsilon,NL}^{\{j-1\}} = \sum_{m,n=-N}^{N} \left[\vec{F}_{\varepsilon,NL}^{\{j-1\}} \right]_{mn} \exp[i(\alpha_{NL,m}x + \gamma_{NL,n}z)], \quad \forall y \qquad (XV.48)$$

does not introduce any Gibb's phenomenon, because $\vec{F}_{\varepsilon,NL}^{\{j-1\}}(\vec{r})$ is a continuous function of the spatial coordinates. Using the matrix $C_{\bar{\bar{\varepsilon}}}^{\{j-2\}}$, which is the inverse of $A_{\bar{\bar{\varepsilon}}}^{\{j-2\}}$, enables us to calculate $\vec{E}_{NL}^{\{j-1\}}(\vec{r})$:

$$\vec{E}_{NL}^{\{j-1\}}(\vec{r}) = C_{\bar{\bar{\varepsilon}}}^{\{j-2\}}(\vec{r})\, \vec{F}_{\varepsilon,NL}^{\{j-1\}}(\vec{r}). \qquad (XV.49)$$

The nonlinear terms in Eqs. (XV.36) can be obtained in the coordinate space by directly using Eqs. (XV.15) or (XV.16) and the fact that the coordinate dependence of $\bar{\bar{\varepsilon}}_L$ and χ^{NL} are known. Eq. (XV.37) gives the coordinate dependence of $\bar{\bar{\varepsilon}}^{\{j-1\}}$ and thus of $C_{\bar{\bar{\varepsilon}}}^{\{j-1\}}$ and $\bar{\bar{\varepsilon}}^{\{j-1\}} \bullet C_{\bar{\bar{\varepsilon}}}^{\{j-1\}}$. Their Fourier components are calculated using numerical Fast Fourier Transformation without factorization. We then obtain the matrices required to compute $Q_{\bar{\bar{\varepsilon}}}^{\{j-1\}}$ given by Eq. (XV.42).

The entire procedure seems rather difficult to explain, but is straightforward for programming, once one has developed the numerical code for the corresponding linear optical problem. The main complication comes from the requirement to calculate and store the field components during the numerical integration.

XV.5.4 Field reconstruction and S-matrix propagation algorithm

At a first glance, there is no problem to obtain the Fourier components of the field at each iteration step, when the amplitudes in the substrate and the superstrate are determined. It can be done by performing one additional integration of the differential set of equations. In order to demonstrate how to do this, let us choose the most simple case of a classical grating with 1-D periodicity along x, because for that case Chapter II contains in great detail the description of the algorithm of finding the solution, including the presentation of the shooting method. Once the amplitudes of the field outside the modulated region are determined, using Eq. (III.48), one integration is required. Instead of using 2N+1 particular solutions, given by Eqs. (II.8) and (II.9) in TE polarization and in Section II.1.2.2 in the TM case, now the field at the bottom is known, so that the integration process can be started with a single *known* solution:

$$\begin{aligned} E_n(0) &= A_n^{(1)} \\ H_n(0) &= -i\beta_n^{(1)} A_n^{(1)} \end{aligned} \tag{XV.50}$$

in the TE case, and

$$\begin{aligned} H_n(0) &= A_n^{(1)} \\ E_n(0) &= -i\frac{\beta_n^{(1)}}{k_1^2} A_n^{(1)} \end{aligned} \tag{XV.51}$$

in the TM case. The field components are determined at each ordinate y during the integration process.

However, numerical problems arise during the integration, due to the growing exponential terms, as described in Chapter III. Eq. (III.28) permits to calculate the unknown field amplitudes at the boundaries of each slices, in which the grating is decomposed when using the S-matrix propagation algorithm. Assuming that there is no incident wave from the substrate side, i.e. $B_n^{(1)} \equiv 0$, Eq. (III.28) with $y_1 = 0$ gives:

$$\tilde{B}'^{(q)} = S_{12}^{(q)} \tilde{A}'^{(q)} \tag{XV.52}$$

$$\tilde{A}^{(1)} = S_{22}^{(q)} \tilde{A}'^{(q)}, \tag{XV.53}$$

where the notations with ~ sign stay for column vectors with components:

$$\left(\tilde{B}'^{(q)}\right)_n = B_n'^{(q)} \exp(i\beta_n^{(M)} y_q) \tag{XV.54}$$

$$\left(\tilde{A}'^{(q)}\right)_n = A_n'^{(q)} \exp(-i\beta_n^{(M)} y_q), \tag{XV.55}$$

$$\left(\tilde{A}^{(1)}\right)_n = A_n^{(1)}. \tag{XV.56}$$

One can assume that Eq. (XV.53) permits to obtain $\tilde{A}'^{(q)}$ form the known amplitudes $\tilde{A}^{(1)}$ by inverting $S_{22}^{(q)}$:

$$\tilde{A}'^{(q)} = \left(S_{22}^{(q)}\right)^{-1} \tilde{A}^{(1)}. \tag{XV.57}$$

However, this block of the matrix S corresponds to transmission from the q-th slide to the substrate and can have very small elements, as discussed in Section X.3.2. To avoid its inversion, it is necessary to use a more complicated procedure in order to obtain $\tilde{A}'^{(q)}$. For that sake, the inverse of $S_{22}^{(q)}$ can be expressed from Eq. (III.46):

$$\begin{aligned}\left(S_{22}^{(q)}\right)^{-1} &= Z^{(q)}\left(S_{22}^{(q+1)}\right)^{-1} = \dots \\ &= Z^{(q)}Z^{(q+1)}\dots Z^{(M-2)}\left(S_{22}^{(M-1)}\right)^{-1}\end{aligned}. \tag{XV.58}$$

This equation can be used together with Eq. (XV.57) to express $\tilde{A}'^{(q)}$:

$$\begin{aligned}\tilde{A}'^{(q)} &= Z^{(q)}Z^{(q+1)}\dots Z^{(M-2)}\left(S_{22}^{(M-1)}\right)^{-1}\tilde{A}^{(1)} \\ &= Z^{(q)}Z^{(q+1)}\dots Z^{(M-2)}\tilde{A}^{(M)}\end{aligned}, \tag{XV.59}$$

where Eq. (XV.57) is used for q = M-1, too. The amplitudes $A_n^{(M)}$ are the field amplitudes incident from the superstrate, which are known, and the matrices $Z^{(q)}$ have been calculated during the integration without numerical difficulties. The other set of amplitudes $\tilde{B}'^{(q)}$ are calculated from $\tilde{A}'^{(q)}$ using Eq. (XV.52).

Finally, the components of the field within each, say, (q+1)-th slice are obtained by integrating the differential set of equations within the slice using as starting values at the bottom the values of $\tilde{A}'^{(q)}$ and $\tilde{B}'^{(q)}$:

$$\begin{aligned}E_n(y_q) &= A_n'^{(q)}\exp(-i\beta_n^{(M)}y_q) + B_n'^{(q)}\exp(i\beta_n^{(M)}y_q) \\ H_n(y_q) &= -i\beta_n^{(M)}A_n'^{(q)}\exp(-i\beta_n^{(M)}y_q) + i\beta_n^{(M)}B_n'^{(q)}\exp(i\beta_n^{(M)}y_q)\end{aligned} \tag{XV.60}$$

in the TE case, and

$$H_n(y_q) = A_n'^{(q)} \exp(-i\beta_n^{(M)} y_q) + B_n'^{(q)} \exp(i\beta_n^{(M)} y_q)$$
$$E_n(y_q) = -i\frac{\beta_n^{(M)}}{k_M^2} A_n'^{(q)} \exp(-i\beta_n^{(M)} y_q) + i\frac{\beta_n^{(M)}}{k_M^2} B_n'^{(q)} \exp(i\beta_n^{(M)} y_q) \qquad (XV.61)$$

in the TM case.

References

XV.1. Y. Kitaoka, K. Urairi, K. Yamamoto, M. Kato, and T. Sasaki, "High-powered green light generation by intracavity frequency doubling using grating feedback optics," Appl. Opt. 34, 5361-5395, 1995

XV.2. D. Miller, D. Chemla, T. Damen, A. Gossard, W. Wiegmann, T. Wood, and C. Burus Jr., "Electric field dependence of optical absorption near the band gap of quantum well structures," Phys. Rev. B 32, 1043-1060, 1985

XV.3. D. Miller, "Quantum wells for optical information processing," Opt. Engineer. 26, 368-372, 1987

XV.4. J. Faist, F. Capasso, C. Sirtori, D. Sivco, A. Hutchinson, and A. Cho, "Quantum cascade laser with a vertical transition and an electron Bragg reflector," Opt. Phot. News 6, p.10, 1995

XV.5a. J. Maserjian, P. Andersson, B. Hancock, J. Iannelli, S. Eng, F. Grunthaner, K. Law, P. Holtz, R. Simes, L. Coldren, A. Gossard, and J. Hertz, "Optically addressed spacial light modulators by MBE-grown *nipi* MQW structures," Appl. Opt. 28, 4801-4807, 1989

XV.5b. P. Koppa, P. Chavel, J. L. Oudar, R. Kuszelewicz, J. P. Schnell, and J. P. Pocholle, "Demonstration of optically controlled data routing with the use of multiple-quantum-well bistable and electro-optical devices," Appl. Opt. 36, 5706-5716, 1997

XV.6a. M. Nevière, J. Flamand, and F. Bonnemason, "TMR – Access to large scale facilities," Report on European Community RTD contract: "Gratings for Ultrabright Lasers," ERBFMGECT#980096

XV.6b. B. Wattellier, "Amélioration des performances des chaînes lasers solides utilisant l'amplification à dérive de fréquence : nouveaux réseaux de diffraction à haute tenue au flux et mise en forme programmable de faisceaux lasers par modulation de la phase spatiale," Ph.D. thesis, Ecole Polytechnique, Paris, 2001

XV.7. J. Webjörn, D. Nam, S. Siala, and R. Waarts, "Nonlinear waveguides on the way to the marketplace," Opt. Phot. News 8, no. 4, 16-20 (April 1997)

XV.8. S. Payne and C. Pollock, eds., *Advanced Solid State Lasers*, v.1 of series Opt. Soc. Am. "Trends in Optics and Photonics," Proc. Topical Meeting, Jan. 31- Feb.2, 1996, San Francisco, CA (Opt. Soc. Am., 1996)

XV.9. N. Bloombergen, *Nonlinear Optics* (Benjamin, New York, 1965)

XV.10. M. Nevière, E. Popov, R. Reinisch, and G. Vitrant, *Electromagnetic Resonances in Nonlinear Optics* (Gordon and Breach Science Publishers, London 2000)

XV.11. P. Vincent, N. Paraire, M. Nevière, A. Koster, and R. Reinisch, "Gratings in nonlinear optics and optical bistability," J. Opt. Soc. Am. B 2, 1106-1116, 1985

XV.12. H. Akhouayri, M.Nevière, P. Vincent, and R. Reinisch, "Second-harmonic generation in a corrugated waveguide," Nonlinear Optics 5, 127-140, 1993

XV.13. E. Popov and M. Nevière, "Surface-enhanced second-harmonic generation in nonlinear corrugated dielectrics: new theoretical approaches," J. Opt. Soc. Am. B 11, 1555-1564, 1994

Appendix I

The z-Dependence of the Total Field in Conical Diffraction

The total field $\vec{E}(x, y, z)$ is the solution of the following boundary-value problem:

(1) $\vec{E}$ verifies Helmholtz equations in each homogeneous medium $\forall z$;

(2) $\vec{N}\times\vec{E}$ and $\vec{N}\times\vec{H}$ are continuous on the grating profile, $\vec{N}$ being the vector normal to the grating surface;

(3) $\vec{E}-\vec{E}^{(i)}$ satisfies an outgoing-wave condition $\forall z$, with $\vec{E}^{(i)}$ denoting the incident field.

As discussed in Section I.2.3, the grating can be associated to a linear operator $\Re$ that transforms the incident field $\vec{E}^{(i)}$ into a total field $\vec{E}$:

$$\vec{E}(x, y, z) = \Re\left[\vec{E}^{(i)}(x, y, z)\right]. \tag{A.I.1}$$

The incident field is a plane wave and its z-dependence is explicitly known:

$$\vec{E}^{(i)}(x, y, z) = \vec{E}^{(i)}(x, y, 0)e^{i\gamma_0 z}. \tag{A.I.2}$$

The linearity of the operator $\Re$ and eq. (A.I.1) thus lead to

$$\vec{E}(x, y, z) == \Re\left[\vec{E}^{(i)}(x, y, 0)e^{i\gamma z}\right] = e^{i\gamma z}\Re\left[\vec{E}^{(i)}(x, y, 0)\right]. \tag{A.I.3}$$

Due to the geometrical invariance with respect to the z coordinate, the grating operator is z-invariant. Thus:

$$\Re\left[\vec{E}^{(i)}(x, y, 0)\right] = \vec{E}(x, y, 0). \tag{A.I.4}$$

As a result:

$$\vec{E}(x, y, z) = \vec{E}(x, y, 0)e^{i\gamma_0 z}, \tag{A.I.5}$$

and the z-dependence of the total (and of the diffracted) field is the same as that of the incident field.

Appendix II

Some Formulas About Toeplitz Matrices

Formula *1 Let I(x) be the unit function: I(x)=1 for any x. Then*

$$[\![\mathrm{I}(x)]\!] = \mathbb{1} \qquad \text{(A.II.1)}$$

is satisfied.

This is obvious because for this function, only the zeroth-order Fourier coefficient does not vanish and is equal to 1.

Formula 2 *Let f(x) and g(x) be piecewise-smooth, bounded, periodic functions and s(x) be their sum:* $s(x) = f(x) + g(x)$. *Then*

$$[\![s(x)]\!] = [\![f(x)]\!] + [\![g(x)]\!] \qquad \text{(A.II.2)}$$

is satisfied.

The proof is a direct concequence of the equality: $s_n = f_n + g_n$, where s_n, f_n and g_n are the nth-order Fourier coefficients of s(x), $f(x)$ and $g(x)$, respectively.

Let us now derive two other useful formulas *which are only valid in the framework of our approximation*, i.e. if the Laurent's rule truncated at order N, as shown in Eq. (IV.9) gives a sufficiently accurate value of the coefficients h_n defined in Eq. (IV.7).

We have therefore to remember that in Formulas 3 and 4, the use of the symbol "=" is abusive from a mathematical point of view. Nevertheless we go on using this symbol for the sake of symplicity.

Formula 3 *Let f(x) and g(x) be two piecewise-smooth, bounded, periodic functions that are never both discontinuous for a given value of x (which is obviously the case if one of them is continuous for all the points of the period), then*

$$[\![f(x)g(x)]\!] = [\![f(x)]\!][\![g(x)]\!] \qquad \text{(A.II.3)}$$

is satisfied.

To prove this "approximated identity," we introduce another continuous function $v(x)$ and consider the product $u(x) = f(x)g(x)v(x)$. First we look at u(x) as a product of f(x)g(x) and v(x). Since v(x) is continuous, f(x)g(x) and v(x) have no concurrent jump discontinuities. Therefore, it is possible to apply the Laurent's rule:

$$[u(x)] = [f(x)g(x)v(x)] = [\![f(x)g(x)]\!][v(x)]. \qquad \text{(A.II.4)}$$

We can also consider $u(x)$ to be the product of $f(x)$ by $g(x)v(x)$ and, due to the continuity of $v(x)$, we can again apply the Laurent's rule, because the two functions g(x) and $g(x)v(x)$ have the same set of discontinuity points, which do not coincide with the points of discontinuity of f(x). Then the Laurent's rule can be again applied to the product g(x)v(x) and we obtain

$$[u(x)] = [f(x)g(x)v(x)] = [\![f(x)]\!][g(x)v(x)] = [\![f(x)]\!][\![g(x)]\!][v(x)]. \qquad \text{(A.II.5)}$$

The equalities (A.II.4) and (A.II.5) hold whatever the column $[v(x)]$ may be, and consequently $[\![f(x)g(x)]\!] = [\![f(x)]\!][\![g(x)]\!]$.

As a consequence, one can establish the commutativity of the product $[\![f(x)]\!][\![g(x)]\!]$ by using the commutativity of the product f(x)g(x):

$$[\![f(x)]\!][\![g(x)]\!] = [\![f(x)g(x)]\!] = [\![g(x)f(x)]\!] = [\![g(x)]\!][\![f(x)]\!] \qquad \text{(A.II.6)}$$

When looking at this equation, a pure mathematician would be probably surprised because he knows that the product of two Toeplitz matrices is not necessarily a Toeplitz matrix; let us repeat that in Eqs. (A.II.3) and (A.II.6) the symbol "=" has not its usual significance. It holds as far as the Laurent's rule is applicable in the truncated Fourier space within a certain precision.

As a direct consequence of the commutativity, Eq. (A.II.6), another useful formula can be established:

Formula 4:

$$[\![f(x)]\!][\![g(x)]\!]^{-1} = [\![g(x)]\!]^{-1}[\![f(x)]\!] \qquad \text{(A.II.7)}$$

It is worth noticing that the inverse of a Toeplitz matrix is not in general, a Toeplitz matrix. An exception is given by diagonal matrices, which fall in the frame of the definition of Toeplitz matrices.

In the case when the inverse rule must be applied instead of the Laurent's rule, on can use the same reasoning as for establishing **Formula 3** in order to obtain the rules of factorization, formulated in

Formula 5 *Let f(x) and g(x) be two piecewise-smooth, bounded, periodic functions which are both discontinuous for a given value of x and have complementary discontinuities, i.e., their product is continuous at the points of their common discontinuities, then*

$$\begin{aligned}[\![f(x)g(x)]\!] &= \left[\!\!\left[\frac{1}{f(x)}\right]\!\!\right]^{-1}[\![g(x)]\!] = \left[\!\!\left[\frac{1}{g(x)}\right]\!\!\right]^{-1}[\![f(x)]\!] = \\ &[\![g(x)]\!]\left[\!\!\left[\frac{1}{f(x)}\right]\!\!\right]^{-1} = [\![f(x)]\!]\left[\!\!\left[\frac{1}{g(x)}\right]\!\!\right]^{-1}\end{aligned} \tag{A.II.8}$$

is satisfied.

The proof is made along the same lines as for **Formula 3**.

Appendix III

Expression of the Transverse Components of the Field in Terms of the Axial Ones

Let us consider an electromagnetic field of the form:

$$\vec{E}(x,y,z) = \vec{e}(x,y)e^{i\gamma_0 z} \qquad \text{(A.III.1)}$$

$$\vec{H}(x,y,z) = \vec{h}(x,y)e^{i\gamma_0 z} \qquad \text{(A.III.2)}$$

and decompose vectors $\vec{e}$ and $\vec{h}$ into what we call the *axial* components (e_z, h_z) and the *transverse* components $(\vec{e}_t, \vec{h}_t)$

$$\vec{e}(x,y) = \vec{e}_t(x,y) + e_z(x,y)\vec{z} \qquad \text{(A.III.3)}$$

and similar notations for $\vec{h}$.

Maxwell equations lead to:

$$\operatorname{curl}\vec{e} + i\gamma_0\vec{z}\times\vec{e} = i\omega\mu\vec{h} \qquad \text{(A.III.4)}$$

$$\operatorname{curl}\vec{h} + i\gamma_0\vec{z}\times\vec{h} = -i\omega\varepsilon\vec{e}\,. \qquad \text{(A.III.5)}$$

When Eq. (A.III.3) is put into Eqs. (A.III.4) and (A.III.5), the result is:

$$\operatorname{curl}\vec{e}_t + i\gamma_0\vec{z}\times\vec{e}_t + \operatorname{grad} e_z \times\vec{z} = i\omega\mu(\vec{h}_t + h_z\vec{z}) \qquad \text{(A.III.6)}$$

$$\operatorname{curl}\vec{h}_t + i\gamma_0\vec{z}\times\vec{h}_t + \operatorname{grad} h_z \times\vec{z} = -i\omega\varepsilon(\vec{e}_t + e_z\vec{z})\,. \qquad \text{(A.III.7)}$$

As can be easily verified, since $\vec{e}_t$ and $\vec{h}_t$ are z-independent, $\operatorname{curl}\vec{e}_t$ and $\operatorname{curl}\vec{h}_t$ are parallel to $\vec{z}$. Projecting Eqs. (A.III.6) and (A.III.7) onto axis Oz and onto the plane perpendicular to $\vec{z}$, gives:

$$\mathrm{curl}\,\vec{e}_t = i\omega\mu h_z \vec{z} \tag{A.III.8}$$

$$i\gamma_0 \vec{z} \times \vec{e}_t + \mathrm{grad}\, e_z \times \vec{z} = i\omega\mu \vec{h}_t \tag{A.III.9}$$

$$\mathrm{curl}\,\vec{h}_t = -i\omega\varepsilon e_z \vec{z} \tag{A.III.10}$$

$$i\gamma_0 \vec{z} \times \vec{h}_t + \mathrm{grad}\, h_z \times \vec{z} = -i\omega\varepsilon \vec{e}_t\,. \tag{A.III.11}$$

Putting the value of $\vec{h}_t$ given by Eq. (A.III.9) inside Eq. (A.III.11) leads to:

$$\vec{e}_t (k^2 - \gamma_0^2) = i\gamma_0 \mathrm{grad}\, e_z - i\omega\mu \vec{z} \times \mathrm{grad}\, h_z \tag{A.III.12}$$

In a similar way, we obtain:

$$\vec{h}_t (k^2 - \gamma_0^2) = i\gamma_0 \mathrm{grad}\, h_z + i\omega\varepsilon \vec{z} \times \mathrm{grad}\, e_z\,. \tag{A.III.13}$$

These equations are useful to express e_x, e_y, h_x and h_y as functions of (e_z, h_z), as well as e_r, e_θ, h_r and h_θ as functions of (e_z, h_z) if cylindrical coordinates are used. And indeed, one can use them to obtain Eqs. (II.26), (IV.36), (IV.37), (IX.16), (XIV.42) and (XIV.43), necessary to match the field components at the boundaries between the different regions.

Appendix IV

The Shooting Method in Matrix Notations

The aim of this appendix is to represent the shooting method, used throughout the book, in matrix notation, for easy programming. To this aim, we first recall the basic ideas of the shooting method. This is a method for solving a boundary-value differential problem, when the field components are *unknown* at interfaces, separated in space. The method can be applied if there exists a *linear* link (operator $\mathbb{L}$) between the unknown amplitudes at the interfaces. This is the case with diffraction problem in linear optics, or with the linearized interactive problem in nonlinear optics (see Chapter XV).

We assume that the geometry of the system allows deriving from Maxwell equations a set of ordinary differential equations with respect to the y coordinate, having the form, presented in Eq. (IV.34):

$$\frac{d}{dy}\begin{pmatrix}[E_x]\\ [E_z]\\ [H_x]\\ [H_z]\end{pmatrix} = M\begin{pmatrix}[E_x]\\ [E_z]\\ [H_x]\\ [H_z]\end{pmatrix}. \tag{A.IV.1}$$

Such cases include classical diffraction gratings, invariant with respect to z and periodical in x, crossed gratings, periodical both in x and z, a stack of classical or crossed gratings forming one-, two- or three-dimensional photonic crystals. The matrix M in each specific case has different components, established for each case throughout the book. Let us consider a stack of gratings (classical or crossed), as presented in Fig. III.2. In each intermediate homogeneous layer with infinitely small thickness, optical index ν_M , and vertical ordinate y_j, the z components of the electric and magnetic field can be represented in the form, given by the Rayleigh expansion:

$$E_z^{(j)}(x,y,z) = \sum_{m,n=-N}^{N}\left[A_{e,m,n}^{\prime(j)} e^{i\left(\alpha_m x-\beta_{m,n}^{(M)} y_j+\gamma_n z\right)} + B_{e,m,n}^{\prime(j)} e^{i\left(\alpha_m x+\beta_{m,n}^{(M)} y_j+\gamma_n z\right)}\right]$$

$$\tag{A.IV.2}$$

$$H_z^{(j)}(x,y,z) = \sum_{m,n=-N}^{N}\left[A_{h,m,n}^{\prime(j)} e^{i\left(\alpha_m x-\beta_{m,n}^{(M)} y_j+\gamma_n z\right)} + B_{h,m,n}^{\prime(j)} e^{i\left(\alpha_m x+\beta_{m,n}^{(M)} y_j+\gamma_n z\right)}\right].$$

Here the amplitudes A' and B' correspond, for the propagating orders, to waves propagating downward and upward, respectively, and their subscripts e and h correspond to the electric or magnetic field. The sum on m appears due to the periodicity in x, the sum on n comes from the periodicity in z. The propagation constants in x and z are given by the incident vector components α_0 and γ_0:

$$\begin{aligned} \alpha_m &= \alpha_0 + m\frac{2\pi}{d_x} \\ \gamma_n &= \gamma_0 + n\frac{2\pi}{d_z} \end{aligned} . \tag{A.IV.3}$$

If the system is invariant with respect to z (classical gratings), the sum in n is canceled to a single term with $\exp(i\gamma_0 z)$ dependence in z, as $d_z = \infty$.

Appendix III enables us to express the x components of the electric and magnetic field as functions of their z components, see Eqs. (IV.36) and (IV.37). When using the quasiperiodicity of E_z and H_z in x (and z, for crossed gratings), one obtains Eqs. (IV.38) in the case of classical gratings. For crossed gratings it is necessary to add a second subscript, corresponding to the subscript of γ_n. Using Eqs. (A.IV.2) and (IV.38), it is possible to express the column vectors $[E_x]$, $[E_z]$, $[H_x]$ and $[H_z]$ containing the Fourier components of the corresponding vectors inside the square brackets in matrix notations as a linear combination of the amplitudes A' and B':

$$F(y_j) \equiv \begin{pmatrix} [E_x(y_j)] \\ [E_z(y_j)] \\ [H_x(y_j)] \\ [H_z(y_j)] \end{pmatrix} = \Psi_{eh}^{(j)} V_{eh}^{(j)}, \tag{A.IV.4}$$

with $V_{eh}^{(j)}$ defined in a form, similar to Eq. (III.12), but containing twice more components, corresponding to electric and magnetic field:

$$
{}'^{(j)}_{eh} = \left(\begin{array}{c} \vdots \\ A'^{(j)}_{e,n} \exp(-i\beta^{(M)}_n y_j) \\ \vdots \\ A'^{(j)}_{h,n} \exp(-i\beta^{(M)}_n y_j) \\ \vdots \\ \hline \vdots \\ B'^{(j)}_{e,n} \exp(i\beta^{(M)}_n y_j) \\ \vdots \\ B'^{(j)}_{h,n} \exp(i\beta^{(M)}_n y_j) \\ \vdots \end{array} \right). \tag{A.IV.5}
$$

The matrix $\Psi^{(j)}_{eh}$ is block-diagonal with 4 x 4 blocks, each one having 2N+1 (for classical gratings) diagonal elements:

$$
\Psi^{(j)}_{eh} = \begin{pmatrix} \mathbf{p} & \mathbf{q_e} & \mathbf{p} & -\mathbf{q_e} \\ \mathbb{1} & \mathbf{0} & \mathbb{1} & \mathbf{0} \\ \mathbf{q_h} & \mathbf{p} & -\mathbf{q_h} & \mathbf{p} \\ \mathbf{0} & \mathbb{1} & \mathbf{0} & \mathbb{1} \end{pmatrix} \tag{A.IV.6}
$$

with

$$
\begin{aligned}
\mathbf{p}_{mn} &= \delta mn \frac{-\gamma_0 \alpha_n}{k_M^2 - \gamma_0^2} \\
\mathbf{q}_{\mathbf{e},mn} &= \delta mn \frac{\omega \mu_0 \beta_n^{(M)}}{k_M^2 - \gamma_0^2} \\
\mathbf{q}_{\mathbf{h},mn} &= -\delta mn \frac{\omega \varepsilon_j \beta_n^{(M)}}{k_M^2 - \gamma_0^2}
\end{aligned} \quad . \tag{A.IV.7}
$$

In-plane incidence $(\gamma_0 = 0)$ decouples the diffraction into two independent cases of polarization, which enables us to treat the pairs (E_x, H_z) and (E_z, H_x) independently of each other.

If we deal with crossed gratings, the matrices **p** and **q** have (2N+1) x (2N+1) diagonal elements, arranged according to the scheme given in Eq. (IX.13).

Numerical integration of the set (A.IV.1) starts with a set of 4(2N+1) linearly independent combinations of the amplitudes $V_{eh}^{(j)} = \hat{V}_p$, $p \in [1, 8N+4]$. They are identical for all slices and can be chosen to form the unit matrix:

$$\left(V_{eh}^{(j-1)}\right) \equiv \left(\hat{V}_p\right) = \mathbb{1}. \qquad \text{(A.IV.8)}$$

The starting vector $\hat{F}$ at the lower interface of the j-th slice (see Fig. III.2) for the Fourier components of the field is then given by Eq. (A.IV.4):

$$\left(\hat{F}(y_{j-1})\right) = \Psi_{eh}^{(j-1)}. \qquad \text{(A.IV.9)}$$

Numerical integration of Eqs. (A.IV.1) then gives the 8N+4 vectors $\hat{F}_{\text{integrated}}(y_j)$ on the top of the j-th slice as set of 8N+4 vectors, which are no more block-diagonal, due to the coupling between the different diffraction components, imposed by the grating. It is now the moment to remember that due to the linearity of the diffraction problem, $\hat{F}_{\text{integrated}}(y_j)$ and $\hat{F}(y_{j-1})$ are linked through a linear operator $\mathbb{L}$, which can be expressed as a matrix $T_F^{(j)}$, representing the transmission matrix for the field components inside the j-th slice:

$$\left(\hat{F}_{\text{integrated}}(y_j)\right) = T_F^{(j)} \left(\hat{F}(y_{j-1})\right). \qquad \text{(A.IV.10)}$$

The 8N+4 sets of 8N+4 amplitudes $\left(V_{eh}^{(j)}\right)_p$ are obtained from the set of $\hat{F}_{\text{integrated}}(y_j)$ by inverting Eq. (A.IV.4):

$$\left(V_{eh}^{(j)}\right) = \left(\Psi_{eh}^{(j)}\right)^{-1} \left(\hat{F}_{\text{integrated}}(y_j)\right). \qquad \text{(A.IV.11)}$$

Taking into account Eqs. (A.IV.8)-(A.IV.11), one can obtain the linear operator that links the amplitudes $\left(V_{eh}\right)$:

$$\begin{aligned} \left(V_{eh}^{(j)}\right) &= \left(\Psi_{eh}^{(j)}\right)^{-1} \left(\hat{F}_{\text{integrated}}(y_j)\right)\left(\hat{V}_p\right) \\ &\equiv \left(\Psi_{eh}^{(j)}\right)^{-1} \left(\hat{F}_{\text{integrated}}(y_j)\right)\left(V_{eh}^{(j-1)}\right). \end{aligned} \qquad \text{(A.IV.12)}$$

On the other hand, this linear operator is nothing but the transmission matrix for the amplitudes in the j-th layer, defined by:

$$\left(V_{eh}^{(j)}\right) = T_{eh}^{(j)} \left(V_{eh}^{(j-1)}\right). \tag{A.IV.13}$$

Thus

$$T_{eh}^{(j)} = \left(\Psi_{eh}^{(j)}\right)^{-1} \left(\hat{F}_{\text{integrated}}(y_j)\right), \tag{A.IV.14}$$

and Eqs. (A.IV.9 and (A.IV.10) lead to

$$T_{eh}^{(j)} = \left(\Psi_{eh}^{(j)}\right)^{-1} T_F^{(j)} \Psi_{eh}^{(j-1)}. \tag{A.IV.14'}$$

In order to reduce the computation time, it is possible to analytically invert the block-diagonal matrix $\Psi_{eh}^{(j)}$, given by Eq. (A.IV.6):

$$\left(\Psi_{eh}^{(j)}\right)^{-1} = \frac{1}{2} \begin{pmatrix} \mathbf{0} & \mathbb{1} & \dfrac{1}{\mathbf{q_h}} & -\dfrac{\mathbf{p}}{\mathbf{q_h}} \\ \dfrac{1}{\mathbf{q_e}} & -\dfrac{\mathbf{p}}{\mathbf{q_e}} & \mathbf{0} & \mathbb{1} \\ \mathbf{0} & \mathbb{1} & -\dfrac{1}{\mathbf{q_h}} & \dfrac{\mathbf{p}}{\mathbf{q_h}} \\ -\dfrac{1}{\mathbf{q_e}} & \dfrac{\mathbf{p}}{\mathbf{q_e}} & \mathbf{0} & \mathbb{1} \end{pmatrix} \tag{A.IV.15}$$

When the matrix $T_{eh}^{(j)}$ is obtained by using Eq. (A.IV.14), the block components of the S-matrix are found by the recursive algorithm, presented in detail in Chapter III. This algorithm involves matrix operations (inversion, sum, product) with matrices having different size, depending on the diffraction problem. Classical gratings, used in conical incidence $(\gamma_0 \neq 0)$, require blocks of size $(4N+2)\times(4N+2)$. In-plane incidence $(\gamma_0 = 0)$ decouples TE and TM polarization, so that the blocks have size $(2N+1)\times(2N+1)$. Crossed gratings require matrices with blocks of size $(4N+2)^2\times(4N+2)^2$.

Equation (A.IV.16), similar to Eq. (III.48), but having twice greater size to account for both electric and magnetic field, finally gives the diffracted amplitudes as a linear combination of the incident amplitudes:

$$\begin{pmatrix} \vdots \\ B_{e,n}^{(M)}\, e^{i\beta_n^{(M)} y_{M-1}} \\ \vdots \\ B_{h,n}^{(M)}\, e^{i\beta_n^{(M)} y_{M-1}} \\ \vdots \\ \hline \vdots \\ A_{e,n}^{(1)} \\ \vdots \\ A_{h,n}^{(1)} \\ \vdots \end{pmatrix} = \begin{pmatrix} S_{eh,11} & S_{eh,12} \\ S_{eh,21} & S_{eh,22} \end{pmatrix} \begin{pmatrix} \vdots \\ B_{e,n}^{(1)} \\ \vdots \\ B_{h,n}^{(1)} \\ \vdots \\ \hline \vdots \\ A_{e,n}^{(M)}\, e^{-i\beta_n^{(M)} y_{M-1}} \\ \vdots \\ A_{h,n}^{(M)}\, e^{-i\beta_n^{(M)} y_{M-1}} \\ \vdots \end{pmatrix}. \quad \text{(A.IV.16)}$$

Thus the outgoing-wave condition ($B_{e,n}^{(1)} = 0 = B_{h,n}^{(1)}, \forall n$) then allows finding the Rayleigh coefficients $B_{e,n}^{(M)}$, $B_{h,n}^{(M)}$, $A_{e,n}^{(1)}$ and $A_{h,n}^{(1)}$ from $A_{e,n}^{(M)}$ and $A_{h,n}^{(M)}$ using equations similar to Eqs. (III.49) and (III.50).

List of Notations

The list contains the notations common to several chapters. Some chapters contain additional 'local' notations, that are only used in a limited part of the text.

General convention:

[ ]	column vector containing the Fourier components of the periodic function inside the brackets
⟦ ⟧	Toeplitz matrix made of the Fourier components of the periodic function inside the brackets
(V)	matrix made from a set of vectors V

subscripts:

i,j	denote coordinate components x,y,z of a given vector or matrix
m,n	denote the Fourier component number
1	attributed to the substrate
2	attributed to the superstrate in case of a single interface
M	attributed to the superstrate in case of multiple interfaces

superscripts:

(i)	attributed to the incident wave, field, direction, etc.
(j)	attributed to the j-th medum
(r)	attributed to refraction
(t)	attributed to transmission
{j}	attributed to the j-th iteration in nonlinear optics
Kerr	optical Kerr effect
NL	nonlinear
SHG	second-harmonic generation

Notation:

$\vec{A}$	polarization vector of an incident plane wave
$A = \Delta [\![c^2]\!]$	
A_ε	matrix linking $\vec{E}$ and F_ε

$A_n^{(j)}$ Rayleigh coefficient of the downgoing or anti-evanescent n-th order in the j-th homogeneous medium

$A_n'^{(j)}$ same as $A_n^{(j)}$, but in the j-th infinitely thin layer

$\tilde{A}_n^{(j)} = A_n'^{(j)} \exp(-i\beta_n^{(M)} y_j)$

a groove depth, height of a bump

a_s, a_r optimal groove depth, given by scalar or electromagnetic approach

A.D. angular deviation (between the incident and the -1st diffracted order)

α_0 x-component of the incident wavevector

α_n x-component of the wavevector of the n-th order $(= \alpha_0 + nK)$

α diagonal matrix with elements α_n

α_q^p q-th pole of matrix S (propagation constant of the homogeneous solution in x-direction)

$\vec{B}$ vector of magnetic induction

$B = \Delta[\![cs]\!]$

$B_n^{(j)}$ Rayleigh coefficient of the upgoing or evanescent n-th order in the j-th homogeneous medium

$B_n'^{(j)}$ same as $B_n^{(j)}$, but in the j-th infinitely thin layer

$\tilde{B}_n^{(j)} = B_n'^{(j)} \exp(i\beta_n^{(M)} y_j)$

B_R branching ratio

b_n, b_n' complex coeeficients of the conformal mapping $z(\mathcal{Z})$ and $dz/d\mathcal{Z}$, respectively

$\beta_0^{(j)}$ y-component of the wavevector in j-th medium

$\beta_n^{(j)}$ y-component of the n-th order wavevector in j-th medium

$\beta_{n,m}^{(j)}$ y-component of the (n,m)-th order wavevector in j-th medium of a crossed grating

$\left(\beta^{(j)}\right)$ diagonal matrix with elements $\beta_n^{(j)}$

$C^{(j)}$ interface matrix

C_ε inverse of A_ε

$C(x)$ coupling coefficient of a grating coupler

c speed of light in vacuum, width of the lamella of a lamellar grating

$c(x)$ — continuation over the entire groove period of the x-component of the unit vector $\vec{T}$ tangential to a given profile

$\vec{D}$ — vector of electric induction

$[\vec{D}]$ — column vector made of three blocks containing the Fourier components of D_x, D_y, and D_z.

$[\vec{\tilde{D}}]$ — column vector made of three blocks containing the Fourier components of D_x and D_z.

D_N — electric induction component normal to a given surface

D — bilayer thickness of multilayers

d — groove spacing

d_x , d_z — period along x- or z-axis

δ — angle (polarization angle in conical diffraction or angle of inclination for slanted lamellar grating)

$\delta(x)$ — Dirac's delta-function of x

δ_{mn} — Krönecker's symbol

$\hat{\delta}_1$, $\hat{\delta}_2$ — round-off errors in the computer calculations ($\sim 10^{-16}$ in double precision arithmetics)

$\Delta = [\![\varepsilon]\!] - \left[\!\left[\frac{1}{\varepsilon}\right]\!\right]^{-1}$

$\vec{E}$ — complex electric field vector

$\vec{\tilde{E}} = \vec{E} / i\omega\mu$

$[\vec{E}]$ — column vector made of three blocks containing the Fourier components of E_x, E_y, and E_z.

E_N, E_T — electric field components, normal or tangential to a given surface

$\hat{E}_{n,p}$ — the n-th component of the p-th starting vector for E_z in the shooting method for TE polarization

$\mathbf{E}_x$ — column vectors containing the two-dimensional set of Fourier components of E_x for a crossed grating

$\mathbf{E}_z$ — same as $\mathbf{E}_x$, but for E_z

$\hat{\mathbf{E}}_{z,r,s}$ — the starting matrix for $\mathbf{E}_z$ in the shooting method for crossed gratings; same for the x-component

$e_n^{(r)}$ — diffraction efficiency in the n-th reflected order

$e^{(r)}_{n,m}$	diffraction efficiency in the (n,m)-th reflected order of a crossed grating
$e^{(t)}_{n}$	diffraction efficiency in the n-th transmitted order
ε	absolute electric permittivity
$\overline{\overline{\varepsilon}}$	tensor of absolute electric permittivity
ε_0	absolute electric permittivity of vacuum
ε_j	absolute electric permittivity of j-th medium
ε^{ijk}	Levi-Civita permutation indicator
F	a column vector containing the Fourier components of the x and z elements of electric and magnetic field vectors
$(\hat{F})$	a matrix containing the starting vectors of F in the shooting method
F_ε	column vector containing the components E_{T_1}, D_N, and E_{T_2}, continuous across each interface
$f(x,y)=0$, $f(x,y,z)=0$	surface equation
$\Phi^{(j)}_n$	flux of $\vec{P}$ of the n-th order in j-th medium
$\Phi(y)$	diagonal matrix with elements $\exp(\rho_m y)$
φ, φ_B	groove angle of an echelette grating
$\tilde{\varphi}$	phase shift
$g(x)$	grating groove shape function
$\mathfrak{G}$	metric tensor
Γ_m	propagation constants in y-direction of 'modes' of photonic crystals
γ_0	z-component of the incident wavevector (out-of-plane mounting)
γ_n	z-component of the wavevector of order n $(=\gamma_0 + n2\pi/d_z)$ of a crossed grating
γ	a diagonal matrix with elements γ_n
$\vec{H}$	complex magnetic field vector
$[\vec{H}]$	column vector made of three blocks containing the Fourier components of H_x, H_y, and H_z.
H_N, H_T	magnetic field components, normal or tangential to a given surface
$\hat{H}_{n,p}$	the n-th component of the p-th starting vector for H_z in the shooting method for TM polarization

$\mathbf{H}_x$ — column vectors containing the two-dimensional set of Fourier components of H_x for crossed gratings

$\mathbf{H}_z$ — same as $\mathbf{H}_x$, but for H_z

$\hat{\mathbf{H}}_{z,r,s}$ — the starting matrix for $\mathbf{H}_z$ in the shooting method for crossed gratings; same for the x-component

H_n^+ — one of the Hankel functions

h_j — thickness of the j-th layer along the y-direction

$\mathbb{1}$ — unit matrix

J_n — n-th Bessel function

j_s — surface current

K — grating spatial circular frequency $(= 2\pi/d)$

K(X, Y) — function defined in the conformal mapping $(= k_2^2 |dz/d\mathcal{Z}|^2)$

$\vec{k}$ — wavevector

$\vec{k}_2$ — wavevector in media 2

k_j — modulus of the wavevector in media j

$k^2(x, y) = \omega^2 \mu \varepsilon(x, y)$

κ — imaginary part of the refractive index

$\mathbb{L}$ — linear operator that tranforms the incident field into the total field

λ — wavelength

λ_j — wavelength in medium j

λ_B — blaze wavelength

M — matrix linking the Fourier components of the (x,z) field components and their y-derivatives

M^{TE}, M^{TM} — particular expression of matrix M in TE or TM polarization

M — the number of the superstrate in a stack of gratings

$\tilde{M}$ — the number of slices in the staircase approximation of an arbitray profile $(\tilde{M} = M - 2)$

$\mathcal{M}$ — the inverse of the scattering matrix (S)

μ — absolute magnetic permittivity

$\bar{\bar{\mu}}$ — tensor of absolute magnetic permittivity

μ_0 — absolute magnetic permittivity of vacuum

N — truncation parameter of the Fourier series

N_p — number of discontinuity points in a single period

N_O	number of propagating order
$\vec{N}$	unit vector normal to a surface
ν_j	refractive index of j-th medium
ω	circular frequency (in time)
$\vec{P}$	complex Poynting vector
$\vec{P}^{(i)}$	complex Poynting vector of incident field
$\vec{P}^{(g)}$	complex Poynting vector of a guided wave
$\vec{P}^{L}(\vec{r},t)$	linear polarization time-dependent vector
$\vec{P}^{NL}(\vec{r},t)$	nonlinear polarization time-dependent vector
Ψ	incident angle in conical diffraction mounting
$\Psi^{(j)}$	a matrix that links the field column vector F and the vector V in the j-th medium, $\Psi^{(j)}_{eh}$ - same vector in conical diffraction, having twice more components
$p(\alpha_0)$	density function of a limited incident beam
Q	matrix containing the eigenvalues of matrix M
Q_ε	matrix linking $[\vec{E}]$ and $[\vec{D}]$ in a truncated Fourier space
Q_μ	matrix linking $[\vec{H}]$ and $[\vec{B}]$ in a truncated Fourier space
$q(x)$	spatial density function of a limited incident beam
$q^{(d)}(x)$	spatial density function of a diffracted beam
R	diagonal matrix with elements ρ_m
r	reflection matrix
ρ_m	eigenvalues of a given matrix
$S^{(j)}$	scattering matrix of a stack of j slices
$s(x)$	continuation over the entire groove period of the y-component of the unit vector $\vec{T}$ tangential to a given profile
$\vec{T},\vec{T}_1,\vec{T}_2$	unit vectors tangential to an interface surface
$T^{(j)}$	transmission matrix of the j-th slice
T_A	matrix linking $V_A^{(1)}$ and $V_A^{(2)}$; $T_A = T_{11}$
T_B	matrix linking $V_B^{(1)}$ and $V_B^{(2)}$; $T_B = T_{22}$
t	time, transmission matrix
$\theta_{j,n}$	angle of propagation with respect to the y-axis of order n in the j-th medium

θ_2	angle of incidence with respect to the y-axis in the superstrate for a single surface
θ_M	angle of incidence with respect to the y-axis in the superstrate for a stack of interfaces
θ	grazing angle of incidence $(= \pi/2 - \theta_2$ or $\pi/2 - \theta_M)$
θ_g	grazing angle of incidence on a multilayer
$U(X,Y)$	field in the coordinate system (X,Y)
$U^{(1)}$	the set of numbers of the orders propagating in the substrate
$U^{(2)}$	the set of numbers of the orders propagating in the superstrate
$u(x,y)$	any component of $\vec{E}$ or $\vec{H}$
$u^{(i)}$	any component of the incident field
u_n	n-th component of u(x,y) in $\exp(i\alpha_n x)$ basis
u^-	downgoing or anti-evanescent field
u^+	upgoing or evanescent field
$V^{(j)}$	column vector containing the pseudo-Fourier components of downgoing and upgoing waves in the j-th medium
$V_A^{(j)}$	column vector containing the pseudo-Fourier components of downgoing in the j-th medium (the upper half of $V^{(j)}$)
$V_B^{(j)}$	column vector containing the pseudo-Fourier components of upgoing in the j-th medium (the lower half of $V^{(j)}$)
$W = \exp(i\mathcal{Z})$	
$w = \exp(iz)$	
$\vec{x}, \vec{y}, \vec{z}$	unit vectors along the three axis
x_j	j-th intersection with a single grating period of the grating profile with the line $y = y_0$
Y_a	matching distance in the conformal mapping method
$Z^{(j)}$	intermediate matrix in the S-matrix propagation algorithm
Z_j	impedance in the j-th medium, Z_0 - in vacuum
$\mathcal{Z}$	complex variable $(= X + iY)$
z	complex variable $(= x + iy)$

Index